2011

Construction Documents & Services

Study Guide

Arthur Kornblut & Lester Wertheimer

* Review :-

1. Lesson 4
2. Construction Contract General Condition

KAPLAN
CONSTRUCTION EDUCATION

President: Dr. Andrew Temte

Chief Learning Officer: Dr. Tim Smaby

Vice President of Engineering Education: Dr. Jeffrey Manzi, PE

Director of Product Management: Brian O'Connor

CONSTRUCTION DOCUMENTS & SERVICES STUDY GUIDE 2011

Published by Kaplan Construction Education

332 Front Street, Suite 501

La Crosse, WI 54601

800-420-1429

www.kaplanarchitecture.com

Printed in the United States of America.

11 12 13 10 9 8 7 6 5 4 3 2 1

ISBN: 978-1-4277-2996-5 / 1-4277-2996-4
PPN: 3200-0583

CONTENTS

INTRODUCTION

WELCOME

Thank you for choosing Kaplan AE Education for your ARE study needs. We offer updates annually to keep abreast of code and exam changes and to address any errors discovered since the previous update was published. We wish you the best of luck in your pursuit of licensure.

ARE OVERVIEW

Since the State of Illinois first pioneered the practice of licensing architects in 1897, architectural licensing has been increasingly adopted as a means to protect the public health, safety, and welfare. Today, the United States and Canadian provinces require licensing for individuals practicing architecture. Licensing requirements vary by jurisdiction; however, the minimum requirements are uniform and in all cases include passing the Architect Registration Examination (ARE). This makes the ARE a required rite of passage for all those entering the profession, and you should be congratulated on undertaking this challenging endeavor.

Developed by the National Council of Architectural Registration Boards (NCARB), the ARE is the only exam by which architecture candidates can become registered in the United States or Canada. The ARE assesses candidates' knowledge, skills, and abilities in seven different areas of professional practice, including a candidate's competency in decision making and knowledge of various areas of the profession. The exam also tests competence in fulfilling an architect's responsibilities and in coordinating the activities of others while working with a team of design and construction specialists. In all jurisdictions, candidates must pass the seven divisions of the exam to become registered.

The ARE is designed and prepared by architects, making it a practice-based exam. It is generally not a test of academic knowledge, but rather a means to test decision-making ability as it relates to the responsibilities of the architectural profession. For example, the exam does not expect candidates to memorize specific details of the building code, but requires them to understand a model code's general requirements, scope, and purpose, and to know the architect's responsibilities related to that code. As such, there is no substitute for a well-rounded internship to help prepare for the ARE.

4.0 Exam Format

The seven ARE 4.0 divisions are outlined in the table below.

DIVISION	QUESTIONS	VIGNETTES
Building Design & Construction Systems	85	Accessibility/ Ramp Roof Plan Stair Design
Building Systems	95	Mechanical & Electrical Plan
Construction Documents & Services	100	Building Section
Programming, Planning & Practice	85	Site Zoning
Schematic Design	-	Building Layout Interior Layout
Site Planning & Design	65	Site Design Site Grading
Structural Systems	125	Structural Layout

ARCHITECTURAL HISTORY

Questions pertaining to the history of architecture appear throughout the ARE. The prominence of historical questions will vary not only by division but also within different versions of the exam for each division. In general, however, history tends to be lightly tested, with approximately three to seven history questions per division, depending upon the total number of questions within the division. One aspect common to all the divisions is that whatever history questions are presented will be related to that division's subject matter. For example, a question regarding Chicago's John Hancock Center and the purpose of its unique exterior cross bracing may appear on the Structural Systems exam.

Though it is difficult to predict how essential your knowledge of architectural history will be to passing any of the multiple-choice divisions, it is recommended that you refer to a primer in this field—such as Kaplan's *Architectural History*—before taking each exam, and that you keep an eye out for topics relevant to the division for which you are studying. It is always better to be overprepared than taken by surprise at the testing center.

The exam presents multiple-choice questions individually. Candidates may answer questions, skip questions, or mark questions for further review. Candidates may also move backward or forward within the exam using simple on-screen icons. The vignettes require candidates to create a graphic solution according to program and code requirements.

Actual appointment times for taking the exam are slightly longer than the actual exam time, allowing candidates to check in and out of the testing center. All ARE candidates are encouraged to review NCARB's *ARE Guidelines* for further detail about the exam format. These guidelines are available via free download at NCARB's Web site (*www.ncarb.org*).

Exam Format

It is important for exam candidates to familiarize themselves not only with exam content, but also with question format. Familiarity with the basic question types found in the ARE will reduce confusion, save time, and help you pass the exam. The ARE contains three basic question types.

The first and most common type is a straightforward multiple-choice question followed by four choices (A, B, C, and D). Candidates are expected to select the correct answer. This type of question is shown in the following example.

Which of the following cities is the capital of the United States?

A. New York

B. Washington, D.C.

C. Chicago

D. Los Angeles

The second type of question is a negatively worded question. In questions such as this, the negative wording is usually highlighted using all caps, as shown below.

Which of the following cities is NOT located on the west coast of the United States?

A. Los Angeles

B. San Diego

C. San Francisco

D. New York

The third type of question is a combination question. In a combination question, more than one choice may be correct; candidates must select from combinations of potentially correct choices. An example of a combination question is shown on the following page.

Which of the following cities are located within the United States?

I. New York

II. Toronto

III. Montreal

IV. Los Angeles

A. I only

B. I and II

C. II and III

D. I and IV

The single most important thing candidates can do to prepare themselves for the vignettes is to learn to proficiently navigate NCARB's graphic software. Practice software can be downloaded free of charge from their Web site. Candidates should download it and become thoroughly familiar with its use.

Recommendations on Exam Division Order

NCARB allows candidates to choose the order in which they take the exams, and the choice is an important one. While only you know what works best for you, the following are some general considerations that many have found to be beneficial:

1. The Building Design & Construction Systems and Programming, Planning & Practice divisions are perhaps the broadest of all the divisions. Although this can make them among the most intimidating, taking these divisions early in the process will give a candidate a broad base of knowledge and may prove helpful in preparing for subsequent divisions. An alternative to this approach is to take these two divisions last, because you will already be familiar with much of their content. This latter approach likely is most beneficial when you take the exam divisions in fairly rapid succession so that details learned while studying for earlier divisions will still be fresh in your mind.
2. The Construction Documents & Services exam covers a broad range of subjects, dealing primarily with the architect's role and responsibilities within the building design and construction team. Because these subjects serve as one of the core foundations of the ARE, it may be advisable to take this division early in the process, as knowledge gained preparing for this exam can help in subsequent divisions.
3. Take exams that particularly concern you early in the process. NCARB rules prohibit retaking an exam for six months. Therefore, failing an exam early in the process will allow the candidate to use the waiting period to prepare for and take other exams.

EXAM PREPARATION

Overview

There is little argument that preparation is key to passing the ARE. With this in mind, Kaplan has developed a complete learning system for each exam division, including study guides, question-and-answer handbooks, mock exams, and flash cards. The study guides offer a condensed course of study and will best prepare you for the exam when utilized along with the other tools in the learning system. The system is designed to provide you with the general background necessary to pass the exam and to provide an indication of specific content areas that demand additional attention.

In addition to the Kaplan learning system, materials from industry-standard documents may prove useful for the various divisions. Several of these sources are noted in the "Supplementary Study Materials" section on page x.

Understanding the Field

Although many find the subject of construction documents and services unexciting, the subject is vital not only in obtaining architectural registration, but also in everyday professional practice. It is via construction documents and services that design and construction professionals translate design concepts into real buildings and the roles and responsibilities of the parties are established. Accordingly, the Construction Documents & Services division of the exam tests candidates' knowledge of the roles, rights, and responsibilities of the architect.

Understanding the Exam

It is important to note that this exam tests a candidate's knowledge of the *contractual* roles, rights, and responsibilities of an architect. Therefore, candidates should answer questions based upon what is contractually or legally correct rather than what may be common practice or real-world experience.

In addition to the material provided within this course, exam candidates should be well versed in the AIA documents B102 *Standard Form of Agreement Between Owner and Architect without a Predefined Scope of Architect's Services,* B201 *Standard Form of Architect's Services: Design and Construction Contract Administration* and A201 *General Conditions of the Contract for Construction.* These documents enumerate the roles and responsibilities of the various parties within the design and construction team. As such, a thorough knowledge of these documents is crucial to success for this exam.

For the Building Section vignette, keep an eye on where the cut line is in relation to the plan. Make sure that all partitions, ducts, and structural elements are shown aligned with those in the plan at the cut line. Also, watch for ducts that change to larger sizes beyond the cut line. If there are any, you will need to accommodate the larger size in the ceiling space.

Preparation Basics

The first step in preparation should be a review of the exam specifications and reference materials published by NCARB. These statements are available for each of the seven ARE divisions to serve as a guide for preparing for the exam. Download these statements and familiarize yourself with their content. This will help you focus your attention on the subjects on which the exam focuses.

Prior CAD knowledge is not necessary to successfully complete vignettes. In fact, it's important for candidates familiar with CAD to realize they will experience significant differences between CAD and the drawing tools used on the exam.

Though no two people will have exactly the same ARE experience, the following are recommended best practices to adopt in your studies and should serve as a guide.

Set aside scheduled study time.

Establish a routine and adopt study strategies that reflect your strengths and mirror your approach in other successful academic pursuits. Most importantly, set aside a definite amount of study time each week, just as if you were taking a lecture course, and carefully read all of the material.

Take—and retake—quizzes.

After studying each lesson in the study guide, take the quiz found at its conclusion. The quiz questions are intended to be straightforward and objective. Answers and explanations can be found at the back of the book. If you answer a question incorrectly, see if you can determine why the correct answer is correct before reading the explanation. Retake the quiz until you answer every question correctly and understand why the correct answers are correct.

Identify areas for improvement.

The quizzes allow you the opportunity to pinpoint areas where you need improvement. Reread and take note of the sections that cover these areas and seek additional information from other sources. Use the question-and-answer handbook and online test bank as a final tune-up for the exam.

Take the final exam.

A final exam designed to simulate the ARE follows the last lesson of each study guide. Answers and explanations can be found on the pages following the exam. As with the lesson quizzes, retake the final exam until you answer every question correctly and understand why the correct answers are correct.

Use the flash cards.

If you've purchased the flash cards, go through them once and set aside any terms you know at first glance. Take the rest to work, reviewing them on the train, over lunch, or before bed. Remove cards as you become familiar with their terms until you know all the terms. Review all the cards a final time before taking the exam.

Practice using the NCARB software.

Work through the practice vignettes contained within the NCARB software. You should work through each vignette repeatedly until you can solve it easily. As your skills develop, track how long it takes to work through a solution for each vignette.

Supplementary Study Materials

In addition to the Kaplan learning system, materials from industry-standard sources may prove useful in your studies. Candidates should consult the list of exam references in the NCARB guidelines for the council's recommendations and pay particular attention to the following publications, which are essential to successfully completing this exam:

- International Code Council (ICC) *International Building Code*
- American Institute of Architects B102-2007 *Standard Form of Agreement Between Owner and Architect without a Predefined Scope of Architect's Services*
- American Institute of Architects B201-2007 *Standard Form of Architect's Services: Design and Construction Contract Administration*
- American Institute of Architects A201-2007 *General Conditions of the Contract for Construction*

Test-Taking Advice

Preparation for the exam should include a review of successful test-taking procedures—especially for those who have been out of the classroom for some time. The following is advice to aid in your success.

Pace yourself.

Each division allows candidates at least one minute per question. You should be able to comfortably read and reread each question and fully understand what is being asked before answering. Each vignette allows candidates ample time to complete a solution within the time allotted.

Read carefully.

Begin each question by reading it carefully and fully reviewing the choices, eliminating those that are obviously incorrect. Interpret language literally, and keep an eye out for negatively worded questions. With vignettes, carefully review instructions and requirements. Quickly make a list of program and code requirements to check your work against as you proceed through the vignette.

Guess.
All unanswered questions are considered incorrect, so answer every question. If you are unsure of the correct answer, select your best guess and/or mark the question for later review. If you continue to be unsure of the answer after returning to the question a second time, it is usually best to stick with your first guess.

Review difficult questions.
The exam allows candidates to review and change answers within the time limit. Utilize this feature to mark troubling questions for review upon completing the rest of the exam.

Reference material.
Some divisions include reference materials accessible through an on-screen icon. These materials include formulas and other reference content that may prove helpful when answering questions in these divisions. Note that candidates may *not* bring reference material with them to the testing center.

Best answer questions.
Many candidates fall victim to questions seeking the "best" answer. In these cases, it may appear at first glance as though several choices are correct. Remember the importance of reviewing the question carefully and interpreting the language literally. Consider the following example.

> Which of these cities is located on the east coast of the United States?
>
> **A.** Boston
>
> **B.** Philadelphia
>
> **C.** Washington, D.C.
>
> **D.** Atlanta

At first glance, it may appear that all of the cities could be correct answers. However, if you interpret the question literally, you'll identify the critical phrase as "on the east coast." Although each of the cities listed is arguably an "eastern" city, only Boston sits on the Atlantic coast. All the other choices are located in the eastern part of the country, but are not coastal cities.

Style doesn't count.
Vignettes are graded on their conformance with program requirements and instructions. Don't waste time creating aesthetically pleasing solutions and adding unnecessary design elements.

ACKNOWLEDGMENTS

Lesson 6, with the exception of the "Time Management" section, was written by Paul D. Spreiregen, FAIA. Mr. Spreiregen is an architect, planner, teacher, lecturer, and author. He began his career as a Fulbright scholar after graduating from the MIT School of Architecture. He is renowned as an urban designer both in the United States and abroad. He currently is in private practice in Washington, D.C.

Lessons 9–12 were written by the late Arthur T. Kornblut, AIA. Mr. Kornblut was a licensed architect and practicing attorney in Washington, D.C. He wrote and lectured extensively on the legal aspects of construction.

Material on the topic of Sustainable Design was written by Jonathan Boyer, AIA. Mr. Boyer is a principal of the firm Boyer Associates Ltd in Chicago. He is a graduate of the University of Pennsylvania (BA) and Yale University (MArch). His firm has focused on sustainable design and environmental planning for more than 30 years, with projects throughout the United States.

Lesson 7, which covers Delivery Methods, was developed by Cayl S. Hollis, an architect with 15 years of professional experience designing and managing architectural projects comprising residential, commercial, institutional, and public works.

Content discussing the Building Section vignette was written and illustrated by Lester Wertheimer, FAIA. Mr. Wertheimer is a licensed architect in private practice in Los Angeles. He has written and lectured on the design aspects of the ARE for many years.

Portions of this edition were revised by Robert Elfer, NCARB. Robert has led ARE seminars since 1999, and currently teaches at the Boston Architectural Center. He also led ARE review seminars at the 2006, 2007, 2008, and 2009 AIA national conventions. Aimee Woodall reviewed and revised portions of this edition.

This introduction was written by John F. Hardt, AIA. Mr. Hardt is vice president and senior project architect with Karlsberger, an architecture, planning, and design firm based in Columbus, Ohio. He is a graduate of Ohio State University (MArch).

ABOUT KAPLAN

Thank you for choosing Kaplan AE Education as your source for ARE preparation materials. Whether helping future professors prepare for the GRE or providing tomorrow's doctors the tools they need to pass the MCAT, Kaplan possesses more than 50 years of experience as a global leader in exam prep and educational publishing. It is that experience and history that Kaplan brings to the world of architectural education, pairing unparalleled resources with acknowledged experts in ARE content areas to bring you the very best in licensure study materials.

Only Kaplan AE offers a complete catalog of individual products and integrated learning systems to help you pass all seven divisions of the ARE. Kaplan's ARE materials include study guides, mock exams, question-and-answer handbooks, video workshops, and flash cards. Products may be purchased individually or in division-specific learning systems to suit your needs. These systems are designed to help you better focus on essential information for each division, provide flexibility in how you study, and save you money.

To order, please visit *www.kaplanarchitecture.com* or call 800-420-1429.

Part I

The Multiple-Choice Exam

LESSON ONE 1

DOCUMENTATION

INTRODUCTION

Construction documents prepared by architects and their consultants consist primarily of drawings and specifications. Architects must understand the purpose of these documents and the conventions used in organizing and producing them.

Drawings and specifications convey the owner's and architect's design intentions to the contractors responsible for bidding and constructing the project. In order to obtain competitive bids from contractors and to enable construction to proceed with a minimum number of changes and conflicts, the construction documents must be complete, concise, correct, clear, and coordinated. If, because of negligence by the architect and/or the architect's consultants in preparing the construction documents, the construction budget is exceeded, or an inordinate number of change orders are required during construction, the architect may be subjected to professional liability claims.

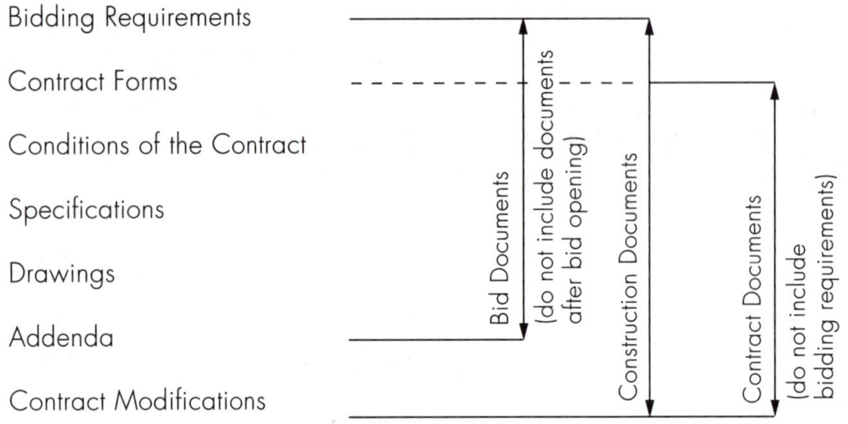

Figure 1.1

The building industry has developed standards for construction documents to achieve these objectives. Architects who have a clear understanding of these standards will be prepared to communicate with consultants and contractors and to objectively review and evaluate their work.

CONSTRUCTION DOCUMENTS

"Construction Documents" is a term used to describe many types of documents used to take a project from design to completed building. Figure 1.1 lists types of construction documents.

Correlation between Drawings and Specifications

Drawings and specifications must be consistent with each other to avoid misinterpretation. Drawings define physical relationships between materials, products, and systems. They indicate physical dimensions and locations of construction elements. Sizes, quantities (implicitly), and configurations are all shown on drawings. Specifications complement the drawings. They express in writing the requirements regarding quality, methods and techniques of installation, and desired performance. For an example of the coordination of drawings and specifications, see Figure 1.2.

Reference noting systems, which use specification section numbers in the drawing notes, can enhance coordination between drawings and specifications.

Construction Drawings

Logic

Although drawings may be organized somewhat differently for each project, they usually communicate their intent most effectively when recognizable and accepted conventions are used, such as standard abbreviations, material designations, graphic symbols, and schedule formats.

The drawings must also be organized logically. Sheets that define symbols and abbreviations or contain other general information are usually placed first in a set of drawings, followed by site and landscape drawings, architectural drawings, structural, plumbing, HVAC, and electrical drawings. Consistent use of this arrangement makes it easier for contractors and others to locate needed information. For an example of the typical sequence of drawings, see Table 1.1.

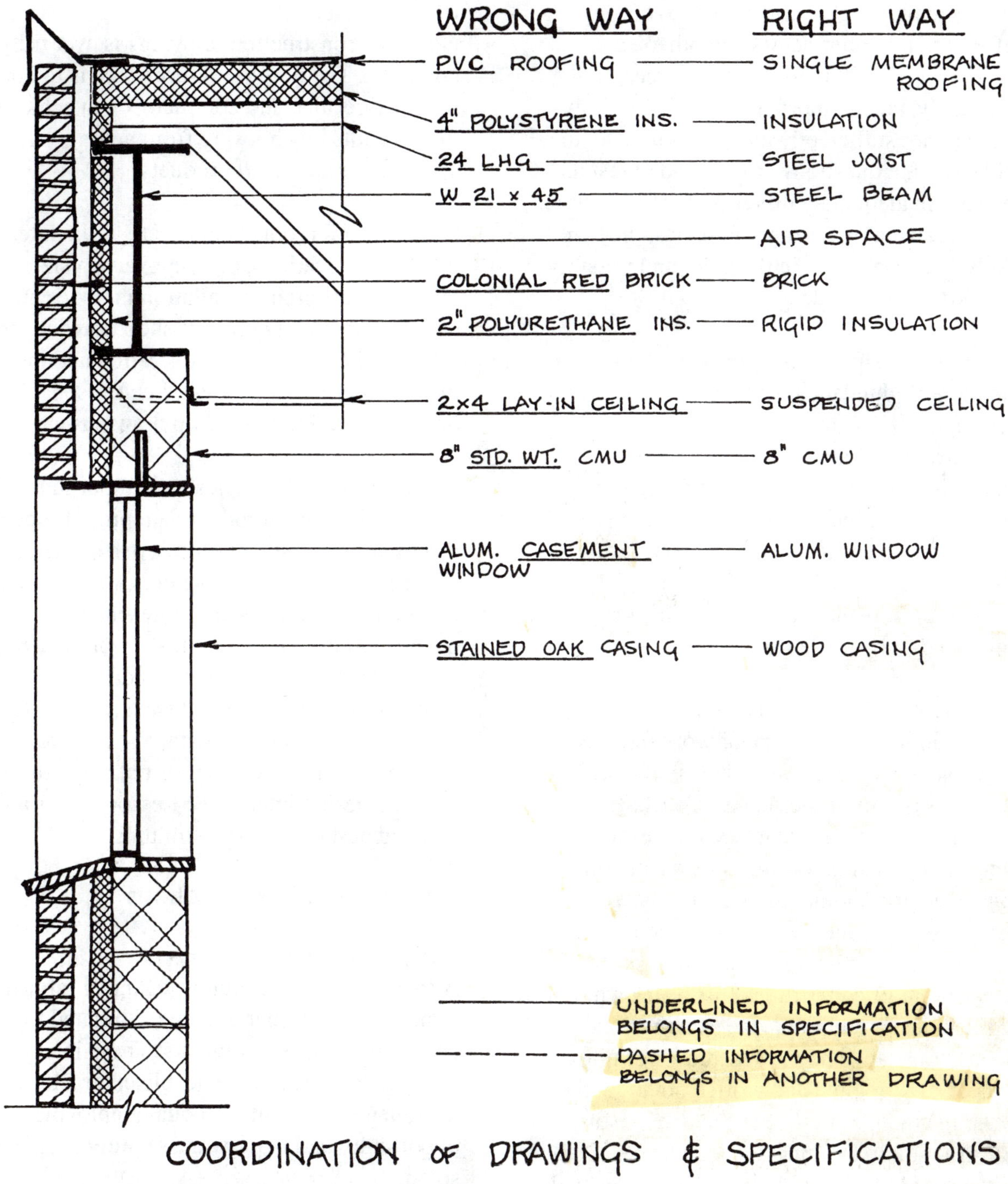
WRONG WAY
RIGHT WAY
PVC ROOFING
SINGLE MEMBRANE ROOFING
4" POLYSTYRENE INS.
INSULATION
24 LHG
STEEL JOIST
W 21 x 45
STEEL BEAM
AIR SPACE
COLONIAL RED BRICK
BRICK
2" POLYURETHANE INS.
RIGID INSULATION
2x4 LAY-IN CEILING
SUSPENDED CEILING
8" STD. WT. CMU
8" CMU
ALUM. CASEMENT WINDOW
ALUM. WINDOW
STAINED OAK CASING
WOOD CASING
UNDERLINED INFORMATION BELONGS IN SPECIFICATION
DASHED INFORMATION BELONGS IN ANOTHER DRAWING
COORDINATION OF DRAWINGS & SPECIFICATIONS

Figure 1.2

Clarity

Since the purpose of drawings is to convey information to contractors and others, architects must present the information clearly, accurately, and at appropriate scales. If drawings are not sufficiently clear or complete, they will generate questions or unintended results. If they are inadequately dimensioned, inaccurate, or are drawn at inappropriate scales, they are likely to cause errors in the field, and possibly additional costs or delays.

Clarity is a result of common sense and coordination. Redundancy, ambiguities, and omissions must be avoided. Each item should be explained and drawn with adequate detail and placed in the most logical location in a set of drawings. Graphic symbols, designations, and cross-references are often necessary. Using cross-references reduces the amount of drawing that must be done, and subsequent revisions only need to be made once.

It is essential that the construction documents clearly delineate the scope of work required for the project. Generally anything shown on the drawings is part of the scope of work. However, there are several important exceptions to this general statement. For example, on site drawings, the surroundings are usually shown. To delineate the scope of work required by the contract, a limit line must be drawn. For utilities, a point of connection must be shown. For renovations or additions, the existing conditions are usually shown. The drawings must indicate, by graphic and written notation, the existing construction to remain, the existing construction to be removed, and the new construction. Equipment or other items furnished by the owner may be shown on the drawings to establish dimensional criteria. These should be noted as not-in-contract (NIC).

Relation to Shop Drawings, Product Data, and Samples

No set of construction drawings is ever truly complete. The purpose of drawings is to convey intended results, and contractors can achieve these results by choosing from among various specified or approved products.

Since several products may satisfy the intent of the specifications, but may vary in size or require different installation methods and details, contractors prepare shop drawings to graphically indicate the fabrication and installation of a particular product. See Lesson 10 for a detailed description of shop drawings.

Product data and samples are similar in purpose to shop drawings. Product data provide specific information about a product's performance in the form of charts, brochures, diagrams, or instructions. Samples are representative of a material's color, texture, finish, workmanship, etc., and establish physical standards for future work. Architects review all such submittals—shop drawings, product data, and samples—but only to determine whether or not they comply with the design intent. See Lesson 10 for a detailed description of submittals.

Construction Specifications

Project Manual

A Project Manual contains technical information, as well as other documents related to legal and procedural requirements. These include bidding documents, applicable contract forms, documents that relate to equal employment opportunity and labor wage requirements, and so on.

TYPICAL SEQUENCE OF DRAWINGS

- Title Sheet, including list of drawings, list of abbreviations, material and symbol schedules, and general notes
- Architectural Drawings
  - Code compliance diagrams
  - Site plan, including vicinity map
  - Site details
  - Landscape plan(s) and details (if required)
  - Demolition plans (if required)
  - Floor plan(s)
  - Exterior elevations
  - Building sections
  - Wall sections
  - Stair and elevator plans, sections, and details
  - Supplementary plan(s) and details of special facilities (if required)
  - Exterior details
  - Window types
  - Window details
  - Interior elevations
  - Interior details
  - Millwork details
  - Finish schedules
  - Door and frame schedule and details
  - Reflected ceiling plan(s)
  - Roof plan(s)
  - Miscellaneous details
- Structural Drawings
  - Foundation plan(s)
  - Floor framing plan(s)
  - Roof framing plan(s)
  - Structural sections
  - Structural details
  - Schedules, including footings, beams, joists, and columns
- Plumbing Drawings
  - Site plan
  - Plumbing plans
  - Fire protection plans (if required)
  - Plumbing details
  - Plumbing schedules, including plumbing fixtures
  - Riser diagrams

(continued)

Table 1.1

11 Yousef 2014

TYPICAL SEQUENCE OF DRAWINGS

Heating, Ventilating, and Air Conditioning (HVAC) Drawings
- HVAC plan(s)
- HVAC details
- HVAC schedules
- Riser diagrams

Electrical Drawings
- Site plan
- Electrical plan(s), including power and lighting
- Single line diagrams
- Electrical details
- Electrical schedules, including lighting fixtures
- Information Technology or Telecommunications Drawings

Table 1.1 (continued)

Technical specifications (Divisions 1 through 16) are prepared by the architect and consultants as part of their basic services, while most other documents are prepared by the owner and the owner's representatives, including attorneys, insurances agents, and consultants. The architect, or construction manager if any, assembles these documents into a Project Manual. For contents of a typical Project Manual, see Table 1.2.

Use of Standard Forms

AIA documents and forms are often included in Project Manuals. They may include standard forms for the Owner-Contractor Agreement, the

TYPICAL PROJECT MANUAL CONTENTS

Title Sheet
Signature Sheet
Table of Contents
Bid Form
Instructions to Bidders
Proposed Owner-Contractor Agreement
General Conditions and Supplementary Conditions
Sample Forms

AIA Document A312	Performance Bond and Payment Bond
AIA Document G702	Application and Certificate for Payment
AIA Document G703	Continuation Sheet
AIA Document G704	Certificate of Substantial Completion
AIA Document G707	Consent of Surety to Final Payment
AIA Document G707A	Consent of Surety to Reduction in or Partial Release of Retainage

List of Drawings
Index to Specifications
Specification Divisions 1 through 16

Table 1.2

General Conditions, Instructions to Bidders, various bond forms, as well as administrative forms relating to payments to the contractor, field administration, and completion of construction. One of the main advantages of using standard AIA forms is that most people in the construction industry are familiar with their contents and how they have previously been interpreted. Since contractors understand their intent, they do not have to provide for contingencies in their bid to allow for potentially unclear documents. Thus, project administration is simplified.

The CSI MasterFormat, developed and published by the Construction Specifications Institute, is widely used in the United States to organize specifications. It is well known in the construction industry and helps architects avoid potential gaps and overlaps in specifications.

The CSI MasterFormat is divided into *Divisions, Sections*, and *Parts*. MasterFormat 1995, currently in use, consists of 16 divisions that comprise the permanent framework of the specifications. This system recently underwent extensive revision, and an expanded MasterFormat was published in 2004. Divisions were added to address the increasing complexity of building systems, existing divisions were modified, and the section numbering system replaced the five-digit system with a six-digit system. Refer to Table 1.5 for a comparison between the two editions. Each Division has a fixed name and number, as shown in the right column. A *Section* describes the basic unit of work, such as a specific product or piece of equipment, and its installation. For example, Division 9, Finishes, might include a Section on Gypsum Drywall, as shown in the right column.

Each Section is further divided into three *Parts*: General, Materials, and Execution. The *General Part* deals with the coverage or scope of a Section. It describes related work, definitions, quality control, submittals, and guarantees/warranties. The *Materials Part* lists and describes the materials, products, and equipment to be used. In shortened or outline specifications, this Part predominates. The *Execution Part* details the manner in which products and materials will be installed and work performed. In addition, coordination with other trades, inspection and acceptance of their work, tests, and other similar items may also be covered. Since contractors usually are permitted to choose among various acceptable products or equipment, it is best not to describe the attributes of particular items too specifically. Affording contractors the opportunity to submit data on products and preferred methods is more practical and, possibly, more economical.

CSI MASTERFORMAT 1995

Division	1	General Requirements
Division	2	Site Work
Division	3	Concrete
Division	4	Masonry
Division	5	Metals
Division	6	Wood and Plastics
Division	7	Thermal and Moisture Protection
Division	8	Doors and Windows
Division	9	Finishes
Division	10	Specialties
Division	11	Equipment
Division	12	Furnishings
Division	13	Special Construction
Division	14	Conveying Systems
Division	15	Mechanical
Division	16	Electrical

Table 1.3

TYPICAL SECTION HEADINGS (MASTERFORMAT 1995)

09250	Gypsum Drywall
09300	Tile
09550	Wood Flooring
09680	Carpeting
09900	Painting

Table 1.4

MASTER FORMAT 2004 EDITION—NUMBERS AND TITLES

DIVISION NUMBERS AND TITLES

Procurement and Contracting Requirements Group

Division 00	**Procurement and Contracting Requirements**

General Requirements Subgroup

Division 01	**General Requirements**

Facility Construction Subgroup

Division 02	**Existing Conditions**
Division 03	**Concrete**
Division 04	**Masonry**
Division 05	**Metals**
Division 06	**Wood, Plastic, and Composites**
Division 07	**Thermal and Moisture Protection**
Division 08	**Openings**
Division 09	**Finishes**
Division 10	**Specialities**
Division 11	**Equipment**
Division 12	**Furnishings**
Division 13	**Special Construction**
Division 14	**Conveying Equipment**
Division 15	*Reserved*
Division 16	*Reserved*
Division 17	*Reserved*
Division 18	*Reserved*
Division 19	*Reserved*

Facility Services Subgroup

Division 20	*Reserved*
Division 21	**Fire Suppression**
Division 22	**Plumbing**
Division 23	**Heating, Ventilating, and Air Conditioning**
Division 24	*Reserved*
Division 25	**Integrated Automation**
Division 26	**Electrical**
Division 27	**Communications**
Division 28	**Electronic Safety and Security**
Division 29	*Reserved*

Site and Infrastructure Subgroup

Division 30	*Reserved*
Division 31	**Earthwork**
Division 32	**Exterior Improvements**
Division 33	**Utilities**
Division 34	**Transportation**
Division 35	**Waterway and Marine Construction**
Division 36	*Reserved*
Division 37	*Reserved*
Division 38	*Reserved*
Division 39	*Reserved*

Process Equipment Subgroup

Division 40	**Process Integration**
Division 41	**Material Processing and Handling Equipment**
Division 42	**Process Heating, Cooling, and Drying Equipment**
Division 43	**Process Gas and Liquid Handling, Purification, and Storage Equipment**
Division 44	**Pollution Control Equipment**
Division 45	**Industry-Specific Manufacturing Equipment**
Division 46	Reserved
Division 47	Reserved
Division 48	**Electrical Power Generation**
Division 49	*Reserved*

Table 1.5

The AIA General Conditions explicitly states that the organization of the specifications into divisions, sections, and articles, and the arrangement of drawings shall *not* control the contractor in dividing the work among subcontractors or in establishing the extent of work to be performed by any trade. General contractors are generally more familiar with labor markets and union labor jurisdiction agreements than architects. Each contractor is also free to utilize a unique procedure for accomplishing the required work and must be responsible for covering potential gaps in the assignments of subcontractors.

Master specifications have been developed using word processing programs that provide automated methods of editing specifications to suit a particular project. Some masters are developed by architectural firms, and some are commercial or proprietary systems. In either case, masters are intended to reduce a specifier's clerical and repetitive work. They allow architects to spend more time on research and make written technical information available to project teams in the early stages of the design process.

Before master specifications were developed, specifiers usually edited sections from previous projects or created new sections based on data from manufacturers and trade associations. The first master specifications were *text-based*, comprising comprehensive data bases of practically all available products and methods. A recent concept is *knowledge-based* specifications, which use a dialog (question-and-answer) method to access the data bases. Text-based systems are reductive, while knowledge-based systems are additive.

However, there are some problems associated with the use of master systems. Since it is easier to edit by deleting unnecessary or inapplicable material than to write new material, a master specification should be applicable to every project. Unfortunately, it is difficult to create a master that is totally comprehensive.

It is also difficult to maintain an accurate and up-to-date master system with rapidly changing technology. For these reasons, many architects subscribe to one of the commercially available proprietary systems. Whichever system is used, it is important to guard against accepting materials and processes merely because they are listed in the master, instead of thorough, independent analysis.

Organization

The relationship between the AIA General Conditions and Division One-General Requirements of the MasterFormat Specifications must be understood. The General Conditions contains contractual provisions that elaborate on elements of the AIA Owner-Contractor Agreement. They are intended to apply to many projects and situations. By contrast, the material in Division One of the specifications describes the administrative rules and work-related provisions for the specific project. A well-written Division One can help a project run smoothly during the construction phase. An example of the typical sections in Division One is shown in Table 1.6.

Types of Technical Specifications

There are several types of technical specifications, and examples of each may be found in any Project Manual. There is nothing inherently wrong with mixing types if they are used correctly.

Proprietary specifications call for desired materials, products, systems, and equipment by their trade names and model numbers. Proprietary specifications are relatively easy to prepare since they rely on commercially available products, which are described in detail in their

DIVISION ONE, GENERAL REQUIREMENTS

CSI Sections MasterFormat 95	MasterFormat 04	Description
01010	01 11 00	Summary of Work
01025	01 20 00	Payment, Modification, and Completion Procedures
01031	01 23 00	Alternates
01033	01 21 00	Allowances
01034	01 22 00	Unit Prices
01200	01 32 00	Progress Documentation and Procedures
01300	01 33 00	Submittals
01400	01 45 00	Quality Control Procedures
01500	01 50 00	Temporary Facilities and Services
01600	01 60 00	Product Requirements and Substitutions
	01 25 00	(Substitution Procedures)
01700	01 70 00	Construction Procedures
01800	01 78 39	Project Record Documents

Table 1.6

manufacturers' literature, not in the specifications. Architects should thoroughly investigate a manufacturer's claims for such products and systems. The track record of the manufacturer and the product, as well as its suitability for a particular application, should be carefully investigated.

There are two kinds of proprietary specifications: *closed* (sole source) and *open* (equal). See Table 1.7. Closed specifications require a particular brand or trade name and do not permit substitution. They are intended for situations where only one product will provide the desired result. For example, in a renovation project where only a few windows need to be replaced, a specific brand and model of window may be required. Closed specifications are not usually permitted on publicly-funded projects, where open, competitive bidding is required.

Open specifications name several (usually three) acceptable materials, products, or systems, and contractors may use any one of them. Alternatively, other approved products that match the capabilities and quality of the named items may be used if the open proprietary specification contains an *approved equal* clause. Open specifications are most often used on publicly-funded projects because they promote competition while avoiding questions of impropriety or favoritism in the selection of materials, products, and systems. Open specifications are also used on private projects because they allow contractors to apply their expertise to the construction process while decreasing costs through open competition.

Frequently, the architect may have to determine whether a material or product proposed to be substituted by the contractor is equal in quality and performance to that specified. Division One, General Requirements, must be specific in describing the administrative procedures for contractors to follow in order to obtain approval

TYPICAL EXAMPLE OF PROPRIETARY OPEN SPECIFICATION

A. Admixtures

1. Water-reducing and air-entraining agents shall be used in all concrete, in strict accordance with the manufacturer's printed instructions. Total air entrained in freshly mixed concrete shall be 5.0 percent plus or minus 1.0 percent of volume of concrete with required strengths maintained.
2. Water-Reducing Agent: "Sonotard WR" by Sonneborn Building Products, "WRDA" by W. R. Grace Company, "Pozzolith 100" by Master Builders Company, or Sika "Plastocrete N." Water-reducing agent shall be by same manufacturer as air-entraining agent.
3. Air-Entraining Agent: "Darex" by W. R. Grace Company, "Aerolith" by Sonneborn Building Products, "MBVR" by Master Builders Company, or Sika "AER."

Table 1.7

TYPICAL EXAMPLE OF PERFORMANCE SPECIFICATION

A. Sprayed-On Fireproofing

Materials, procedures for application, dry densities, and thicknesses necessary to provide the required protection shall have been tested in accordance with ASTM E-119 and approved by UL for the uses indicated.

B. Structural Steel Members and Roof Deck

All structural steel members and roof deck shall be protected under this Section with adequate fireproofing thicknesses and densities to provide the following fire resistance ratings.

Steel columns and beams supporting more than one (1) floor .... 3 hours
Steel columns supporting roof deck .... 2 hours
Metal roof deck and supporting steel members .... 1 hour
Steel members supporting one (1) floor .... 2 hours

Load: In addition, beams shall have sustained the applied load during the ASTM E-119 fire endurance test; and the transmission of heat through the beam protection during the period of fire exposure for the specified rating shall not have raised the average (arithmetical) temperature of the steel at any one of four sections above 1,200°F (648.8°C) and shall not have raised the temperature above 1,400°F (760.0°C) at any one of the measured points.

Thickness and Density: Where the thickness of fire protection material for the specified fire resistance rating is given as an average thickness, the minimum thickness shall be that given as average thickness. Acceptable minimum thickness of applied material shall be that measured at specified dry density. Minimum applied dry density per cubic foot shall be 18 pounds.

Fire ratings interpolated or extrapolated from actual test data will not be accepted. Provide evidence prior to application that proposed materials and installation methods and materials have been approved by all authorities having jurisdiction.

Table 1.8

for such substitutions. The contractor should be responsible for submitting complete technical data to the architect for evaluation. Without the architect's review and approval, a substitution cannot be considered to be an *approved equal.* In evaluating proposed substitutions, architects must determine whether the aesthetic intent

TYPICAL EXAMPLE OF REFERENCE SPECIFICATION
A. Steel Stud Shear Connectors shall conform to the requirements of Articles 4.26 and 4.27 of "Structural Welding Code" AWS D1.1-77 of the American Welding Society.
B. Bolts, Nuts, and Washers shall comply with ASTM A325. Bolt dimensions shall comply with requirements of ANSI Standard B18.2 for structural bolts, except that the radii of the filler under the bolt head shall not be less than 1/32" for bolts up to 1" in diameter. Nut dimensions shall comply with requirements of ANSI B18.2 for heavy semi-finished hexagonal nuts. Circular washers shall be flat and smooth and bevel washers square or rectangular. All washers shall comply with requirements of ANSI B27.2 for Type A washers. Where clipping of washers is necessary, clip one side only and not closer than 7/8 of the bolt diameter from the center of the washer.

Table 1.9

TYPICAL EXAMPLE OF DESCRIPTIVE SPECIFICATION
Solid Wood Door Construction
A. Except as otherwise indicated, all flush wood doors (except UL doors specified hereinafter) shall be wood solid core doors 3'-0" x 7'-0", 1-3/4" thick of 5-ply construction with face veneers bonded to both faces. Cores shall be solid stave low density wood blocks bonded together under heat and pressure. Cross bands shall be thoroughly kiln-dried hardwood, 1/10" thick, extending full width of door. Core construction shall be AWI Type "SLC," non-resinous wood.
B. Face veneers shall be standard thickness (1/28") paint grade veneer. Vertical stile edges and top and bottom rail edges shall be hardwood. Vertical edges shall be 5/8" minimum, top and bottom 2 rails shall be 1-1/4" minimum. Face veneers shall be AWI Type "1." Doors shall be completely sanded, ready to receive paint finish in the field under PAINTING Section.
C. Solid core doors shall meet or exceed the requirements of U.S. Department of Commerce Commercial Standard CS-171, and shall be equal to DSC-1 manufactured by Weyerhaeuser Company, or equal as approved by the Architect from manufacturer specified hereinbefore. Except as otherwise indicated, doors shall be AWI "Custom" Grade.

Table 1.10

will be met. Furthermore, they must consider value, quality, warranties, the manufacturer's reputation, compliance with code requirements and regulations, operating and maintenance costs, size and weight, ease of construction, construction labor and equipment requirements, and operational characteristics.

Substitute products or systems do not have to be identical to those specified, since not all the features of a specified product may be required. If a product can meet desired results and has the most important features, substitutions may be acceptable.

Performance specifications define products or systems by describing desired end results that are performance oriented. In such specifications, the precise composition of individual components or systems is not described. This method allows contractors and manufacturers to apply their unique expertise and encourages broad competition and maximum creative input. Performance specifications are most appropriate when new or unusual products or systems are required or when innovation is necessary.

TYPICAL EXAMPLE OF CASH ALLOWANCE SPECIFICATION

A "Schedule of Allowances," showing amounts included in Contract Sum, is included at the end of this Section. Coordinate allowance Work with related Work, to ensure that each selection is completely integrated and interfaced with related Work. Requirements for the Work of allowances are shown and specified, to extent established by date of Contract Documents; additional requirements are established by Change Order. At earliest possible date, advise Architect of date each final allowance selection must be completed. Submit proposals for allowance Work as directed, and in the manner specified for Change Orders. Indicate quantities, unit costs, total purchase amounts, taxes, delivery charges and trade discounts. Where requested, furnish detailed breakdown of quantity survey. Contractor mark-up on overrun of allowance purchases will be permitted where purchase amount exceeds established allowance by more than 15 percent; otherwise, and except as otherwise indicated, amount of Change Order on each allowance will be difference between purchase amount and allowance. Deliver excess materials of allowance Work to Owner's storage space, or dispose of by other means as directed.

Schedule of Allowances

Allowance No. 1	A lump sum of $3,000 for purchase of finish hardware, as defined by and specified in Specification sections of Division 8.
Allowance No. 2	A lump sum of $5,000 for purchase of carpet, as defined by and specified in Specification sections of Division 9.

Table 1.11

It can be challenging to prepare performance specifications. Describing the problems or the conditions in which products or systems must operate, and the parameters for acceptable solutions, is difficult. Performance specifications must explicitly define required testing methods and procedures for evaluating performance. Energy consumption costs, aesthetics, and similar factors may be especially difficult to specify.

Reference specifications refer to quality standards established by recognized testing authorities or by the federal government. They are typically used in conjunction with other types of specifications. It should be understood, however, that the quality and performance described in the referenced specifications may only be a minimum level and not appropriate or sufficient for the specific application. An architect's specifications must clearly state which parts of the referenced specifications are meant to apply. Standard reference specifications are also dated, and the latest version should be researched before it is cited.

Descriptive specifications are the most detailed of all specifications. They describe all components of products, their arrangement and methods of assembly, physical and chemical properties, arrangement and relationship of parts, and numerous other details and requirements. In descriptive specifications, the architect assumes total responsibility for the function and performance of a product. Unless the architect is certain the assembled product will function properly, the use of this type of specification should be avoided. See Table 1.10.

Cash allowance specifications are used in lieu of specifying a particular portion of the work.

Under this method, an architect directs bidders to set aside a specified amount of money to be applied to the construction work at the architect's direction. Cash allowance specifications are used when full information on levels of quality has not been determined or is not available at the time bids are solicited. Hardware and carpeting are often handled in this manner. Types and quantities of hardware may be determined on the basis of the drawings, while levels of quality may have to be determined later. Similarly, the extent of carpeting may be known, but not the type or quality. These determinations are sometimes delayed in order to meet a project's budget limitations.

Cash allowances may be used for the purchase and delivery of the product only, in which case the installation is indicated in the construction documents and included in the base bid. Alternatively, a cash allowance may be used for both furnishing and installing the product.

In some cases, products are indicated as *owner furnished-contractor installed* on the construction documents. Although not a cash allowance, an owner-furnished product may be installed as part of the contractor's scope of work. A product indicated as *NIC (not-in-contract)* is neither furnished nor installed by the contractor.

If cash allowances are used, specifications should include information on installation methods, the dollar amount of the allowance, and methods of measuring costs to be applied against allowance amounts. When installations are complete, costs can be compared with allowance amounts and the difference credited to the owner or contractor as appropriate. The example in Table 1.11 is taken from Division One, General Requirements. The installation requirements are located in Divisions 8 and 9 as noted.

Effect of Multiple Prime Contracts

When multiple prime contracts are used, rather than one general contract, specification sections may have to be written for individual construction trades. This situation, which often occurs on public projects, requires increased effort on the part of the architect. First, architects must understand local trade union work rules and jurisdictional requirements. Second, many parts of the specifications must be duplicated, increasing the coordination required of architects. For example, each section must include its own agreement, general conditions, and general requirements documents. If there are gaps in assignment of construction work, the architect may be held responsible for such omissions.

Interpretation

Where construction documents are inconsistent or ambiguous, or have gaps or overlaps in coverage, they may be open to interpretation.

When two clauses in the specifications conflict, the more specific clause will usually prevail over the more general clause. Handwritten provisions will usually prevail over typewritten provisions, which in turn take precedence over pre-printed provisions. These cases illustrate the principle that individual and personal attention to an item more likely reflects the author's intent.

When two drawings conflict or are inconsistent, the more recent drawing will usually prevail. Dates of all revisions should appear on drawings, and the items involved in each revision should be clearly indicated. When different drawings are prepared at the same time, large scale detailed drawings will usually prevail over small scale general drawings, such as floor plans and elevations. This type of hierarchy can be made part of the construction contract by incorporation in the Supplementary Conditions.

Specifications sometimes indicate one requirement and drawings another. Subparagraph 1.2.1 of the AIA General Conditions, Document A201, states: "The Contract Documents are complementary, and what is required by one shall be as binding as if required by all…" Inconsistencies are not resolved by an arbitrary order of precedence, but must be brought to the architect's attention for appropriate resolution. The AIA General Conditions is clear in regard to the interpretation of the contract documents. Subparagraph 4.2.11 states: "The Architect will interpret and decide matters concerning performance under, and requirements of, the Contract Documents..."

DOCUMENT COORDINATION

Compliance with Code Requirements and Regulations

A coordinated and detailed response to code requirements from the entire design team is essential to the success of a project.

Consider, for example, energy requirements. Siting, preliminary selection of materials, and schematic organization of programmatic elements are largely within an architect's control. These energy considerations must be balanced against other requirements more closely controlled by others, including structural requirements.

Fire protection also requires building team coordination. The incorporation of interior courtyards or atriums, for example, may require engineering for fire protection. Mechanical, electrical, and plumbing equipment are often critical elements in a fire protection plan. When there are no physical barriers to the spread of potential fires, protection depends upon sensing devices, sprinkler systems, and air handling equipment. These systems and building components are likely to be designed or selected by the engineering and fire protection consultants, rather than the architect.

The mechanical, plumbing, and electrical codes often have provisions that are the same as or that complement the building and life safety codes. These common provisions are generally understood by most design professionals. Architects, however, cannot always be certain that engineers and other consultants have complied with all code provisions. As a practical matter, architects of complex projects may simply inform consultants about which codes are applicable, and ask them to research the detailed requirements. This does not relieve architects, however, of responsibility to meet code requirements. As leader of the design team and the party contracting with the owner for professional design services, the architect has prime responsibility for code compliance. However, each engineering consultant must sign the drawings submitted for plan review by the code official and thereby also becomes responsible for compliance. Moreover, the AIA Architect-Consultant Agreement (Document C401) states that the consultant is responsible for code compliance in the same manner and extent that the architect is responsible to the owner.

Initially, architects should verify that each member of the project team is working from the same set of code requirements. Consultants should inform the architect about significant aspects of their work that are required by code. Although codes generally allow several responses to requirements, they occasionally require specific design features. Consequently, architects must know which design elements may change and which may not.

Architects are responsible to notify their consultants of design decisions that have code implications. For example, fire walls must be clearly identified, so that air handling ducts passing through them include fire and smoke dampers. See Figure 1.3. Alternatively, the duct work could be arranged to avoid fire walls altogether. Ceiling appearance is affected by the type and location of sprinkler heads. If ceilings are required to be fire rated, light fixtures and air handling grilles must be properly accommodated.

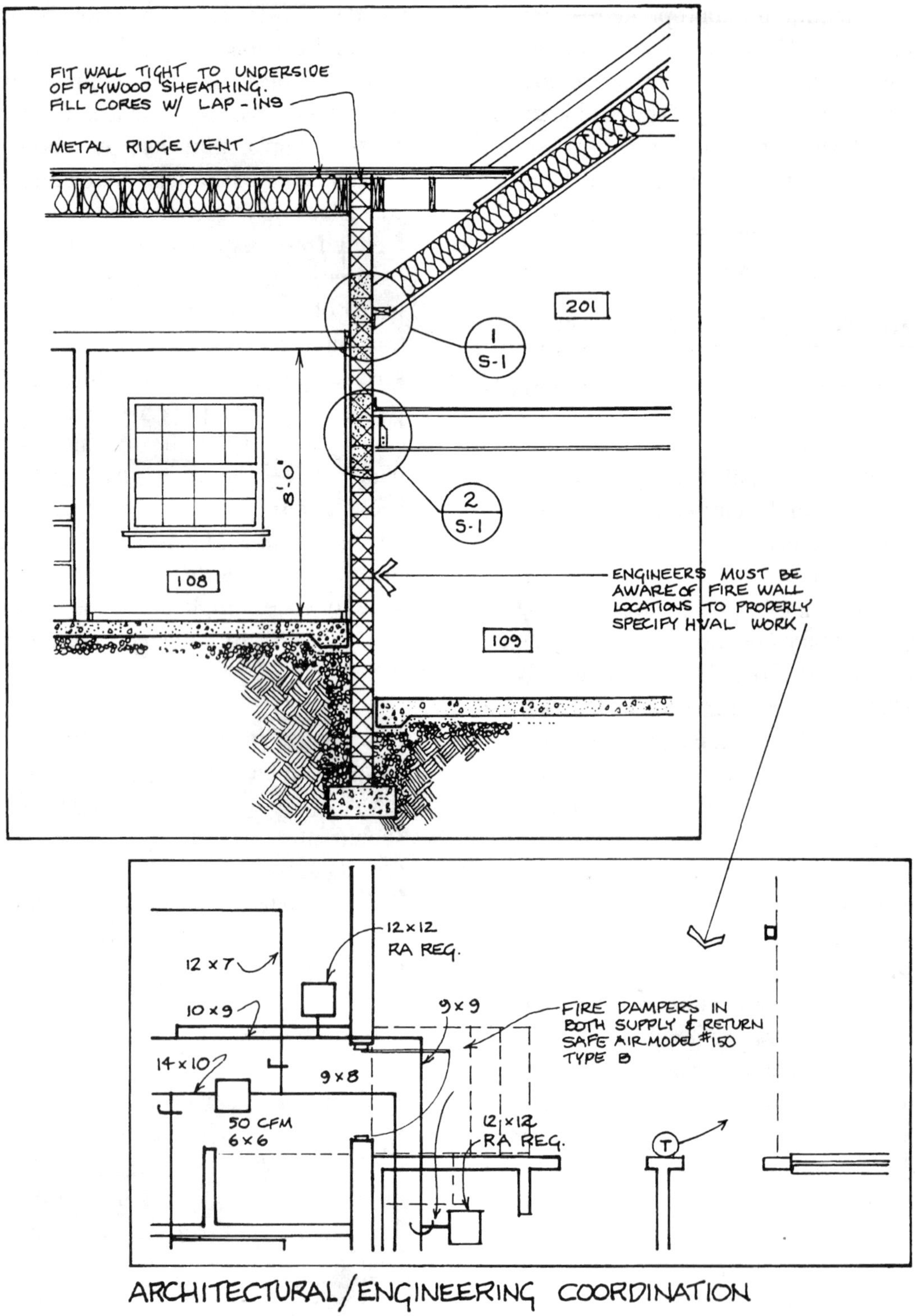
FIT WALL TIGHT TO UNDERSIDE OF PLYWOOD SHEATHING. FILL CORES W/ LAP-INS
METAL RIDGE VENT
201
1
S-1
2
S-1
8'-0"
108
109
ENGINEERS MUST BE AWARE OF FIRE WALL LOCATIONS TO PROPERLY SPECIFY HVAC WORK
12×12 RA REG.
12×7
10×9
9×9
FIRE DAMPERS IN BOTH SUPPLY & RETURN SAFE AIR MODEL #150 TYPE B
14×10
9×8
50 CFM 6×6
12×12 RA REG.
T
ARCHITECTURAL/ENGINEERING COORDINATION

Figure 1.3

Compliance with Design Criteria

Aesthetics

Consultants can significantly influence the aesthetic character of a project. Structural expression, for instance, is an important element in many architectural designs. Structural engineers often collaborate with architectural designers to achieve such aesthetic goals. The structural design of the cross-braced frame of the John Hancock Building and the bundled tube design of the Sears Tower, both in Chicago, are good examples of positive aesthetic qualities achieved through the mutual efforts of architects and their structural engineers. On a smaller scale, the structural design of framing members influences floor-to-floor height, and thus overall building height, by establishing the floor structure's depth. The relationship of spandrels to window openings is often critical to the proportions of a building's facade. In many instances, the basic character of a building is a result of its structural expression, as in a domed structure or an air-supported roof.

Mechanical engineers may influence wall treatments by their response to energy considerations. Their work can affect the character of the building's envelope, including its fenestration in relation to solar orientation. On a smaller scale, the location and design of air diffusers can affect the aesthetic appearance of interior spaces. Where mechanical equipment is exposed to view, architects normally ask to review and approve illustrations showing the equipment's physical appearance. Unsightly fans on rooftops can seriously detract from an architect's design.

Electrical engineers, through selection and placement of light fixtures, can affect the aesthetic quality of spaces and ceilings. With the development of open plan office design and the use of task lighting, electrical engineers may also influence the design and placement of partitions, furniture, and equipment. Offices commonly contain video display equipment, computers, communications equipment, and electronic sensing devices for security and fire protection. These items of equipment are generally selected by the electrical engineer, in consultation with the architect.

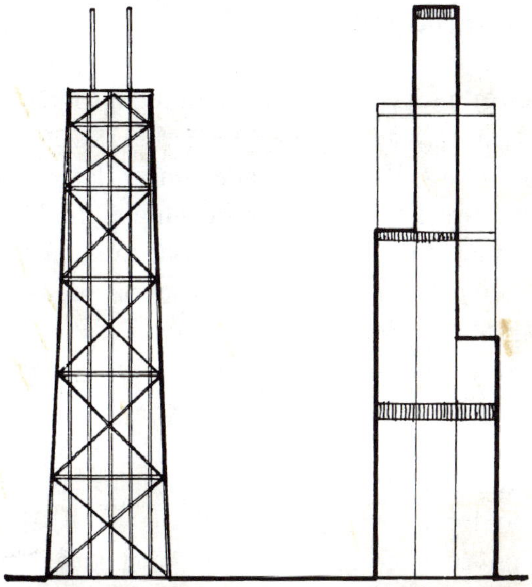

STRUCTURAL EXPRESSION IN ARCHITECTURE

THE DESIGN OF THESE TWO WELL-KNOWN BUILDINGS HAS BEEN STRONGLY INFLUENCED BY CONSULTANTS TO THE ARCHITECT.

Figure 1.4

Food service consultants, lighting consultants, acoustical consultants, and art advisors may also influence a building's aesthetic qualities. Architects must always inform their consultants of design criteria and the aesthetic effects they are trying to achieve. Product data, study models, and photographs may be used to assess intermediate design progress, and are subject to the architect's final approval. Architects must know enough about the details of their consultants' work to maintain design control. At times, they may suggest alternate approaches or solutions more compatible with the desired aesthetic character of the project.

Quality Control

Since many of the construction documents prepared by consultants are based on calculations, quality control is relatively easy to achieve. Parameters are well defined and solutions can easily be checked.

Details that are shown on the drawings must be in conformance with engineering design assumptions. If a structural engineer designs a moment-resisting frame, for example, the joint details must reflect that condition. Architects may review consultants' construction documents to confirm that designs, details, and specifications are compatible with the consultants' calculations and assumptions.

An architect can support a consultant's quality control by informing her of all relevant design criteria to which the consultant must adhere, and by requiring the consultant to schedule periodic reviews by experienced senior staff members or *peer review* by others in the consultant's profession.

Although architects can check for internal consistency and for apparent compliance with standards, consultants are primarily responsible for quality control of their own work.

Cost Control

Estimating *initial costs* is an essential part of a consultant's work. The percentage of the total budget allocated to each discipline varies with building type and project scope. Architects often establish budgets for the major elements of construction work for incorporation into an overall project budget. Once the budget is established, consultants are expected to design within its limits. Consultants must, therefore, be accurate in predicting initial costs so that the architect can prepare a reliable overall project estimate.

Because operating costs tend to vary inversely with initial costs, a relatively low construction budget may imply that life cycle costs will be relatively high. Consultants must evaluate conflicting considerations in order to produce optimum design solutions. Likewise, architects must review each alternative to be sure that a consultant's decision serves not only the architect's particular discipline, but the project as a whole.

Operational costs may be difficult to calculate. Calculations involve more than the characteristics of specified equipment; they can involve the operating characteristics of the owner's organization and other factors affecting a facility, such as changing climatic conditions. Engineering calculations may, in some instances, be based on assumptions different from actual conditions. For example, a facility may be operated differently than anticipated by its program; calculations may be based on average conditions, in spite of the fact that extreme weather conditions may have been experienced in recent years; or fuel prices may have increased suddenly and unexpectedly.

It is important that basic design assumptions are realistic. Architects should understand the operating characteristics of facilities, and they must ensure that design assumptions are accurate and that designed elements and systems will be appropriate.

Maintenance is an important aspect in the selection of products and systems. Some mechanical and electrical systems are complicated, sophisticated, and sensitive. If properly balanced, they can be efficient and economical. But, they can also be troublesome and more difficult to maintain than simpler, less technically advanced systems. Equipment maintenance costs vary with the size and skill of maintenance staffs. Some design professionals have expanded their practices to include facilities management services, including the preparation of detailed operational and maintenance programs.

Specified systems must be properly installed, reliable, and receive scheduled maintenance to be successful. The architect should determine that such systems are appropriate to the contractor's and building manager's degree of sophistication.

Compatibility with Other Elements

The *size and weight* of equipment is another design consideration. Engineering drawings are largely diagrammatic, making it difficult to verify that design criteria have been met. For example, a large pipe or duct may be represented by a single line on a drawing, but its actual size determines the clearances which must be provided and maintained. Unfortunately, these considerations are sometimes ignored. Architects can create similar problems by providing insufficient space for equipment and services during design development phases. Allotted spaces might prove to be too small, and increasing the building's gross area may be difficult without disrupting the overall architectural solution and budget.

Operational characteristics of mechanical and electrical equipment must be considered by the design team before final selections and placement are made. In critical cases, a special consultant, such as an acoustical engineer, may be retained to advise the design team on the placement, isolation, and construction of large air handling equipment. Electrical distribution equipment can interfere with the operation of sensitive laboratory or hospital equipment. In this case, the architect may ask the owner to provide the services of a special consultant to advise the design team on the placement, selection, and isolation of certain equipment.

Ease of Construction

Labor and Equipment Requirements

Architects and their consultants should determine that the systems they design can actually be built, considering the space, equipment, and labor required. For example, if a floor system utilizes precast concrete T-beams, there must be sufficient room on the site to position the cranes required to erect these units. If construction access is available from one side of a site only, construction must be able to proceed in only one direction. Post-tensioned structures require accurate placement of tension cables and hydraulic jacks to stress tendons properly. The availability of the skilled and experienced labor necessary for these operations influences the decision to utilize such systems.

Large air-conditioning chillers and cooling towers are often placed on the upper stories or roofs of multistory buildings. If they cannot be disassembled and installed in sections, they must be lifted intact to their final locations. Once in place, equipment and systems may require sophisticated pneumatic and electrical controls and precise balance in order to operate properly. The installation of sensitive equipment requires the availability of skilled technicians.

In making design decisions, the architect's consultants must consider the limitations of local labor and the availability of special equipment. They must be aware of the implications of applicable union rules. Although contractors must determine the appropriate trade for each part of the work, both architects and consultants should follow established and generally accepted operating procedures, and understand their impact on design decisions.

Sequencing

Engineers and other consultants must see their drawings in terms of the construction sequence as well as the final product. Very large components of mechanical equipment must be brought up to, and placed into, equipment penthouses after they are manufactured. Buildings must remain structurally stable during construction. Once installed, equipment must be accessible for servicing or to remove and replace malfunctioning units.

Architects should review consultants' construction documents with the construction process in mind. The sequence of construction and workability of the scheme throughout the construction process must be considered. Major building elements must fit into place at the appropriate time and without disrupting other ongoing activities.

Scheduling

It may be desirable to order certain components of a building well in advance of their installation. Major HVAC components, large electrical transformers or switchgear, and curtain wall systems are frequently custom made for a particular project. These elements are not generally in a warehouse waiting to be purchased. Even standard catalog items are often manufactured only when specifically ordered and require a significant amount of lead time before delivery.

Architects' consultants must be involved in scheduling to enable major items to be available when needed. Contractors are often selected too late to order long-lead time equipment in a timely manner. One solution is for the owner, on the advice of the architect and consultants, to order equipment directly. When a contractor is subsequently selected, purchase orders are assigned from owner to contractor. Upon delivery, the items are received and installed in the same way as if the contractor had been involved from the beginning.

Fast-track delivery procedures work generally the same way. A project is divided into packages or stages of work, each of which represents a separate prime contract. Starting construction and ordering items before all the construction drawings are completed helps to ensure the availability of products when needed, and tends to control costs during periods of rapid inflation.

Architects must be sure that consultants specify and package items according to proper criteria. Information about a project's ultimate character and configuration may be limited when ordering. Circumstances may change between the time orders are placed, or a construction package let, and the time an item is received or final drawings completed. An architect must work with consultants to determine important features, while leaving other aspects open to inevitable change. This may result in excess capacity in equipment or the need to alter designs to integrate with equipment or items already ordered.

Consultants must also be aware of overall construction schedules and, within these schedules, pertinent installation periods. If a new chiller or cooling tower is required before summer, or a new boiler or heating plant before winter, engineering designs must allow equipment to be built and installed in time. Or, if construction must occur during winter months, structural engineers may want to avoid the use of reinforced masonry, which requires special measures to protect mortar from freezing.

These concerns are especially applicable to renovation projects. An old system may have to be changed to a new one, or an owner may require that a new wing or suite be ready before the old one is abandoned. Some considerations will be apparent from construction documents, while others will not. Architects must be certain that timing has been considered and is realistic.

An owner may rely upon the architect and the architect's consultants for pre-construction services such as cost estimating, scheduling and sequencing, and reviewing ease of construction.

Construction Management

With the advent of fast-track and other sophisticated methods of procurement, some owners have retained *construction managers (CMs)* to provide these pre-construction services.

Construction management may be defined as *activities over and above normal architectural and engineering services, conducted during the pre-design, design, and construction phases, which contribute to the control of time and cost.*

Despite this simple definition, the scope of the CM's functions vary widely from project to project. The CM often joins the project team during the design phases and either remains as an adviser or becomes the constructor as well.

If the CM is an adviser, it acts as the owner's agent and provides the owner with impartial technical advice. The appropriate AIA form is the Owner-Construction Manager Agreement (Document C132).

If the CM is the constructor, there are two appropriate AIA forms: either Document A133 or A134.

Internal Coordination of Consultants' Documents

The architect is the prime professional under contract to the owner, and as such, liable for the consultants' work. Prudent architects, therefore, try to make certain that their consultants provide appropriate levels of professional service. There are some practical limits, however.

One limit is that architects cannot check each consultant's documents for internal consistency and coordination. That is the responsibility of each consultant. If an electrical engineer specifies one type of lighting fixture, the drawings should not show another. Dimensions should be accurate, and drawings and specifications should be coordinated. The AIA Architect-Consultant Agreement (Document C401) specifically requires the consultant to be responsible for coordinating her own work.

When a consulting firm combines more than one engineering discipline, coordination becomes more complicated. For example, structural, HVAC, plumbing, and electrical work may all be done in different departments of the same consulting firm. Generally, a consultant's documents must be made internally consistent by that consultant. Structural and mechanical documents must be checked against each other for conflicts prior to being sent to the architect. Someone in the consulting firm must be responsible for this interdisciplinary checking.

Overall Coordination of Consultants' Documents

Format for Specifications

Specifications prepared by an architect and the consultants are bound together into a Project Manual. All the work of the individual parties must be coordinated to produce a unified document, not a collection of individual parts. To accomplish this, architects establish formats for consultants to follow.

Coordination extends from simple considerations, such as the color of the paper on which the specifications of different consultants is printed, to the format and numbering system used. The consultants' input to bid forms, including instructions to bidders, and to Division One, the general requirements of the specifications, must be established. Overall, each consultant's work must be coordinated with that of the architect and the other consultants.

The architect must require that the consultants participate in the preparation of the requirements of Division One, so that their individual specification sections are appropriately coordinated. The architect is the one professional on a project team with the required perspective to coordinate the many diverse elements of a Project Manual.

Diagrammatic Mechanical and Electrical Drawings

Most construction documents prepared by mechanical and electrical consultants are diagrams or schedules. HVAC drawings show dimensions of ducts. Major pieces of equipment are shown, but other physical conditions are not represented. Duct dimensions may not include the thickness of required insulation. Electrical documents are more diagrammatic. Typically, wiring is indicated in floors or in ceilings, as are home runs to panelboards. Actual conduit locations, however, are usually determined by contractors in the field. Plumbing drawings are less diagrammatic than HVAC and electrical drawings, but pipes and fittings are not drawn to scale. The exact location of piping may be determined by the contractor in the field.

While these different methods of representation are logical, checking and coordination is difficult. Architects can overlay drawings of the various consultants to spot potential conflicts. Even where lines do not cross in such overlays, this does not guarantee adequate clearances, since the diagrams may not be precise enough. Overlay drafting and CAD make it easier for architects to identify and resolve conflicts before they become construction problems.

Serious construction problems may be caused by uncoordinated drawings. Contractors may have problems installing mechanical ducts and electrical conduits within the space actually provided. For example, walls may be framed without adequate space for plumbing lines. Architects must address such potential problems when checking the consultants' documents.

LESSON 1 QUIZ

1. When preparing specifications relating to moisture control, an architect should rely on

   A. manufacturers' literature.

   B. the local building code.

   C. personal judgment and experience.

   D. all of the above.

2. Construction work is divided among the construction trades in accordance with

   I. the specifications.

   II. the drawings.

   III. trade union rules.

   IV. the general contractor's judgment.

   A. IV only **C.** III and IV

   B. I, II, III, and IV **D.** I, II, and III

3. Which type of information is NOT normally contained in construction drawings?

   A. Dimensions

   B. Level of quality

   C. Quantities

   D. Configurations

4. Shop drawings provide the architect with information on which of the following? Check all that apply.

   A. Installation details

   B. Equipment operating data

   C. Color and texture

   D. Standard of workmanship

5. A Project Manual contains

   I. drawings.

   II. general conditions of the contract for construction.

   III. instructions to bidders.

   IV. sample contract administration forms.

   V. the bid form.

   A. II and IV

   B. I, III, and V

   C. II, III, IV, and V

   D. I, II, III, IV, and V

6. All of the following statements about descriptive specifications are true EXCEPT

   A. they describe desired end results.

   B. they make the architect responsible for proper performance of the specified items.

   C. they explain all components of the specified items in detail.

   D. they describe the arrangement and assembly of the components of the specified items.

7. Cash allowance specifications

A. require the contractor to set aside money in the bid to be applied to the cost of an item of work once the level of quality is known.

B. provide that the owner will obtain a discount for construction materials purchased with cash.

C. provide that the owner will be rebated a cash amount for each specified item that the contractor can buy at a discount.

D. establish a fixed price for each unit of material so that a final bid amount can be calculated once quantities are known.

8. When selecting mechanical systems for a project, the architect and mechanical engineering consultant should consider which of the following factors?

I. Skill of the owner's maintenance staff

II. Weight of the equipment

III. Noise characteristics of the equipment

IV. Operating clearances

A. II and IV

B. II, III, and IV

C. III only

D. I, II, III, and IV

9. Reference specifications are

A. comprehensive specification checklists to which an architect refers at the start of each new job.

B. used by architects to refer contractors to federal or other standard specifications that are to apply to work on the project.

C. proprietary systems like MasterSpec® to which architects may subscribe.

D. useful because they do not have to be updated once the initial research is done.

10. All of the following statements about performance specifications are true EXCEPT

A. they include test parameters for the items specified.

B. they make the contractor responsible for proper performance of the specified items.

C. they explain all components of the project.

D. the specified items are in detail.

E. they are best used for new or unusual situations.

11. Which type of information is NOT normally contained in construction specifications?

A. Level of quality

B. Quantities

C. Desired performance

D. Installation methods

12. When evaluating substitutions proposed by the contractors, an architect should consider all of the following factors EXCEPT

A. the terms of the warranty.

B. code compliance.

C. projected maintenance costs.

D. equipment required for installation.

13. Which of the following statements concerning master specifications is FALSE?

A. They tend to reduce repetitive clerical work.

B. They are usually edited by adding appropriate sections.

C. They can be difficult to keep accurate and up-to-date.

D. They make initial draft specifications available early in a project's development.

14. Proprietary specifications

A. are supplied by product manufacturers on loose sheets of paper for architects to bind into project manuals.

B. are usually *open*, but may be *closed* on private projects.

C. contain full technical data on the products specified.

D. contain no trade names.

15. Plumbing drawings have all of the following characteristics EXCEPT

A. they are diagrammatic.

B. they are frequently superimposed on blank architectural floor plans.

C. they graphically show the physical dimensions of pipes.

D. they indicate connections between pipes.

LESSON TWO

QUALITY MANAGEMENT AND COORDINATION

INTRODUCTION

This lesson provides a general overview of issues related to quality. Quality management begins with careful coordination of documents and work with consultants, as discussed in the previous lesson. Here, subheadings create a suggested checklist for developing and coordinating working drawings. More details on these topics may be found in the *Building Design & Construction Systems* study guide.

QUALITY MANAGEMENT

General Concepts

In recent years, manufacturing companies in the United States have become increasingly aware of the need for strict quality control. Similarly, quality management in the service sector has also improved.

In response to an unacceptable number of errors in building design, documentation, and professional practice, many architects have started to implement *quality control* procedures, which involve carefully checking the work (a contract, a set of drawings, a design sketch) before it is distributed to the user (an owner, a contractor, a drafter). However, quality control identifies errors late in the process, when they are costly to correct.

In response to this limitation, *quality assurance* has been implemented to supplement quality control. Basically, quality assurance requires that in designing, documenting, and constructing a building, the proper resources and scrutiny are applied to each part of the process in order to prevent errors before they are made, or at least to correct errors early, before they are compounded. The limitation of quality assurance is that it is segmented rather than holistic.

Because of this limitation, *total quality management* (TQM) was developed. TQM incorporates quality control and quality assurance, but also includes all aspects of service to achieve the goal of customer satisfaction, where the term *customer* is broadly defined to include the client, the user, the public, and the profession.

Structural Integrity

Vertical Forces

In order to design safe buildings, architects must understand the effects of gravity loads on structures. These loads, which are caused by the weight of a building and its contents, act vertically and are transmitted through the structure to the underlying earth or rock. Usually, a geotechnical engineer determines the pertinent soil characteristics and recommends appropriate design criteria and bearing capacities. However, the architect must verify that the required investigations have been made and the construction documents incorporate the geotechnical engineer's recommendations.

One relevant characteristic is the presence and amount of water in the soil. If the hydrostatic pressure exerted upward on a building exceeds the building's weight, it will tend to float. In this case, either the weight of the building must be increased or the hydrostatic pressure reduced.

Soil characteristics may vary within the area bounded by the perimeter of a building, which could result in differential settlement problems. Although a reasonable amount of uniform settlement may be acceptable, differential settlement can cause structural damage.

When wet soils expand, they exert an upward force on the foundation. The geotechnical engineer may recommend various ways to deal with this situation. Frost may create similar problems, requiring foundations to be located below the frost line in order to avoid upward heaving of the structure.

Winds also create vertical loads on a building. They exert uplift forces on the roof, which may detach the roof from the structure. Winds may have a similar effect on eaves, overhangs, and porch roofs, particularly in parts of the country where hurricanes occur. In those areas, building code provisions require continuous structural connections from foundations to roof.

In summary, consider

- potential differential settlement caused by nonuniform soil conditions;
- expansive soils, frost action, or a high water table exerting upward pressure on foundations; and
- wind forces creating uplift on roofs and overhangs.

Lateral Forces

In addition to vertical loads, building structures must also be designed to resist forces acting horizontally, usually called lateral forces. Below the surface of the ground, soils exert lateral pressures on buildings, and subsurface walls must be designed to withstand these pressures. Such walls act like vertical slabs that span between the first floor and the foundation and resist the pressures applied from outside the building. Water in the soil, or the presence of expansive soils, may increase the pressure on subsurface walls. Proper subsurface drainage may alleviate or eliminate the increased horizontal pressures created by water or expansive soils.

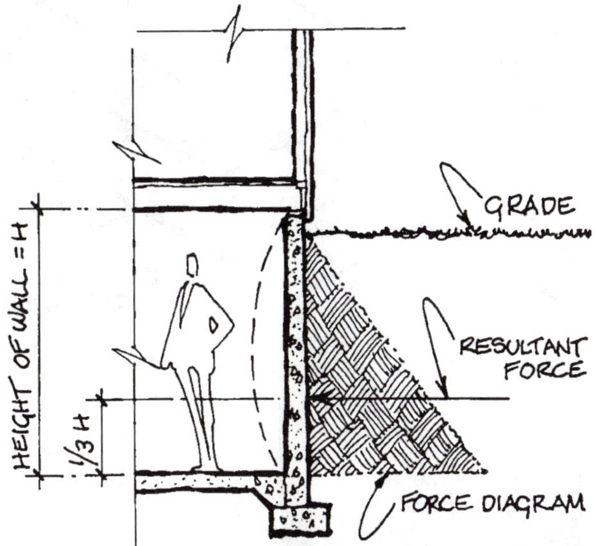

LATERAL FORCE ON SUBSURFACE WALL

Figure 2.1

Structures must also be designed to resist lateral forces acting above the surface of the ground, such as wind or earthquake. Wind forces vary, depending on the building's geographical location as well as its location on a site (surrounded by other structures versus out in the open). The height of a building also affects the intensity of wind forces. Wind may create large overturning moments. However, proper structural design considers such forces and the methods to resist them.

There are both inward and outward pressures on building surfaces exposed to wind, which are not uniform. Thus, the corners of structures must be sufficiently rigid to resist complex racking and twisting forces. Windows and other openings must also be able to resist wind forces. For example, glazing must be detailed to prevent inward pressures from pushing the glass from its frame.

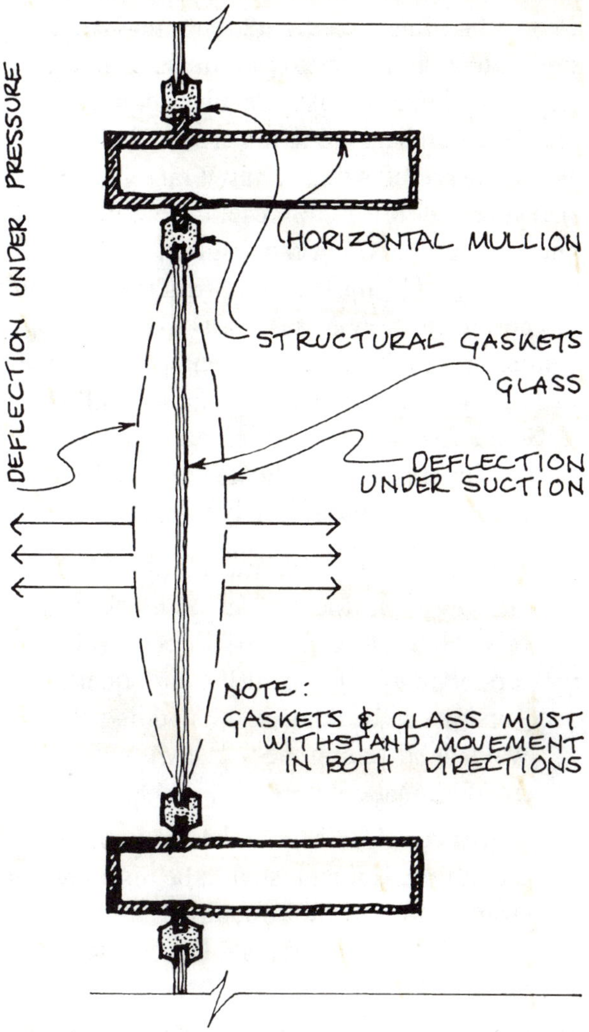

FLUCTUATING FORCES ON EXTERIOR ASSEMBLIES

Figure 2.2

Winds create suction on most roof surfaces and on leeward walls. Proper mechanical anchorage or sufficiently heavy ballast is necessary to hold the roof membrane in place. Window frames must also be capable of holding glass against forces of suction. Furthermore, since the wind direction can change, each wall element must be able to withstand both inward pressure and outward suction.

Earthquakes also create lateral forces in buildings that vary in direction, intensity, and duration. A building's configuration affects its ability to resist seismic (earthquake) forces. Lightweight buildings of regular geometric shape and stiffness tend to perform well. Portions of structures that are not homogeneous in shape or construction react differently to seismic forces, which can result in stress concentrations. The most careful structural design cannot make an inherently weak building configuration safe from seismic damage. However, even proper configurations contain building design details that must be carefully considered during the preparation of construction documents. It is essential to verify the following:

- There are no extreme variations in the strength or stiffness of the perimeter frame or wall structure. Variations tend to create torsion in a building during earthquakes. Torsion can be alleviated by adding shear walls or stiffening the frame with bracing or additional columns.
- A stiff core is not located asymmetrically in a building. If it is, it should be disconnected from the rest of the structure to prevent it from inducing torsional forces into the structure.
- There are no notches or reentrant corners to concentrate stresses. If there are, the building should be structurally separated into two or more uniform elements.
- No *soft story* is created by elevating the structures on *pilotis* at the first story. Bracing columns at the first story or increasing the number of columns to stiffen that level will tend to alleviate the problem.

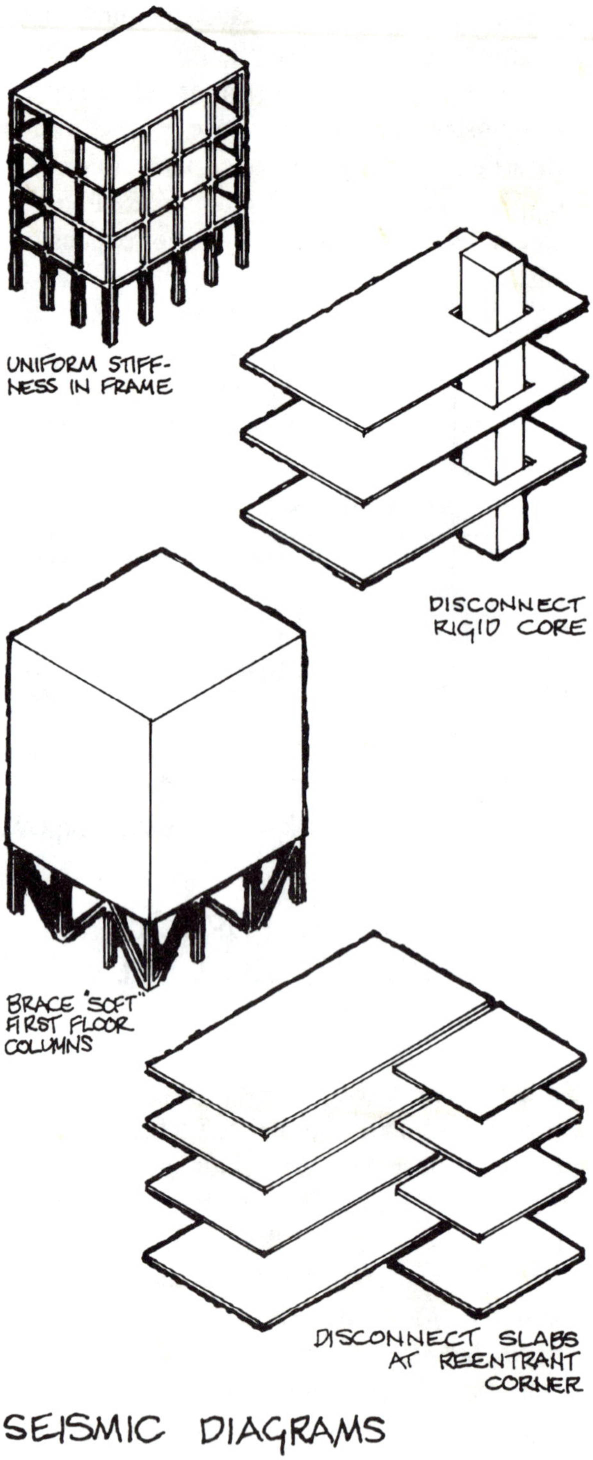

Figure 2.3

In general, for seismic design, buildings should be as uniform in configuration and stiffness as possible. Furthermore, connections should allow a structure to act as a unit as much as possible.

In summary, consider

- providing reinforcing in underground walls to resist lateral forces exerted by soils;
- providing drains to prevent or control hydrostatic pressure against underground walls and slabs;
- detailing windows and other openings to resist both inward and outward wind pressures;
- providing mechanical connections or weight on the roof membrane to resist upward forces created by winds; and
- shaping buildings uniformly with relatively uniform stiffness and strength to resist seismic forces.

Water and Moisture Control

Buildings are exposed to water from various sources. When water penetrates a building, it may damage building materials, destroy the contents, and inconvenience occupants. Consequently, architectural design and detailing must control the effects of water on buildings.

Floodproofing

Buildings must be protected from floods by locating them outside of the 100-year flood plain if possible. However, if that is impossible, buildings can be floodproofed in other ways.

Buildings can be elevated by placing them on fill or stilts, or a combination of the two, to raise floor levels above anticipated flood levels. Dry and wet methods of floodproofing may also be used. Dry floodproofing methods utilize water resistant materials and sealants to prevent flood waters from entering the building and are thus effective in keeping building interiors dry. Wet floodproofing methods allow water to penetrate buildings. When flood waters recede, standing water is pumped out. This method requires mechanical equipment and other water-sensitive materials to be located above anticipated flood levels. The wet method does not require the structural system to resist the force of flood waters. Both dry and wet methods, however, require building superstructures to be firmly anchored to solid ground.

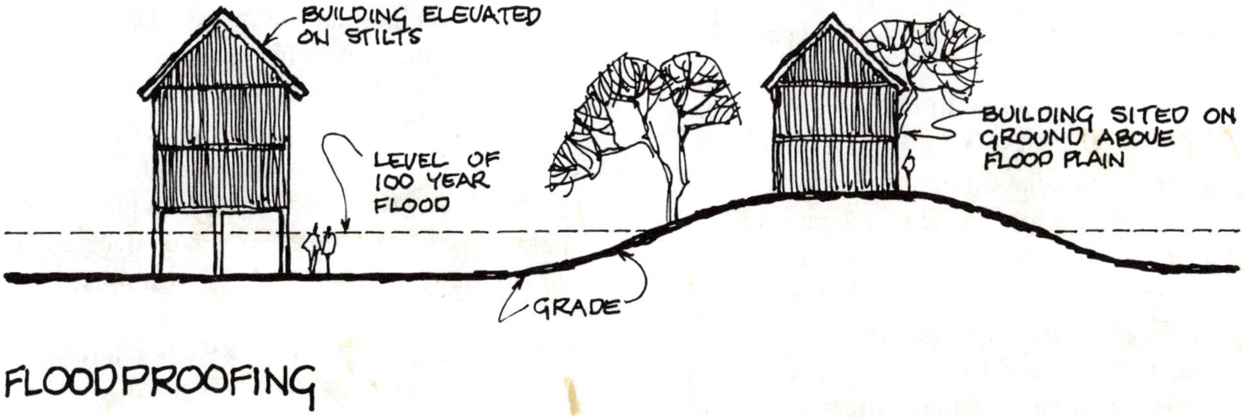

Figure 2.4

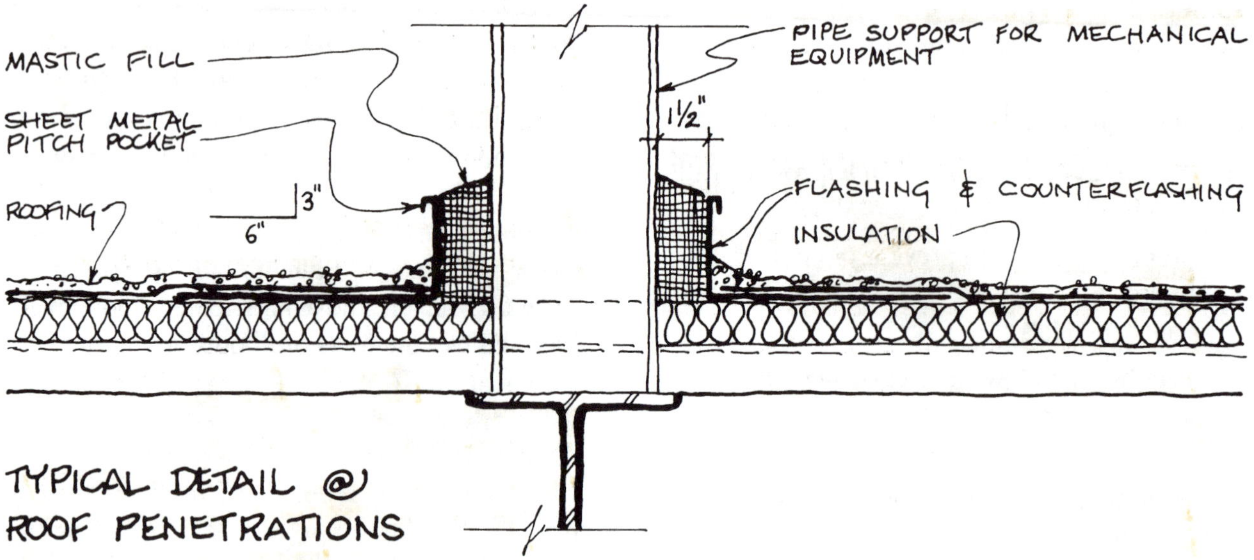

Figure 2.5

Horizontal Surfaces

Roofs are subject to water and moisture penetration. Sloped roofs, where the pitch is 4:12 or greater, have fewer problems than flat, or slightly sloped, roofs. Shingles, tile, or metal may be applied to sloped roofs to produce attractive and generally leakproof installations, since most water flows down the slope. Critical areas where roof planes form valleys, at penetrations of vents and equipment supports, and intersections between walls and roof surfaces, require flashing to prevent water penetration.

Flat surfaces are more difficult to waterproof. All flat roofs, decks, and other horizontal surfaces should have a minimum slope to drain of 1/4" per foot to avoid standing ponds of water. Details at interruptions or terminations of roof membranes are critical. Particular care is required where roof membranes terminate at parapet walls and roof edges. Details should allow water to run off, incorporating physical barriers, like flashing, to prevent water from penetrating walls. Joints between monolithic roof membranes, as well as penetrations or terminations of the membrane, must be at least as waterproof as the membrane itself. Flashing and mastic are often used for that purpose.

Many manufacturers of roofing systems offer a guarantee, or bond, for a specific time period. They may, however, reserve the right to approve the roofing contractor, and to provide a manufacturer's representative to inspect the installation and provide technical assistance. However, manufacturers may disclaim responsibility for parts of roof systems supplied by others, such as flashing materials. Such disclaimers may result in problems because the integrity of the joint between roof membrane and flashing is sensitive and critical to a successful installation. However, many roof systems include all components, including flashing, enabling architects and owners to benefit from a single source of responsibility.

Costly roof failures can be prevented through the use of proper materials and detailing. The cost of roofing is relatively minor compared to overall project costs. Therefore, it is best to design and specify quality roofing systems.

Vertical Surfaces

Exterior walls and openings for windows and doors must also be designed to resist water intrusion. Chemical sealers may be applied to the surface of masonry materials to repel water. Alternatively, impermeable backup materials, such as sheet metal, elastomeric membranes, or building paper, may be used. If moisture does penetrate exterior walls, the backup material will direct it through weep holes back to the exterior. Manufactured window and door assemblies are usually designed to work this way. The component parts resist water penetration and provide positive drainage if water intrudes into the assembly.

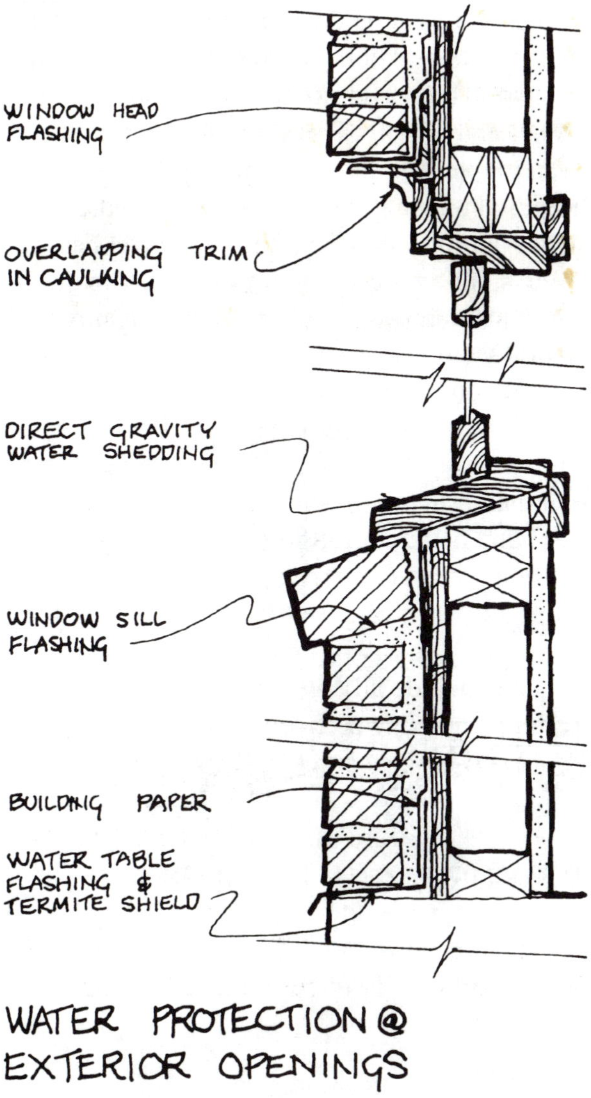

Figure 2.6

Most problems with vertical surfaces occur at joints where wall materials meet and between walls and window or door assemblies. Water driven by gravity, wind, or capillary action may enter a building at these points. Flashing, overlapping materials and trim, and the application of sealants, are used to prevent such intrusion. Sealants may be used alone or in combination with flashing and overlapped joints to increase the degree of protection.

Sealants

Sealants must be specified for their particular application, and such joints must be carefully detailed. This may include anything from setting a door threshold in two beads of sealant to specifying a sealant for complex joints between exterior materials. In the latter case, the joint width must be adequate to accommodate anticipated movement of the materials. Such movement may occur because of thermal expansion or structural deflection. Initial joint widths help to determine the type of sealant required because sealants must be sufficiently elastic to keep joints sealed under all circumstances. The width of a joint also indirectly determines its depth. Joint fillers, such as backer rods, may be used if the depth of a joint is greater than that required for proper sealant performance. The consequences of joint failures are great, and the relative cost of sealants is small. Therefore, it is not advisable to economize when detailing exterior joints and specifying sealants.

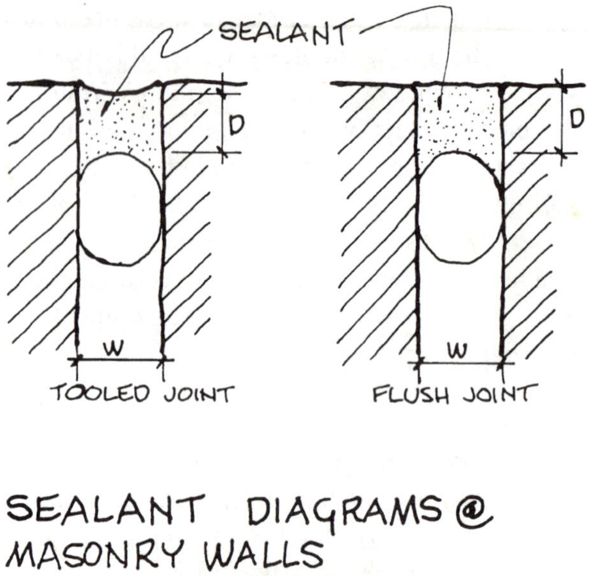

Figure 2.7

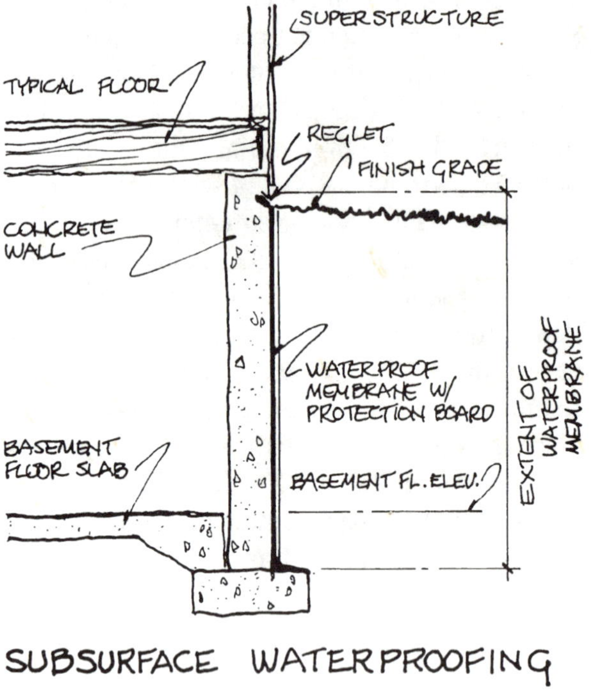

Figure 2.8

Subsurface Water

Water may also enter buildings at, or below, the surface of the ground. The floor elevation of a building should be set high enough so that the adjacent grade can be sloped to provide positive drainage away from the building. Most common building materials, including concrete, which are used below grade are permeable. Consequently, a waterproof membrane must be applied to the outside of a wall below the surface of the ground to prevent water penetration. Membranes must be continuous and penetrations by pipes at footings and other junctures may require mechanical waterstops. In any case, membranes should extend below the level of the lowest floor of a building.

Foundation drains may be required to alleviate hydrostatic pressure created by ground water and to minimize the potential for water penetration. They are made of perforated pipe and are placed in gravel fill areas outside foundation walls. Drains should be placed at the perimeter of a building to collect ground water and carry it away. The gravel fill helps keep pipes free of silt and facilitates water flow to them. If the water table is high, a full gravel base may be required under a basement floor, with drain pipes appropriately spaced. The gravel base course not only aids the flow of water to drains, but it also prevents capillary action against slabs.

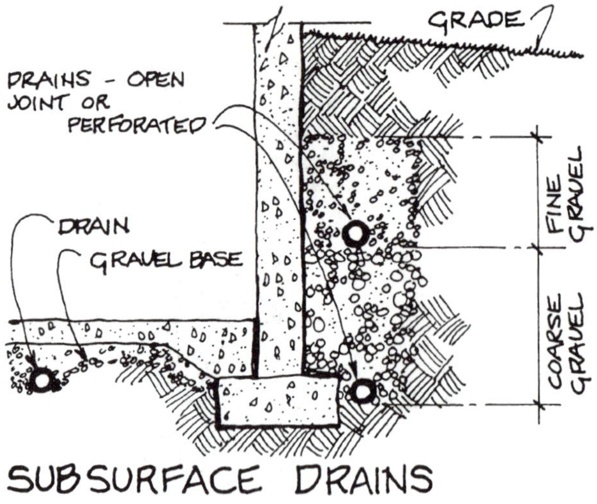

Figure 2.9

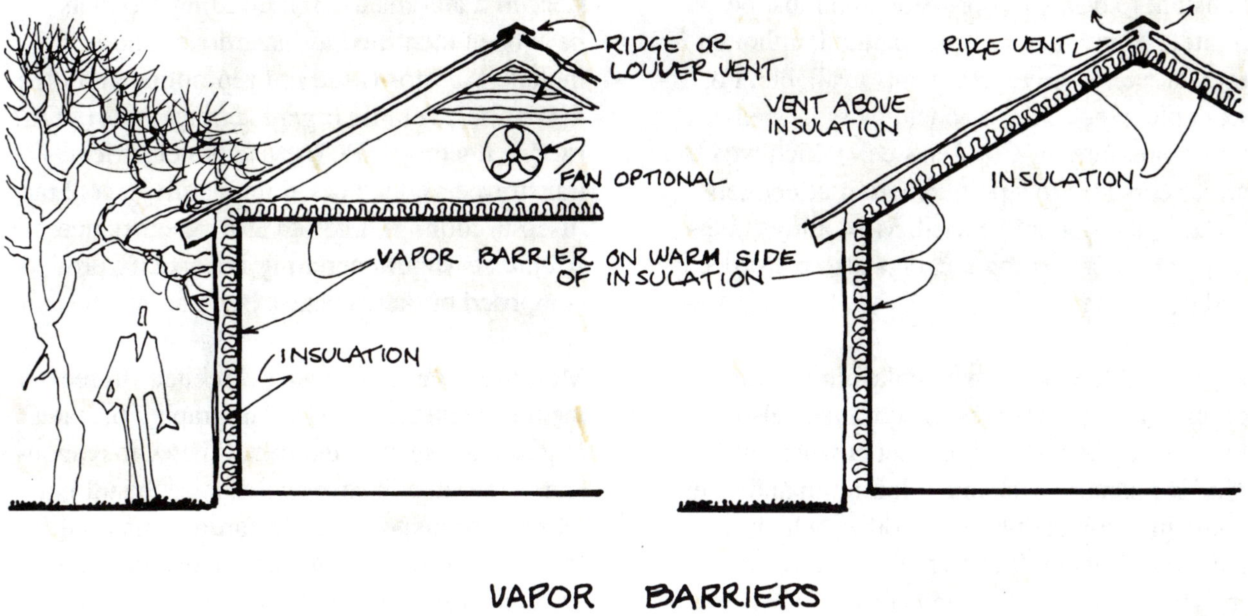

Figure 2.10

Condensation

Moisture that originates within a building must also be considered. Water vapor passing through an interior surface of a wall or ceiling will condense to liquid upon contact with a sufficiently cold exterior wall or roof surface. The resulting condensation may drip onto ceilings or along the inside of walls, damaging interior finishes, blistering paint, and causing mold growth and decay.

A vapor barrier on the warm (interior) side of a wall or roof space usually prevents condensation, since it stops water vapor before it comes in contact with a cold surface or space. Mechanical or natural ventilation, particularly in attic spaces, is used to evaporate moisture that may occur in roof areas.

In summary, consider

- floodproofing by siting structures outside of flood plains, elevating buildings, or using methods of dry or wet floodproofing;
- providing proper flashing details at all roof penetrations, junctions, and terminations;
- using sealants or backing materials at junctions of vertical surfaces;
- using flashing, overlapping materials, and sealants at openings in vertical surfaces;
- selecting appropriate sealants with proper width-to-depth ratios at joints;
- sloping ground surfaces away from buildings for positive drainage;
- providing waterproof membranes and subsurface drains at building perimeters and, in some cases, under floor slabs; and
- providing vapor barriers on the warm side of walls and roofs, combined with adequate ventilation to control condensation.

Environmental Considerations

Architects must select materials that are resistant to decay and corrosion. For example, redwood, cedar, and concrete are naturally resistant to decay and may be left untreated. Most other materials, however, are not naturally resistant to decay or corrosion and must be treated. Many materials are painted, although there are alternative treatments available. For example, wood may be chemically treated under pressure, or Cor-Ten steel, which weathers or corrodes to a point where it effectively protects itself, may be used. Most alloys, plastics, and fluorocarbon coatings also resist decay and corrosion.

Corrosion may occur where flashing meets other building materials. Electrolysis, also known as galvanic action, occurs when two reactive materials, such as aluminum and steel, come into contact with each other, which may lead to corrosion. It is critical to avoid electrolytic galvanic action and corrosion because flashings are vital to making a building watertight. Corrosion may occur even where materials do not touch. Where water passes over one material onto another, electrolysis may occur. Therefore, reactive materials should be separated by mastic, building paper, or other material, to prevent water from creating a galvanic connection between materials.

Insects, particularly termites, present an additional hazard. Termites attack the wood parts of structures. Metal shields are installed to block termites trying to reach those parts. These termite shields must create a continuous barrier to the movement of insects in order to be effective. Added protection may be provided by introducing poison into the soil at the perimeter of a building, under floor slabs, in crawl spaces, and in voids within concrete masonry units. New research is being done to test particle barriers, typically non-toxic particles that are too big for termites to move and too small for termites to move between. Since termites prefer warm, damp, poorly ventilated spaces, they may be controlled by minimizing such conditions.

Certain materials used in building products have been identified as hazardous. These include asbestos (used in fireproofing and pipe insulation), volatile organic compounds (VOCs, used in coatings), PCBs (used in electrical transformers), lead (used in coatings), coal tar (used in roofing), and other toxic substances. Architects should certainly not specify products composed of these materials.

Moreover, the AIA Owner-Architect Agreement (Document B102), Paragraph 7.6, states in part that the architect "shall have no responsibility for the discovery, presence, handling, removal or disposal of...hazardous materials." Professional liability insurance excludes coverage for services performed related to these materials.

The phrase *sick building syndrome* is now used primarily to describe poor indoor air quality. This phenomenon can result from inadequate mechanical ventilation, germ-breeding filters and stagnant condensate reservoirs, off-gassing from fabrics, coatings and backing materials, tobacco smoke, and other toxic substances. Some regulations exist, while others are being developed, to remedy this situation.

In addition, architects are placing greater emphasis on conservation when specifying building materials. Products made from renewable resources, recycled and recyclable materials, and products that require less energy to manufacture, are being specified in response to environmental awareness.

Thermal and Acoustical Control

Thermal Control

Siting, landscaping, the location of uses within a structure, the amount and location of openings in exterior walls, and the selection of building materials, all affect the thermal performance of a building. Certain basic design

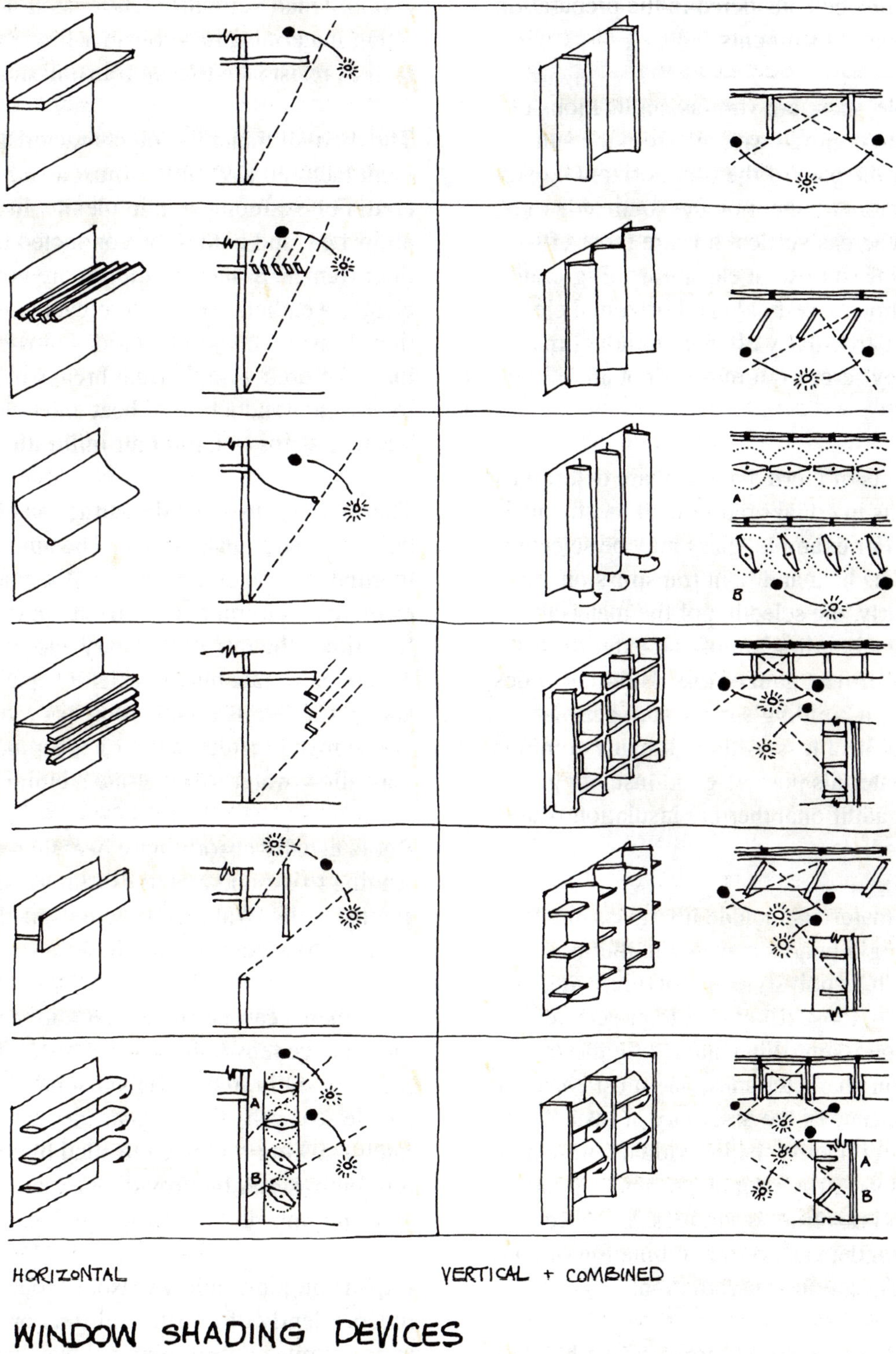
A
B
A
B
A
B
HORIZONTAL
VERTICAL + COMBINED
WINDOW SHADING DEVICES

Figure 2.11

elements must be considered in the preparation of construction documents.

For example, there are various applications of sunshades to control heat gain from the sun. Because of the path of the sun, horizontal overhangs are more efficient on the south side of a building, whereas vertical fins are more efficient on the east or west elevations. Egg-crates, a combination of vertical and horizontal elements, are particularly efficient in hot climates because they screen out most, if not all, of the sun's rays.

Many aspects of thermal control are determined by architects in collaboration with mechanical engineers. For example, glass may be selected because of its heat and light transmission qualities. Similarly, the selection of the materials comprising a building's roof and exterior walls must consider heat gain or loss. Building codes may specify a building's maximum permissible annual heat loss. Although many common building materials possess good insulating properties, additional thermal insulation is usually required.

Insulation materials include fiberglass blankets and batts, rigid boards made from thermoplastics such as polystyrene, polyurethane, or isocyanurate, loose fill of perlite or vermiculite, and spray-on foam. When analyzing thermal transmission characteristics, one must consider the performance of the assembly of insulating materials, not merely the individual component parts. Air space, for example, provides good insulation in material assemblies since heat transmission depends on a combination of conduction, convection, and radiation.

Some insulating materials are combustible. For example, certain rigid insulation boards may create a fire hazard if they are exposed to air. Others may be more flammable installed horizontally than vertically. The selection of insulating materials must consider fire resistance, as well as resistance to heat transmission.

The thermal transmission characteristics of manufactured assemblies must also be considered. For example, certain metal windows may allow heat and cold to be conducted through their frames. Some assemblies are not sufficiently weather-stripped to prevent air infiltration. Most metal window today, however, are manufactured with thermal breaks in the frames to prevent conduction of heat and cold and with weather-strips to inhibit air infiltration.

Thermal expansion and contraction cause building materials to move. The amount of thermal expansion depends on the material. For example, aluminum expands at approximately four times the rate of masonry and twice that of concrete. Metals may buckle or separate from anchors if details do not allow for movement. Loose metal results in poor appearance and may allow water to penetrate a building.

Roofs usually absorb more heat than walls, tending to expand and to exert horizontal forces against walls. Walls must be designed and detailed to accommodate these forces.

Movements caused by temperature changes may induce tensile forces that result in cracking, especially in materials that have low tensile strength, such as masonry and concrete. Such cracking can be controlled by using steel reinforcing, and by providing expansion and control joints at the appropriate locations.

Expansion joints allow masonry to move independently of a structural steel or concrete frame, while control joints relieve forces that may build up within individual masonry elements. Expansion joints should be located at regular intervals along walls, at major changes in wall height or thickness, at columns and pilasters, at wall openings, near wall intersections, and near junctions of walls in L-, T-, or U-shaped buildings. Masonry must remain structurally stable and watertight at control and expansion joints.

Concrete and plaster are similar to masonry and react in much the same way. Some cracking will inevitably occur, most of which does not affect the building's structural integrity. However, cracks are unsightly and may allow water to enter the building. Therefore, architects attempt to control cracking by providing weakened joints (control joints) in which the cracks are likely to occur, in preference to the random cracking that would otherwise take place.

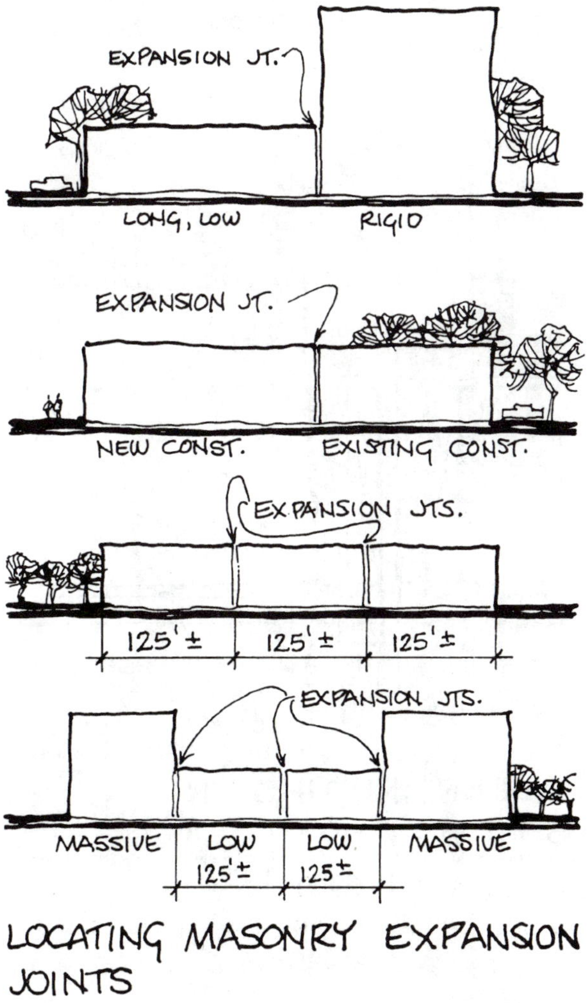

Figure 2.12

In summary, consider

- sunshades appropriate to a building's solar orientation;
- insulating materials that are integral parts of construction assemblies;
- manufactured assemblies with appropriate thermal control features;
- details that accommodate thermal expansion and contraction; and
- expansion and control joints for masonry, concrete, and plaster.

Acoustical Control

In analyzing the transmission of sound, we must consider three factors: source, path, and receiver. Each factor may be modified or controlled in the design of buildings.

The level of outside noise which a *receiver* can tolerate may be expressed as the Preferred Noise Criteria (PNC), for which applicable standards are published. Building construction must reduce noise generated by outside sources to a tolerable level. There are two basic ratings of assemblies of materials: STC and IIC. Sound Transmission Class (STC) is the method of rating the acoustic efficiency of various wall and floor systems in isolating airborne sound transmission. The higher the STC rating, the more efficient the construction. STC ratings are not appropriate in considering the effect of impact noises on floor assemblies caused by foot traffic or machinery. For that purpose, Impact Isolation Class (IIC), also referred to as Impact Noise Rating, is used. This is a simple numerical rating developed by the Federal Housing Administration to estimate the impact sound isolation performance of floor/ceiling systems.

In selecting assemblies and developing details, an architect must be familiar with the following principles of acoustical design: lightweight, porous materials are effective for sound *absorption*, while heavy, impervious materials are effective for sound *isolation*; they cannot be used interchangeably. Penetrations at walls, floors, and ceilings must be minimized, and openings, including those at the tops and bottoms of partitions, should be airtight. Impact noises, such as those cause by foot traffic on floors, can be minimized by the use of resilient flooring. Finally, sound waves are transmitted over long distances through structural frames. Discontinuities at junctions of walls, partitions, floors, and ceilings, therefore, are valuable sound control devices.

SOUND ABSORPTION COEFFICIENTS OF MATERIALS @ 500 HERTZ

Acoustical plaster (average)	0.50
Brick, exposed and unpainted	0.03
Carpet, heavy, on solid surface	0.14
Concrete block, unpainted	0.31
Painted	0.06
Concrete, unpainted	0.02
Wood floor	0.10
Glass, ordinary window	0.10
Gypsum board, 1/2" on 2 × 4s @ 16" oc	0.05
Plaster on lath	0.06

Table 2.1

Sound absorption within spaces must also be considered. The Noise Reduction Coefficient (NRC) is a rating that compares absorptive capabilities of acoustical materials. See Table 2.1. Echo and reverberation control is critical in the design of spaces. For a given use, there is an optimum balance of absorption and reflection achieved largely through the selection of appropriate materials. Porous, fibrous finish materials are more effective in absorbing sound than hard, reflective surfaces. Carpets are used to reduce impact noise from foot traffic and also absorb airborne sounds. Sound absorbing materials may be applied to walls, although less effectively than when applied to ceilings. Acoustical tile is an effective absorptive material for use at ceilings. Light fixtures and air diffusers, which reflect sound, must also be considered in the design of ceilings.

In summary, consider

- sound source, path, and receiver;
- Sound Transmission Class (STC) and Impact Isolation Class (IIC) ratings of walls and floor assemblies that equal or exceed required Noise Reduction (NR) criteria;
- sealing joints, openings, and other penetrations in walls and ceilings/floors between spaces; and
- sound absorption within spaces.

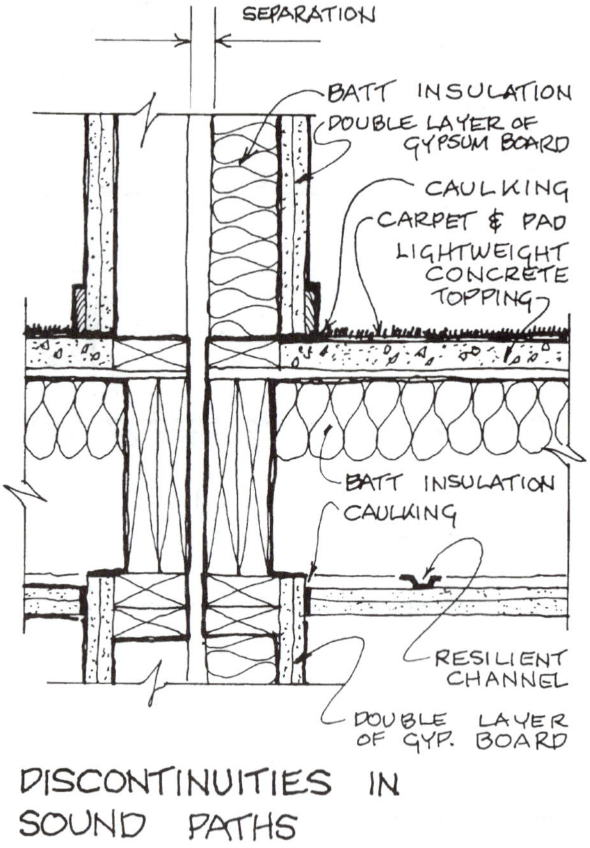

Figure 2.13

Durability and Maintenance

Building materials should be both durable and easily maintained. The appropriate level of durability or required maintenance may vary, depending on the building's occupancy (use), its required life span or that of its component parts, its available level of required maintenance, accessibility, and other factors. Trade-offs between initial and long-term costs may have to be made. For example, it may be more economical to replace a system or component every 10 years than to purchase and install one that will last for 20 years. This type of analysis is referred to as *life cycle costing*. Sustainability concerns should also be taken into account.

Dimensional and Finish Tolerances

Architectural details must be able to accommodate construction tolerances and imprecision of workmanship. Consequently, details must allow for some dimensional adjustments, such as slotted holes at connections, shim spaces, and trim applied over imperfect joints. The problem is to determine how much dimensional tolerance is necessary.

Dimensional tolerance depends on both the quality of materials used and the stage in the construction process. Less expensive or bulky materials have poorer dimensional quality and stability. For example, concealed concrete and rough carpentry work do not require the same level of precision as tile work or cabinetry.

In some cases, a very high level of precision may be required. For example, walls at property or setback lines must be located very accurately. Spaces or recesses to receive prefabricated or manufactured equipment of a specific size must be precisely constructed. Precision may be so critical that the contractor may have to make measurements of the completed construction before ordering required equipment. High levels of precision and low tolerances, however, usually result in higher costs and increased construction time.

Trade associations, such as the American Institute of Steel Construction (AISC) and the American Concrete Institute (ACI), publish construction standards applicable to their materials. An example is in Table 2.2. These standards provide useful rules of thumb.

However, if an architect refers to these standards in specifications, the necessary relevant tolerances must also appear in the detail drawings. Otherwise, conflicts may result. For example, steel construction that is meant to be at the same elevation as adjacent concrete construction may not align exactly, even if both materials are within the requirements of their individual tolerances.

In summary, consider

- there are dimensional variances within materials;
- the accuracy and precision of building construction is limited by the ability of materials and workers to perform within their inherent limitations;
- high levels of precision may be costly and unnecessary; and
- where great precision with minimum tolerances is mandatory, it should be specified.

Aesthetics

Specifications may require a contractor to submit shop drawings, product data, or samples. An architect can control aesthetics, in part, by reviewing these submittals. Shop drawings show precisely how a contractor intends to install a specific item of equipment or material. Product data are standard information sheets that define and describe the physical and operational characteristics of items of equipment. Samples may show the proposed color, texture, or finish of materials. Samples approved by an architect become the standard against which actual installations are compared.

Where an installation is the first use of an untested material or system, or where many repetitive elements occur, a mock-up of that element may be required. Full scale mock-ups are generally not within the architect's or owner's budget during the preconstruction phases. On occasion, a material supplier or fabricator may be willing to produce such a mock-up. However, more commonly, the specifications may require the successful bidder to construct a full scale mock-up at the job site or at the factory. Visual mock-ups, such as brick panels, may be used to select brick shades and mortar colors, and to demonstrate the contractor's ability to construct the brickwork as detailed and specified. Performance mock-ups, such as aluminum and glass curtain walls, may be tested to verify that the assembly can withstand specified wind loads, seismic loads, and rain and wind infiltration criteria. The results of these mock-ups may lead to refinements in the detailing and methods and sequence of construction. However, during the construction phase, major design changes are not feasible without the owner's agreement to issue a change order, if requested by the contractor.

A mock-up may cost ten times as much as the same amount of work on the job. Since contractors include such expenses in the contract sum, architects must be selective, requiring mock-ups only when necessary.

The AIA General Conditions (Document A201) gives the architect the right to reject construction work that does not comply with the intent expressed in the construction documents. Usually such rejection must be based on objective standards, such as field tests. For decisions concerning aesthetics, however, the architect is the final authority.

Compliance with Codes and Regulations

Fire Protection

The purpose of licensing architects is to assure that architects are competent to protect the public's health, safety, and welfare. An architect must be able to design buildings that are structurally safe and meet the requirements for fire and life safety. The safety requirements specified in building and fire protection codes are minimum standards only, and do not necessarily provide the best solution in every case.

Architects should understand the following basic concepts of fire protection in building codes:

- A fire in one building should not damage another building.
- A building's construction should be able to withstand the effects of fire for a specified period of time.
- The more hazardous the use, the more protection the building's construction must provide.
- As the size of a building or the number of its occupants increases, the fire protection, fire separation, and required means of egress also increase.
- Occupants should be provided with a direct and safe means of escape from a building in the event of fire.

TYPICAL DIMENSIONAL TOLERANCE IN CONCRETE CONSTRUCTION

Formwork Tolerances for standard reinforced concrete buildings up to approximately 100 feet tall.

Variation from plumb

1. *In the lines and surfaces of columns, piers, and walls:*
   1/4" per 10', but not more than 1".
2. *Exposed corner columns, control joint grooves, and other conspicuous lines:*
   In any bay, or 20' max .... 1/4".
   In 40' or more .... 1/2".

Variation from level

1. *In slab soffits, ceilings, and beam soffits:*
   In 10' .... 1/4".
   In any bay, or 20' max .... 3/8".
   In 40' or more .... 3/4".
2. *For exposed lintels, sills, parapets, horizontal grooves, and other conspicuous lines:*
   In any bay, or 20' max .... 1/4".
   In 40' or more .... 1/2".

***Variation of linear building lines** (from established position in plan and related position of columns, walls, and partitions):*
In any bay, or 20' max .... 1/4".
In 40' or more .... 1/2".

Variation of sizes and locations
Sleeves, floor openings, and wall openings .... 1/4".

Variation in
Cross-sectional dimensions of columns and beams; thickness of slabs and walls:
Minus .... 1/4".
Plus .... 1/2".

Footings

1. *Variation in dimensions in plan:*
   Minus .... 1/2".
   Plus .... 2".*
2. *Misplacement or eccentricity:*
   2% of the footing width in the direction of misplacement but not more than 2".*
3. *Reduction in thickness:*
   Minus .... 5%.

**Concrete only, not reinforcing bars or dowels.*

Variation in steps

1. *In a flight of stairs:*
   Rise .... 1/8".
   Tread .... 1/4".
2. *In consecutive steps:*
   Rise .... 1/16".
   Tread .... 1/8".

Table 2.2

Building codes regulate how these basic concepts are applied to building design. There were three model building codes used in the United States: the BOCA code, used primarily in the east; the SBCCI (Southern) code, used primarily in the south; and the Uniform Building Code (UBC), used primarily in the west. In 2000, the International Code Council began publishing the International Code Series, intended to replace the separate codes used in various regions. The International Building Code has been or is in the process of being adopted nationwide. A model code must be adopted by a state or local jurisdiction in order to be enforced. Some jurisdictions amend the model codes and several write their own.

An outline of the basic steps in determining code compliance for a building design follows:

1. **Construction documents:** Determine compliance with the requirements for construction documents.
2. **Use group:** Determine the appropriate use group classification of the building.
3. **Height and area:** Determine the type of construction required based on the building use group and the height and area limitations.
4. **Type of construction:** Determine compliance with the required type of construction of the building by the building materials used and the fire resistance rating of the building elements.
5. **Siting:** Determine the location of the building on the site, including separation distances from lot lines and other buildings. Determine exterior wall and wall opening requirements based on proximity to lot lines and adjacent buildings.
6. **Fire performance:** Determine compliance with detailed requirements for fire resistance and fire protection systems.
7. **Interior environment and design:** Determine compliance with special use and occupancy requirements, *means of egress* requirements, accessibility requirements, and interior environment requirements.
8. **Exterior envelope:** Determine compliance with exterior envelope requirements, as well as energy conservation.
9. **Structural performance:** Determine compliance with structural requirements and building material requirements.
10. **Building service system:** Determine compliance with various building service system requirements.

Once the use, massing, and location on the site of a building have been determined, the architect can consider options for the type of construction. For example, an office building could be constructed of a type of construction that uses a combination of unprotected noncombustible and combustible construction. The same building could also be constructed of a more restrictive type of construction, using only protected noncombustible materials. Note that *protected* means having fire-resistive protection, such as sprayed fire-proofing or concrete around structural steel construction, while *combustible* refers to a material that can ignite and burn, such as wood. Conversely, for a similar building, but which houses a more restrictive use, such as a hospital, a more restrictive construction type would be required. In general, the higher the construction type, the more costly the building. The basic allowable areas, heights, and number of stories can be increased by providing other features to increase the building's fire resistance, such as an automatic fire sprinkler system, setbacks from adjacent buildings and streets, and fire separation assemblies within the building.

Once the use group, allowable area, height, number of stories, and type of construction have been determined, the next major safety consideration is egress. The building code includes the requirements for calculating the number of occupants, the number of exits for each room, floor, and building, and the arrangement and width of each component of the *means of egress* (aisle, door, corridor, stair). Dimensional requirements for stair treads and risers, guardrails and handrails, and vertical clearances are also found in the code.

Additionally, specific drawings should be checked as follows:

1. **Site Plan**
   a) Is there proper access to streets?
   b) Are building separations adequate? All objects, such as other structures, that might expose a building to fire must be located and separation distances correlated with fire resistance ratings of exterior walls. Walls with windows and other unprotected openings may require additional treatment.
2. **Plans and Sections: Height, Area, Construction**
   a) Verify that ceiling heights are permissible. Check heights of habitable spaces below grade.
   b) Verify that the height of the building, in feet or number of stories above grade, is within allowable limits.
   c) Verify that the floor area of each floor is within permissible limits. Area limitations may be increased under certain conditions.
   d) Verify fire resistance of all building components.
3. **Plans, Sections, and Elevations**
   a) Verify that openings are protected where they are exposed to property lines, adjacent buildings, interior courts, or other openings in the same wall, above and below.
4. **Plans and Sections: Egress**
   a) Confirm the number of occupants anticipated to occupy each floor of a building. This number may be based on an actual count, such as the number of seats in an auditorium, or on an assumed number of occupants per unit of floor area.
   b) Verify the correct number, width, and arrangement of exits.
   c) Verify that distances to exits are within allowable limits.
   d) Verify that required exit door hardware is specified.
   e) Confirm the adequacy of aisles, corridors, and horizontal exits.
   f) Check that all stairways are properly located and enclosed, and that they provide a legal means of egress.
5. **Plans**
   a) Verify that different occupancy types are separated by properly rated fire walls, etc.
   b) Verify that elevator, escalator, and vertical shaft enclosures are properly fire rated.
6. **Specifications**
   a) Determine that finishes and trim meet requirements for proper classifications. Installation methods must meet code requirements for control of smoke emission and limitations on smoke toxicity.
7. **Miscellaneous**
   a) Verify that roof coverings have required ratings.

Life Safety

The primary objective of building codes is the protection of life, health, and property. Codes published by the National Fire Protection Association (NFPA) are designed to protect human life. The NFPA code that most directly affects the architectural design of buildings is NFPA 101, the Life Safety Code, which is also a model code. It may be adopted and enforced by state or local officials. In many cases, the egress portions of building codes are based on NFPA 101.

NFPA's Life Safety Code Handbook, the companion to NFPA 101, cites the Code's goals, which form a good checklist in preparing or reviewing a set of construction documents:

- To provide for adequate exits without dependence on any one safeguard
- To ensure that construction is sufficient to provide structural integrity during a fire while occupants are exiting
- To provide exits that have been designed to the size, shape, and nature of the occupancy
- To ensure that the exits are clear, unobstructed, and unlocked
- To ensure that the exits and routes of escape are clearly marked so that there is no confusion in reaching an exit
- To provide adequate lighting
- To ensure early warning of fire
- To provide for backup or redundant exit arrangements
- To ensure the suitable enclosure of vertical openings
- To make allowances for those design criteria that go beyond code provisions and are tailored to the normal use and needs of the occupancy in question

Barrier-Free Provisions

For the past 20 years or so, federal and state regulations concerning accessibility to buildings by persons with disabilities have been developed. These regulations are typically based on the American National Standards Institute (ANSI) Standard 117.1. Once adopted, these regulations are enforced like any other code requirement; that is, plans must be reviewed by a code official before a building permit is issued, and the work must be inspected before a certificate of occupancy is issued.

In 1990, the Americans with Disabilities Act (ADA) was enacted. In contrast to building codes, the ADA is civil rights legislation. A person denied equal access to a building may sue the owner of the building if the owner does not make the building accessible. Code officials do not enforce the ADA. However, architects are exposed to professional liability if they ignore the accessibility guidelines included in the ADA. State and local code officials, as well as the model code organizations, are currently studying the feasibility of modifying their codes to conform to the ADA.

Design criteria for persons with disabilities usually include requirements for

- walks, ramps, stairs, and circulation routes;
- parking space size, number, and location;
- site furniture;
- tactile, visual, and audible warning signals;
- overhanging and projecting objects;
- entries, doorways, and hardware;
- elevators;
- floor surfaces;
- water closets, drinking fountains, toilet stalls, urinals, lavatories and mirrors, dispensers, and receptacles; and
- electrical systems, outlets, switches, and telephones.

Architects must determine which of the various codes and standards apply to a particular project. As with other aspects of architectural practice, written records in this regard should be kept and copies distributed to the various interested parties.

In summary, consider the following:

- Codes are intended to provide minimum standards of building design for the protection of persons and property.
- The greater the risk, the greater the level of protection required.
- In order for codes to be enforced, they must be adopted by an enforcing agency of the federal, state, or local government.
- The Americans with Disabilities Act (ADA) is federal civil rights legislation, not a building code.

LESSON 2 QUIZ

1. Select the correct statement.

   A. Caulking, when properly applied, may be relied upon to prevent water penetration.

   B. Durability is generally the primary consideration when selecting finish materials.

   C. Samples are not usually required if the product is manufactured.

   D. As a last resort, an architect has the right to reject work that does not conform to the contract documents.

2. In general, construction items may be manufactured, shop fabricated, or job fabricated. Match each item with its most common method of fabrication.

   A. Concrete—job fabrication
   Elevators—shop fabrication
   Light fixtures—manufactured

   B. Plumbing fixtures—manufactured
   Cabinets—job manufactured
   Structural steel—shop fabricated

   C. Plumbing—manufactured
   Stairways—manufactured
   Electrical equipment—manufactured

   D. Air-conditioning equipment—shop fabricated
   Air-conditioning ductwork—shop fabricated
   Railings—shop fabricated

3. All of the following require details which allow for movement caused by thermal changes, EXCEPT

   A. concrete paving.

   B. wood framing.

   C. sheet metal rain gutters.

   D. metal storefronts and curtain walls.

4. The principal consideration in acoustic control is the sound absorptive quality of the

   A. floor.

   B. walls.

   C. ceiling.

   D. roof.

5. Select the correct statement concerning shop drawings.

   A. They are prepared exclusively by the general contractor and submitted to the architect for review.

   B. In case of discrepancy between the shop drawings and the contract documents, the shop drawings will prevail.

   C. They are usually prepared by fabricators and submitted to the general contractor and architect for review.

   D. An architect must carefully check dimensions and quantities shown on shop drawings, because *approving* the shop drawings implies that dimensions and quantities are accurate.

6. Which of the following tend to reduce the transmission of vibration and/or sound?

   I. Installing isolators under air-conditioning equipment
   II. Increasing the mass of the base under air-conditioning equipment
   III. Increasing the cross-sectional area of air-conditioning ducts
   IV. Lining the inside of air-conditioning ducts
   V. Using a resilient suspension system for ceilings

   A. I, II, IV, and V
   B. I, II, III, IV, and V
   C. II, III, and V
   D. I, III, and IV

7. List the following in order of required accuracy, from most to least accurate.

   1 I. Location of property lines
   2 II. Location of column center lines
   5 III. Dimensions of concealed concrete
   4 IV. Dimensions of exposed concrete
   3 V. Cabinet work

   A. I, II, V, IV, III
   B. II, I, V, IV, III
   C. I, V, II, III, IV
   D. V, I, II, IV, III

8. Which of the following are created by winds acting on buildings? Check all that apply.

   A. Moments
   B. Lateral forces
   C. Vertical forces
   D. Torsion

9. One of the acoustical ratings given to building materials and assemblies is Sound Transmission Class (STC). This rating is concerned with

   A. impact noises.
   B. speech isolation.
   C. amplification.
   D. reverberation.

10. To prevent electrolysis, an architect should

   A. specify only galvanized materials.
   B. specify ground fault interrupters (GFI) on all convenience outlets.
   C. separate all reactive metals with building felt or other non-conductive material.
   D. avoid the use of aluminum that will be exposed to the weather.

11. Life safety codes

   A. are primarily intended to protect property from fire damage.
   B. are used in some jurisdictions instead of building codes.
   C. specify the location and number of exits in a building.
   D. consider all factors necessary to produce safe buildings.

12. Mock-ups are most appropriately specified when which of the following conditions exists?

I. The project will incorporate the first installation of a new system or assembly.

II. The system, assembly, or method of installation is very innovative.

III. The time available for construction is short.

IV. The system or assembly is to be used repetitively.

V. The project has a relatively tight budget.

A. III only

B. IV only

C. I, II, and IV

D. I, II, III, and IV

13. Select the most correct statement. Building codes

A. suggest ideal solutions to technical construction problems.

B. are intended to insure that buildings will be structurally safe until the arrival of the fire department.

C. are concerned only with assemblies of construction materials and systems, whereas zoning codes deal with the uses of a building.

D. are concerned with containing fires so that a fire in one building will not damage other buildings.

LESSON THREE

3

DOCUMENTATION AND SUSTAINABILITY

INTRODUCTION

Architectural Process

After the planning process has been concluded, and the site has been selected, the architectural team will begin to focus on the project, including the project's buildings and related infrastructure.

Traditionally, the architect is faced with four components to every design decision:

1. Cost
2. Function
3. Aesthetics
4. Time

The new, sustainable ecological paradigm adds one additional component to form a pentagon of concerns.

5. Sustainability

The ingredients of the normal process have been discussed previously, but the new ingredient, sustainability, changes the meaning of all these pieces of this architectural process.

1. *Cost*

As architects put together budgets for their clients, they are always concerned with the first costs of the design components—the initial cost to purchase and install the design element.

Sustainable design has made the economic decision process more holistic. The decision to select a design element (such as a window, door, flooring, exterior cladding, or mechanical system) is now concerned with the "life-cycle" cost of the design.

1.1 *Life-Cycle Costing*

Life-cycle costing is concerned not only with the first cost, but the operating, maintenance, periodic replacement, and residual value of the design element.

For example, two light fixtures (A and B) might have different first cost: Fixture A has a 10 percent more expensive first cost than B. But when the cost of operation (the lamps use far less energy per lumen output) and the cost of replacement (the bulbs of A last 50 percent longer than the bulbs of Fixture B) is evaluated, Fixture A has a far better life-cycle cost and should be selected.

In this kind of comparison, the life-cycle cost may be persuasive; the extra cost of Fixture A may be recovered in less than two years due to more efficient operation and replacement savings.

In this situation the architect justified Fixture A to the owner, who benefits from more energy efficient lighting that continues to save the owner operating costs for the life of the building.

1.2 *Matrix Costing*

While designing a typical project, the architect faces numerous alternate decisions, a process that may be both intriguing and complex.

In nearly all projects, there is an established budget and program (including all the owner's functional requirements). The architect must balance the functional issues with the budgetary and aesthetic issues.

Sustainable design adds an ingredient to this matrix of decisions that may actually help the composition.

For example, decisions that allow the improved efficiency of the building envelope, light fixtures, and equipment may permit the architect to allow the engineer to reduce the size of the HVAC system, resulting in a budgetary trade-off. The extra cost of the improved envelope may be economically balanced by the diminished cost of the mechanical system.

This type of economic analysis, which evaluates cost elements in a broad matrix of interaction, is a very valuable architectural skill. The ability to understand the interaction between different building systems in a creative and organized fashion can differentiate an excellent architectural design from a simply adequate one.

2. Function

Functionality is one of the primary standards of architectural design. If the building doesn't perform according to the client's needs, then the building design has failed.

Sustainability adds a facet to functionality that even the owner may not initially appreciate.

As previously mentioned, life-cycle costing will affect the decisions in which elements are finally selected to form the final design. However, the search for sustainability may increase the dimensions of functionality.

Years ago, the design element could perform at the highest level regardless of its impact on the environment or energy use.

The fact that many industrial and residential buildings are operating much more efficiently now than in 1960 is evidence that the building design and construction profession is learning how to tune buildings to a higher degree of energy operation. But, with diminishing natural resources and increasing pollution of the environment, even more efficient design is necessary.

Today, architects will include sustainability in the selection of optimal functional design components.

For example, a roof system must be able to withstand a variety of weather conditions, be warranted to be durable a minimum of years, be able to be applied in a range of weather conditions, and have a surface with reflectivity that does not add to the urban heat effect.

3. Time

The schedule of a project is always a difficult reality of the design process. Time is a constraint that forces a systematic and progressive evaluation of the design components.

The sustainable component of the architectural process may add to the amount of time the architect will spend on the research for the project.

The architect may spend more time on a sustainable design with the result being a more integrated, sustainable project.

4. *Aesthetics*

The aesthetic of a project is the combination of the artistry of the architect and the requirements of the project.

Sustainable design has the reputation of emphasizing function and cost over beauty and appeal.

It is the architect's responsibility to keep all the design tools in balance. A project without aesthetic consideration will fail the client, its user, and the potential client that may be deciding between the normal design process and one that considers a broader, integrated, sustainable approach.

5. *Sustainability*

The fifth point in the calculus of the architect is a new component that leads to a new, holistic evaluation of the design process. Because a piece of any living element must be part of the cycle of nature in order to survive, all manmade elements should now consider the mantra, "do no harm and be designed to be integrated within the cycle of all living things."

Architectural designs should create by-products, that can be recycled with other natural elements and not cause depletion of natural resources necessary for the health of future generations.

Sustainable designs should have six goals:

1. Designs that use less
2. Designs that recycle components
3. Designs that have components that are easily recyclable
4. Designs that have components that are fully biodegradable
5. Designs that are efficient
6. Designs that give back to the earth, in the form of energy

Evaluation Standards for Sustainable Design

How can we objectively evaluate the quality of a sustainable project?

The architect is faced with responding to many standards and regulations in the course of assembling a design. Building codes, life safety standards, fire code, zoning regulations, and health and sanitary regulations are some of the many municipal, state, and federal standards that an architect must evaluate in the course of any project.

Sustainability is a new filter for the design process and there are several organizations that have offered checklists for evaluating the inclusion of environmentally sensitive elements into the project.

One of the measures of performance is Leadership in Energy and Environmental Design (LEED)®, which is sponsored by the U.S. Green Building Council (USGBC). This standard was developed in the 1990s by a consortium of building owners, architects, suppliers, engineers, contractors, and governmental agencies.

The goal of LEED and similar environmental design standards is to introduce new sustainable approaches and technologies to the construction industry. LEED is a voluntary environmental rating system that is organized into six categories:

1. Sustainable Sites
2. Water Efficiency
3. Energy & Atmosphere
4. Materials & Resources
5. Indoor Air Quality
6. Innovation & Design Practice

LEED covers the range of architectural decisions: including site design, water usage, energy conservation and production, indoor air quality, building materials, natural lighting, views of the outdoors, and innovative design components.

The LEED point award matrix is a mixture of teaching, persuasion, example, and incentive. It is good checklist for the entire project team to evaluate the quality of sustainable design decisions for the complete project—from initial planning through final construction, maintenance, and training procedures.

These categories combine *prerequisites* (basic sustainable practices such as building commissioning, plans for erosion control, or meeting minimum Indoor Air Quality Standards) with optional *credits* (water use reduction, heat island reduction, or measures of material recycled content).

Most of the credits are performance based—solutions based on system performances against an established standard such as American Society of Heating, Refrigeration and Air Conditioning Engineers (ASHRAE). ASHRAE has created one of most widely recognized standards of energy design that is used by mechanical engineers and architects.

For example, one credit (under the Energy & Resources category) is "Optimize Energy Performance."

The number of points for this credit depend on how the architectural and engineering team can optimize the design of the building's energy systems against the ASHRAE 90.1 standards.

The possible design solutions include optimizing the heating, cooling, fans, pumps, water, and interior lighting systems.

In the graduated point matrix for a new building, if the team improves the performance (against ASHRAE standards) by 15 percent they receive one point and if they manage to improve by 60 percent they would receive 10 points.

LEED describes suggested results but allows the architectural team to find a variety of solutions. The LEED certification awards range from Bronze to Platinum. The LEED certification is innovative and rigorous.

RESEARCH AND EDUCATION FOR SUSTAINABLE DESIGN

Is there additional education and research necessary for a sustainable project?

Yes. Innovative HVAC systems, durable yet non-toxic materials, recycled materials, recyclable materials, native plant material, energy efficient lighting, and controls are examples of design components that are not normally designed and installed by general contractors and architectural consultants on typical projects.

In order to serve the client, the architect of sustainable projects has to be involved in many research and education occasions.

1. *Education of the Client*

Sustainable design requires a new way of examining the architectural design process. Concepts such as life-cycle costing, recycled versus recyclable materials; non-VOC (volatile organic compounds) substances; daylighting; and alternate energy sources are among the several new concepts that the architect should discuss with the client before the design process commences.

It is critical that the client understands the sustainable process and is sympathetic to its potential economic and environmental benefits.

2. *Education of the Project Team*

Once the project has been assigned to an architect, but before the design process begins, the project team (architect, engineer, contractor, consultants, and owner) should assemble and discuss the project scope and objectives with all the project team members.

2.1 *Establishing Project Goals*

Among the many items included in the scope of work (including the extent of work, program elements, budget, and schedule) are the objectives for sustainable design.

For example, the architect and owner might establish goals for several environmental areas such as:

- X percent reduction of energy usage from the established norm;
- improved lighting (less energy used and more efficient dispersal of indirect light with less glare);
- non-toxic and low VOC paint and finishes;
- increased recycled content in materials such as carpeting, gypsum wallboard, ceiling tiles, metal studs, and millwork;
- high-efficiency (ENERGY STAR®) appliances;
- wood elements are all certified wood products; or
- daylighting in all work/occupied spaces.

As the leader of the project team, it is the architect's responsibility to include sustainable goals with the rest of the project scope of work.

A detailed explanation of the benefits of these sustainable design elements to all of the project team will ensure that they fully understand the design potential and economic implications of these concepts.

2.2 *Verify Extent of Work*

Sustainable design involves a more comprehensive approach to pre-project planning.

The LEED certification process will require record keeping and verification of the source of materials—a process that is beyond the normal design and construction work. For purposes of selecting a contractor and consultants, the team should be briefed on these additional obligations.

For example, the demolition process (if LEED certified) will require verification that materials have been sorted and delivered to an approved recycling organization. By contrast, the normal demolition process does not require recycling nor verification that each material is sorted by type.

Clearly establishing the extent and type of effort required for each member of the sustainable design team is critical. The extent and type of effort will affect each member's ability to participate and their fees for services and construction work.

3. Energy and Optimization Modeling

Building shape, orientation, fenestration location, roof color, envelope configuration, and HVAC system efficiency are some of the variables in sustainable design projects that can be fine tuned with DOE-2 (U.S. Department of Energy's building analysis program) and other computer energy modeling programs.

The "fine-tuning" of a project's energy components are one of the elements in the architect's design matrix that affects the final appearance, cost, and performance of the final design.

Energy modeling will not govern the final design. Issues such as compatible scale, color, texture, and functionality are still part of the architect's palette. But energy modeling is one additional factor that the architect will employ as part of the "best practices" approach to architecture.

In addition, modeling can assist in the cost analysis of a project. The fact that the modeling program is interactive helps the architect simultaneously adjust design elements to demonstrate alternate energy efficient solutions.

For example, energy modeling might allow the architect to demonstrate to the team that a more durable, aesthetically pleasing, and energy-efficient building skin could be economically justified by reducing the size and cost of the mechanical system.

The ability to visually and numerically quantify the efficacy of trading certain design elements may be an effective tool for the architect when discussing the building design with the consultants and owner.

4. The Bid and Specification Process

The requirements of a sustainable design will often vary from a normal project.

For example, the millwork section of bid documents will normally specify the finish material, configuration of the design and methods of attachment, delivery, and installation. But the requirement of non-VOC glues and non-VOC substrate may confuse a potential bidder and cause him to unnecessarily increase the bid price.

To facilitate the bidding and construction process, the architect should include:

- Simple definitions of sustainable elements:

  For example, what does "VOC," "certified" wood product, or "daylighting" mean.

- Explanations of specific characteristics of sustainable elements:

  For example, specifically state the standard that must be met [e.g., Green Label Testing Program Limits, carpet's total VOC limit: formaldehyde 0.05 (mg/m^2)].

- References of specific regulatory agency's information (name, address, e-mail, phone, etc.)

  For example, the Carpet and Rug Institute, www.carpet-rug.com, 800-882-8846.

- Examples of suppliers that could meet the sustainable standards indicated. In the case of sustainable products, there are at least two approaches to a list of suppliers for products:

  1. Limit the installer to three to five suppliers of a product that are known to satisfy the sustainable design specifications.

     This approach assures the architect that the product will meet specified standards.

But, with the constantly changing nature of the emerging sustainable design market, a limited list could limit competition and the diversity of creative alternatives.

2. Identify a list of qualified suppliers, but permit the bidder/contractor to submit alternative suppliers who satisfied the sustainable design criteria.

   This approach creates a more competitive environment, but it will require more effort of the architect to properly review and qualify the bids.

4.1 *Changes and Substitutions*

Every project is faced with the reality of time and budgetary pressures. And, in those instances, there may be situations when one product or design element may not be available in the form originally specified.

Sustainable designed projects require more stringent architectural supervision to ensure that original design standards are met.

For example, in the rush to project completion the installer may claim that paints used for "touch-up" of damaged areas are so small that they may be installed with normal, higher VOC paints. The minor transgression might jeopardize the integrity of the project and the ability to receive certification for LEED credits in certain areas.

LESSON 3 QUIZ

1. Life-cycle costing includes which of the following? Check all that apply.
   - **A.** Construction cost
   - **B.** Maintenance cost
   - **C.** Operating cost
   - **D.** Replacement cost

2. Which of the following is NOT a primary goal of sustainable design?
   - **A.** Designs that have components that are easily recyclable
   - **B.** Designs that have lower initial costs
   - **C.** Designs that use less
   - **D.** Designs that have components that are easily biodegradeable

3. Which of the following are recommended approaches to listing suppliers for sustainable products?
   - **I.** Describe product requirements but do not list any suppliers by name.
   - **II.** Limit the installer to three to five suppliers that meet specification requirements.
   - **III.** List sustainable reference standards and elements only.
   - **IV.** Identify a list of qualified suppliers, but allow the bidder/contractor to purpose substitution that meet specifications.
   - **A.** I only
   - **B.** III only
   - **C.** II only
   - **D.** II and IV

4. LEED stands for
   - **A.** Low Energy Environmental Developer.
   - **B.** Leadership in Energy Efficient Design.
   - **C.** Leadership in Energy and Environmental Design.
   - **D.** Lighting, Electricity, and Environmental Dynamics.

5. LEED is sponsored by
   - **A.** NIBS.
   - **B.** USGBC.
   - **C.** ASHRAE.
   - **D.** CSI.

6. Which of the following is NOT a category in the LEED rating system?
   - **A.** Innovation and design practice
   - **B.** Water efficiency
   - **C.** Recycling
   - **D.** Materials and resources

7. VOC stands for
   - **A.** volatile organic compounds.
   - **B.** volatile odiferous chemicals.
   - **C.** ventilation-ordered chemistry.
   - **D.** violent organic compound.

8. Sustainability affects which of the following traditional components to every design decision?
   - **I.** Cost
   - **II.** Function
   - **III.** Aesthetics
   - **IV.** Time
   - **A.** I and III
   - **B.** I and II
   - **C.** I, II, and IV
   - **D.** I, II, III, and IV

1 A, B, C, D
2 B
3 D
4 C
5 B
6 C
7 A
8 D

LESSON FOUR

CONDITIONS OF THE CONTRACT FOR CONSTRUCTION

INTRODUCTION

Application of these principles are covered in Lesson 9, "Construction Contract Administration."

The general conditions of a construction contract between an owner and contractor establish the legal basis for constructing a project. AIA Document A201, General Conditions of the Contract for Construction, is the most widely used form of this kind for private construction projects. It is a complex and extensive document, and often provides the basis for a number of questions on the exam.

Remember two basic things about AIA A201:

1. It is general and covers items that are common to every project. It must, therefore, be tailored to meet the needs of a specific project by supplementary or special conditions.
2. It forms a part of the owner-contractor agreement, which is a contract between those parties only. The architect is not a party to that contract and is not permitted to prepare it. Only an attorney is authorized to prepare such a contract.

The architect should forward all legal contract forms to the owner. These documents must be reviewed by the owner's attorney and incorporated into the Project Manual after written instructions have been received from the owner. Architects should never act unilaterally concerning general, supplementary, or special conditions of the construction contract.

PRACTICE OF ARCHITECTURE VS. PRACTICE OF LAW

Many owners and their attorneys seek advice from architects on construction contracts. Architects must restrict such advice to technical matters within their training and experience. An architect should not prepare construction contracts or apply the law to specific contractual matters. These are duties of the owner's attorney.

The AIA has produced a series of standard documents intended for use on construction projects. Although architects frequently advise clients that AIA documents should be used, only an owner and his or her legal counsel can decide if the AIA documents are appropriate for a specific project.

CONDITIONS OF THE CONTRACT

Standard Form of General Conditions

Building construction is a complex undertaking involving many parties who often have not previously worked together. These parties join to build a structure on a unique site under environmental conditions that are sometimes difficult. They need a fair and comprehensive set of rules to establish and govern their relationship.

The conditions of the contract establish rules that are consistent from project to project. They deal with contractual matters, not with procedural requirements necessary to administer a contract. Although the general conditions are extensive, there may be supplementary and special conditions required to cover unique situations. Together, all of the conditions expand upon the basic agreement between owner and contractor.

A number of organizations individually publish standard forms of general conditions. AIA Document A201, the General Conditions, has the longest history. It was first published in 1911, and has since undergone many revisions. It represents the careful work of many knowledgeable and experienced people. No one individual can be expected to create a set of general conditions that matches the collective wisdom contained in this standard document. It also has a history of interpretation. Its language has been tested and its meaning determined by the courts. Experience has shown the AIA General Conditions to be generally fair and equitable.

People in the construction industry are familiar with the format of A201. Contractors understand the meaning of the terms in this document and the consequences of working under it. Standard documents inspire confidence, which is often reflected in bid prices that are not inflated to cover unknown contingencies. These advantages can be significantly diluted if the documents are printed in a new format for each project, even if copied verbatim from standard forms.

Supplementary General Conditions

Standard documents are general and therefore cannot possibly address the exact requirements of each project. Supplementary general conditions are used to modify the standard forms. They accommodate the legal, physical, or climatic conditions of the specific project, usually following the format of the general conditions and paralleling its provisions. New items can be inserted, and other items revised or deleted according to the requirements of the project.

Why not reformat the general conditions with all modifications integrated into the text? This might be practical if standard forms were not so widely known and accepted. But since they are, it is much easier and more efficient for a contractor to read the supplementary conditions to determine what is unique about the project.

Supplementary conditions are particularly important on projects with multiple prime contracts, phased or fast-track construction, or a contract price based on cost plus fee or any method other than a stipulated sum. Standard general conditions alone are not intended to cover such situations.

Special Conditions

Special conditions are used when supplementary conditions must be further extended. For example, when governmental agencies require that their standard conditions be incorporated into a contract, special conditions may be used. Local laws or customs may also require special conditions. See pages 251 though 254.

RELATION TO DIVISION ONE—GENERAL REQUIREMENTS OF THE SPECIFICATIONS

The general, supplementary, and special conditions of the construction contract have a legal and contractual purpose. They establish rights and responsibilities of the parties, which are consistent from project to project.

The conditions do not establish administrative procedures. Such procedures are found in the General Requirements division of the specifications. For example, the AIA General Conditions (Document A201) establishes the contractor's legal obligation to review shop drawings prepared by subcontractors. The contractor must approve such shop drawings and point out any deviations from the contract documents prior to submitting them to the architect. The General Requirements division of the specifications defines the specific procedures that a contractor must follow with regard to that obligation. For example, the General Requirements might state that six sets of prints and one reproducible copy of each shop drawing must be submitted. Likewise, if shop drawings must be sent to the attention of a particular person or department in the architect's office, this requirement is also stated. A duty is established in the General Conditions, and a means for discharging that duty is defined in the General Requirements division of the specifications.

RELATION TO LAWS, CODES, AND STANDARDS

Construction contracts are subject to the laws in effect at a project's location, which may supersede specific contractual provisions. While contracting parties establish legal provisions that deal strictly with their own relationship, they cannot establish or modify provisions that involve the government or other third parties.

For example, an owner and contractor may agree to limit the amount of damages that the contractor will pay the owner if construction is not completed on time, but they cannot agree to limit their legal liability to a pedestrian who is injured while walking past the construction site. Building codes and zoning ordinances also regulate aspects of construction processes. Contracting parties do not have the right to agree between themselves to do less than the minimum required by such codes and ordinances.

RIGHTS AND OBLIGATIONS OF THE OWNER

An owner has both implicit and explicit rights and obligations relative to construction projects. Under the law, owners implicitly promise that contractors will have access to the construction site. Contractors are also entitled to assume that owners will cooperate and not interfere with construction work.

Other rights and obligations of owners are explicitly stated in the general conditions. An owner's principal obligation is to pay the contractor. The AIA General Conditions requires an owner, upon request, to provide evidence that money is available to make payments.

Owners also have obligations related to the property on which a project will be constructed. For example, an owner must provide required surveys, legal descriptions of the site, and easements if necessary.

An owner is required to provide the contractor with as many copies of the drawings and specifications as are reasonably necessary during the construction period. According to AIA standard contracts, the architect owns such documents, and the contractor must return them or suitably account for all sets given during this time.

Article 6 of the AIA General Conditions discusses an owner's right to perform construction work on a project or to hire more than one contractor at the same time. The article also contains provisions by which multiple prime contractors can perform work simultaneously.

Under certain circumstances, an owner has the right to order a contractor to stop the work. This may occur if construction is determined to be defective or not being performed according to the contract documents. If a contractor does not correct mistakes or does not make proper progress on a project, an owner has the right to have the work corrected or completed by others.

OWNER'S RELATIONSHIP TO SUBCONTRACTORS

The AIA General Conditions assumes that there will be one prime, or general, contractor on a project. While much of the construction work is typically subcontracted, the construction contract is between the owner and the general contractor. In this regard, subparagraph 1.1.2 of the AIA General Conditions (Document A201) clearly states that nothing in the contract documents creates a contractual relationship between an owner and any subcontractor or sub-subcontractor.

Although there is no contractual relationship, there is an implied relationship between an owner and subcontractors. The General Conditions requires a general contractor to bind all subcontractors to the same terms and conditions that bind the contractor to the owner. It also requires that subcontractors have the same rights relative to the contractor that the contractor has relative to the owner. The General Conditions extends rights and responsibilities from owner to contractor to subcontractor and creates the context for all contractual relationships on a project.

SEPARATE PRIME CONTRACTS

Special attention must be paid to supplementary conditions on projects where there will be more than one *prime contractor*. A prime contract is one executed between owner and contractor. *Separate prime contracts* may be used if an owner wants to, or is required by law to, contract directly with several contractors. Separate prime contracts may be written for structural, HVAC, plumbing, electrical, or other major portions of the work.

In this event, each set of contract documents requires its own agreement, conditions, drawings, specifications, and so forth, and there is no single contractor responsible for the whole project. Each contractor is responsible for his portion of the construction work and to coordinate with the others.

Problems frequently arise when one contractor delays or damages another contractor's work. Other problems of a minor nature may also occur. For example, general clean-up may be neglected while other construction details may be needlessly duplicated. Many contractors could, for example, use the same scaffolding if there were adequate coordination. Such problems may be solved in several different ways: a *construction manager* may be retained to provide coordination services, the owner may have a staff capable of the task, or administrative and supervisory responsibilities may be assigned to one of the prime contractors.

MAJOR ELEMENTS OF THE CONDITIONS

Bonds

The AIA General Conditions (A201), Paragraph 11.4, gives owners the right to require the contractor to provide bonds assuring that contract work will be performed in accordance with the contract documents. These bonds also guarantee that obligations arising out of the contract will be paid or satisfied. Bonds reduce an owner's risk of financial loss if a contractor defaults in performance or does not pay the bills.

Bonds referred to in the General Conditions include the Performance Bond and Payment Bond. Refer to AIA Document A312. They are usually written for the full amount of the construction contract.

Bonds are provided by *surety* companies that guarantee that the contractor (the *principal* or *obligor*) will fulfill the contractual duties to the owner (the *beneficiary* or *obligee*).

Contractors are usually required to furnish a bid bond, Document A310, on most competitively bid projects. Bid bonds are usually written for 5 or 10 percent of the bid amount. If the low bidder refuses to sign a contract for the amount bid, the surety partially or fully compensates the owner for the difference between the low bid and the actual contract price of the substituted contractor.

Insurance

Owner's

Paragraph 11.3 of the AIA General Conditions states that an owner must purchase property insurance for the full insurable value of construction work. The insurable value of a project is usually less than the construction contract sum because certain elements of the contract, such as earthwork and grading, are not included. Owners must also purchase boiler and machinery coverage if required.

Property insurance is commonly referred to as *builder's risk* insurance. It includes fire and property damage insurance for a project during construction. It also covers temporary structures, materials, equipment, and supplies, and usually covers tools and equipment if they are located within 100 feet of the project site, as well as property in transit or stored off-site.

The General Conditions usually requires that builder's risk coverage be the *all risk* type, which provides broader coverage than standard *named peril* forms. All risk covers everything but specified exceptions, while named peril insurance only covers risks that are specifically identified.

Property insurance must include riders for extended coverage, including theft and vandalism and malicious mischief. None of these riders includes coverage for earthquake or flood damage. These coverages must be purchased separately.

Since title to construction work passes to the owner upon incorporation of that work into the project or upon payment for it, the AIA General Conditions provides that owners, not contractors, purchase property insurance. If the owner wants the contractor to purchase property insurance, the General Conditions may be modified by appropriate provisions in the supplementary conditions.

Property insurance will be cancelled if an owner occupies all or part of a project prior to substantial completion without the insurance company's consent. Partial occupancy of a project is relatively common and insurance companies must therefore be notified if coverage is to remain in effect.

All coverages mentioned must include the financial interests of contractors and subcontractors, as well as those of owners. For example, a contractor may have paid for materials that are on-site, but which have not been incorporated into the project. If these materials are destroyed, the contractor will have an interest in money paid by the insurance company to cover the loss. The General Conditions provides that the owner will act as trustee for the proceeds paid by insurance companies.

Finally, an owner may purchase *loss of use* insurance that protects against losses arising out of delays or other events that prevent an owner from using a project when and as intended.

Contractor's

The AIA General Conditions also requires contractors to purchase certain types of insurance. Paragraph 11.1 lists required coverages that include the following:

1. *Workers' compensation* insurance. Covers job-related injuries and is usually required by law, which also sets required limits.
2. *Liability* insurance. Covers claims for damages for bodily injury, sickness, disease, or death of the contractor's employees or any other person.
3. *Personal injury* insurance. Covers libel, slander, false arrest, and defamation of character.
4. *Property damage* insurance. Intended to cover property, other than construction work itself, which may be damaged by construction activities. Usually, explosion, collapse, and underground damages are excluded and must be added by endorsement if required.

5. *Automobile liability* insurance. Covers claims for damages arising out of the use of an owned, non-owned, or hired automobile.
6. *Contractual liability* insurance. Covers liability assumed by contract. Under Paragraph 3.18 of the AIA General Conditions, this is primarily *indemnification*, wherein contractors agree to hold owners and architects harmless from damages arising out of specified events.

AIA Document G715, Certificate of Insurance, summarizes the insurance coverage required under the general conditions of the contract.

Contractors are usually required to obtain *products and completed operations* coverage. This insurance covers claims for bodily injury or property damage arising from accidents that may occur after the construction work has been completed and turned over to the owner. However, completed operations insurance does not apply to damage of the completed work itself.

All required coverages, in amounts specified in the supplementary conditions, must be purchased prior to starting construction work, and certificates of insurance must be provided to the owner and the contractor. The certificates must state that the insurance company will notify all parties in advance if insurance is going to be cancelled for any reason.

Liens

Mechanics' liens exist by statute in each state to provide security for payment for labor and materials supplied to construct buildings.

Such liens apply only to the real property in question, buildings and land, and not against other assets of the owner. Liens give the worker, contractor, or material supplier the right to force the sale of an owner's property in order to satisfy a claim for payment. Lien rights give contractors leverage over an owner to force payment of legal claims because valid liens encumber an owner's title to the property. Without clear title, an owner may be prevented from transferring the property through sale or other means.

Liens were devised to encourage people to supply labor and materials for construction projects. Lien laws may seem unfair from an owner's point of view because subcontractors and suppliers who have not been paid by a general contractor may file liens against an owner even if the owner has paid the general contractor for their work.

However, liens are basically fair and equitable since, in most cases, it is impossible to repossess labor and materials that have been incorporated into a construction project.

The AIA General Conditions addresses liens in two places. In subparagraph 9.3.3, the contractor warrants that all labor, materials, and equipment for which payment is requested are free of liens and claims of every kind. Subparagraph 9.10.2 states that final payment will not be made until the contractor submits an affidavit that all debts arising out of the project for which the owner or the owner's property may be liable have been satisfied. The owner may also ask for releases and waivers of liens from the contractor and subcontractors, though sometimes a bond is acceptable in lieu of such waivers.

Liens are not permitted on publicly owned projects. Instead, labor and material payment bonds provide the necessary protection. The Federal law requiring such bonds is known as the *Miller Act*, and many states have similar laws, referred to as *little Miller Acts*.

Details related to these matters vary considerably, depending on the jurisdiction in which the project is located, and the owner's attorneys should respond to questions about liens. Architects should not advise owners on the legal sufficiency of required affidavits, waivers, or releases of liens.

Shop Drawings and Submittals

Improper processing of submittals by the contractor and architect may result in controversy, delays, additional costs, and even litigation.

Responsibility of the Contractor

Paragraph 3.12 of the AIA General Conditions defines a contractor's responsibilities concerning shop drawings. The contractor must review, approve, and promptly submit shop drawings required by the contract documents.

A contractor's stamp of approval on shop drawings means that the contractor will or has *determined and verified materials, field measurements, and field construction criteria related thereto, or will do so, and has checked and coordinated the information contained within such submittals.*... Because contractors have this obligation, architects should *not* review shop drawings unless they contain such a stamp of approval.

The General Conditions prescribes that the contractor is responsible for errors or omissions in shop drawings. Deviations from the requirements of the contract documents are not considered approved by the architect simply because they appear in approved shop drawings. If there are proposed deviations, they must be called to the architect's attention in writing and approved in writing before they become effective. Similarly, if there are revisions to resubmitted shop drawings, other than those revisions requested by the architect on the prior submittal, they must specifically be called to the architect's attention or they will not be effective.

Responsibility of the Architect

The architect's responsibilities are primarily described in the owner-architect agreement. Those described in the AIA Owner-Architect Agreement (Document B102 or B201) are restated in the AIA General Conditions (Document A201) for the contractor's information. If Document B102 or B201 is not used, contractors may not know what services the architect is obligated to provide.

Architects must promptly *review and approve or take other appropriate action* on shop drawings that contractors submit. The General Conditions clearly states that review is strictly to determine if shop drawings conform to the design concept and information given in the contract documents. Matters outside the scope of the design concept and the contract documents are not reviewed.

An architect may be reluctant to use the word *approved* when acting on a submittal because some professional liability exposure may be perceived. However, the AIA General Conditions and the AIA Owner-Architect Agreement both limit the architect's duty with regard to submittals, which results in limited liability exposure. As with other duties, if the architect has performed services in accordance with these AIA documents and within a *standard of care*, there should be no concern about professional liability.

In addition to *approved*, other appropriate action that an architect can take includes *approved as noted, revise and resubmit*, and *not approved.*

Time Limits, Schedule, and Delay

The General Conditions, particularly Article 8, contains many provisions related to timely completion of the work. For example, contractors must complete construction work in the allotted number of calendar days. Calendar days (7 days per week), as opposed to working days, are used to preclude questions about how to count weekends and holidays.

Contract time may start on the date of signing the owner-contractor agreement, upon the issuance of a formal notice to proceed, or on another date stated in the owner-contractor agreement. Contract time ends when construction work is substantially complete. If an owner anticipates a financial loss if the project is not completed on time, and if the owner's attorney recommends the inclusion of a provision for *liquidated damages*, such a provision is included in the owner-contractor agreement. Liquidated damages is an amount of money stipulated in the contract that is chargeable against the contractor as reimbursement for damages suffered by the owner because of the contractor's failure to fulfill contractual obligations.

According to subparagraph 8.2.1 of the *AIA* General Conditions, time limits are *of the essence of the contract*. This term has a special legal meaning. When timing is not crucial, there is no breach of contract as long as work or services are completed reasonably close to anticipated deadlines. However, when *time is of the essence*, even a slight delay may constitute a breach of contract.

The contractor is required to prepare and submit a progress schedule, showing how the work is to be completed within the contract time. On small projects, a bar chart, also known as a Gantt chart, is often used. On large or complex projects, a *critical path method (CPM)* schedule is usually required. The CPM schedule is computerized and differs from a bar chart because it shows relationships between activities, not just durations and sequences.

The General Conditions also provides for extensions to the contract time if an owner's actions or inactions delay a contractor. Changes to required work may also justify time extensions. The General Conditions provides for events of *force majeure*, or acts of God, that are beyond the control of either party. If any delays of this kind occur, the contractor must notify the architect within 21 days or the claim for time extension is waived. The contract time may only be changed by change order or construction change directive.

All of the delays mentioned above may result in an extension of time to complete the contract work. Without this extension, a contractor might be assessed monetary damages for late completion. On the other hand, the contractor may claim monetary damages because of delays allegedly caused by the owner or architect. Time extensions, however, do not automatically entitle the contractor to monetary damages. Standard AIA documents provide means by which a contractor may make a claim for time extensions, monetary damages, or both. Supplementary conditions often contain detailed provisions related to scheduling and the consequences of delays.

Payment Procedures

Prior to submitting the first *application for payment*, the contractor is required to submit a *schedule of values* for all parts of the construction work. This schedule is the basis for the contractor's subsequent applications for payment.

When reviewing the contractor's proposed schedule of values, the architect must determine that appropriate values are assigned to each portion of the work. Usually the contractor is entitled to initial payment for site mobilization, preparation of submittals, and other work that is not actual construction. However, the contractor may *front load* the schedule of values by overstating the value of materials and understating the value of labor. In that event, after the materials are delivered and paid for, the payment for the labor required to install the materials may be inadequate. Of course, the contractor must still pay the labor charges. In the worst case, if the contractor defaults and has unpaid bills from subcontractors or suppliers, the owner may be exposed to financial loss. The architect can protect the owner from such exposure, and save the owner from unnecessarily advancing funds to the contractor, by carefully reviewing the contractor's schedule of values.

Prior to each progress payment, the contractor must submit an application for payment for the architect's review. On the basis of the application, the contractor will generally be paid for all materials and equipment incorporated into a project and for all materials and equipment stored on-site. Special provisions must be made if the contractor requests payment for materials and equipment stored off-site or in transit.

The contractor must also warrant that the owner will obtain title to materials and equipment free and clear of liens and claims. This guarantee becomes effective either upon their incorporation into the project, or payment by the owner in the case of stored materials and equipment.

The architect reviews the contractor's applications for payment and assesses the progress of the construction work. Each application must be based on the approved schedule of values. For example, a typical entry on an application for concrete work with a total value of $60,000 might be for 50 percent completion less 10 percent retainage, for a partial payment of $27,000 ($60,000 × 50% × 90%), less amounts previously paid. In that case, the architect must determine if 50 percent of the concrete work is actually complete, and that the retainage figure is correct. Each line item must be similarly analyzed.

Architects have the right to certify that *no* payment, or an amount less than that requested by a contractor, is due. This action might be necessary if the work has not progressed as claimed by the contractor or if it is not completely acceptable.

Architects also have the right to nullify all or part of previous certificates for payment if it is necessary to protect an owner against

- defective work that has not been corrected;
- claims, or the likelihood of claims, by third parties;
- a contractor's failure to pay subcontractors;
- evidence that the contract work cannot be completed for the unpaid balance of the contract sum;
- damage to the owner or another contractor;
- evidence that the contract work cannot be completed on time; or
- a contractor's persistent failure to comply with contract documents.

When an architect issues a certificate for payment, an owner must pay the contractor as stipulated in the contract documents.

Once a contractor has received payment, all subcontractors must be paid proportionately. Similarly, the amounts of money retained by the contractor from payments to the subcontractors must be in proportion to the amounts retained by an owner. Subparagraph 9.6.3 of the AIA General Conditions states: "The Architect will, on request, furnish to a Subcontractor, if practicable, information regarding percentages of completion or amounts applied for by the Contractor and action taken thereon by the Architect and Owner on account of portions of the Work done by such Subcontractor."

It is solely the contractor's responsibility to pay subcontractors. Neither the owner nor the architect has an obligation to insure that subcontractors are paid. The architect is not responsible for the contractor's disposition of funds paid as a result of applications and certificates for payment.

Finally, the General Conditions states that nothing in the payment process shall be construed as acceptance of work not performed in accordance with the contract documents. Acceptance is determined by provisions for final payment, correction of work, and warranty.

Safety

Contractors are responsible for safety precautions and programs relative to construction work. This includes the safety of the contractor's employees, the construction work itself, and adjacent property that is not part of the construction work.

If damage is caused by a contractor, subcontractor, or other entity for which a contractor is responsible, the contractor must repair or pay for it. Normally, a contractor's superintendent is the individual responsible for construction site safety. In case of emergency, the superintendent may act in any reasonable manner to prevent loss of life or damage to property. If such action or inaction delays work or increases costs, the contract time and contract sum may be changed appropriately by change order.

Since safety is the contractor's obligation, an architect should not review or enforce safety programs or procedures. If an architect notices an unsafe condition, however, the condition should promptly be reported to the contractor's superintendent, and a written record of the circumstances should be prepared. Contractors are the appropriate parties to decide what actions, if any, are required to remedy unsafe conditions. It is improper for an architect to approve or order changes to a contractor's safety programs. Such involvement may create liability for the architect should an accident occur on the site.

Substantial Completion

Legal Effect

In general, contracts must be *completely* performed. Anything less than completeness may be interpreted as a breach of contract. In fact, however, neither the owner nor the contractor would benefit from such a strict definition of completion. Therefore, the law recognizes *substantial completion*, which is sometimes called *substantial performance*. Once a construction contract is substantially complete, a contractor is entitled to the contract sum, less the value of incomplete work and retainage. Upon *final completion*, the contractor is entitled to final payment.

Under the AIA General Conditions, the contract time is the period allotted for substantial completion, including authorized adjustments. Warranty periods begin upon substantial completion. In cases where there are separate prime contracts or phases of construction, there may be more than one date of substantial completion, and more than one warranty period.

Paragraph 9.8 of the AIA General Conditions explicitly defines the *date of substantial completion* and describes its significance. Although the date on which the code official issues a certificate of occupancy is often the same as the date of substantial completion, this is not necessarily the case. For example, a given contract might entail a great deal of finish work that is not required for a C of O but essential to substantial completion. All of the requirements of the contract must be substantially performed to achieve substantial completion, mere occupancy is not sufficient.

The AIA Certificate of Substantial Completion (Document G704), when signed by the owner, architect, and contractor, establishes the date of substantial completion and the responsibilities of each party.

Role of the Contractor

Paragraph 9.8 of the AIA General Conditions discusses the requirements for the date of substantial completion. It is the contractor's responsibility to decide when a project is substantially complete. According to the contract documents, this occurs when the owner can occupy or use the project for its intended purpose. At that time, the contractor prepares a list of items to be corrected and those that must still be completed. That list is submitted to the architect, and a date is set for the architect's inspection. The contractor's list of items may not be definitive, and it is of no legal consequence if items are omitted. The contractor is still required to comply with the contract documents. However, such a list is helpful to the architect performing an inspection.

Role of the Architect

According to the AIA Owner-Architect Agreement, an architect *inspects* a project at only two points in the construction process—substantial completion and final completion. All other site visits are "to become generally familiar with and to keep the Owner informed about the progress and quality of the Work…" Such visits are less detailed and comprehensive than *inspections*. Upon being notified by the contractor, an architect schedules an inspection to determine if a project is substantially complete.

The list of items to be completed or corrected is provided by the contractor and is usually expanded by the architect as a result of a detailed inspection. This list is commonly referred to as a *punch list*. At times, additional work is required before an architect is able to certify that a project is substantially complete.

Many architects include provisions in the supplementary conditions requiring the contractor to pay the architect's fee and expenses for subsequent inspection trips if the project is not substantially complete when first inspected.

The certificate of substantial completion establishes the date of substantial completion and states the time within which the contractor must perform the work described on the *punch list*. The certificate also states the responsibilities of both owner and contractor for "… security, maintenance, heat, utilities, damage to the Work, and insurance.…"

These items are important because the contractor must continue to work on-site, which usually requires access to substantially completed and occupied portions of a project. Construction workers may damage completed portions of the project, use power and water, and possibly injure themselves or others between substantial and final completion. All such events must be anticipated and responsibility for the consequences predetermined.

Role of the Owner

The owner, the contractor, and the architect must sign the certificate of substantial completion. By doing so, they indicate their acceptance of the responsibilities assigned in it.

At substantial completion, the full contract sum, less the value of incomplete work and retained amounts, is normally due the contractor. The amount of retainage may be reduced at this time to reflect the actual value of the work to be completed and corrected.

Insurance

Owners usually occupy projects immediately after the issuance of a certificate of substantial completion. Occupancy invalidates property insurance coverages that were in effect during the construction period. It is important, therefore, for owners to inform the insurance company and make necessary insurance changeover arrangements prior to occupying any portion of a project.

A contractor's surety has a financial interest in money retained by an owner from amounts due the contractor, since that money can be used by the surety if the contractor defaults. If any portion of the retainage is refunded to a contractor, the surety's risk of loss is theoretically increased. Applicable bonds are voided if retainage is released without the surety's consent.

It is essential, therefore, for an owner to obtain the surety's written consent before any portion of the retainage is released to a contractor. The AIA Consent of Surety to Reduction in or Partial Release of Retainage (Document G707A) is intended for this purpose.

Warranties and Guarantees

Paragraph 3.5 of the AIA General Conditions contains a general warranty for performance from the contractor to the owner and the architect. It states that all materials and equipment are to be new, unless otherwise specifically required, and that completed construction, including workmanship, is to be of good quality and free from faults and defects.

In subparagraph 9.3.3, the contractor warrants that legal title to all materials and equipment will pass to the owner upon its incorporation into construction or upon the contractor's receipt of payment, whichever is earlier. Title must be free and clear of liens, claims, or other security interests. The owner's attorney, not the architect, must determine if that requirement is met.

Since these warranties are governed by state law, their terms may vary. They are generally in force for several years from their effective date. The owner may recover monetary damages from the contractor during the full time the warranty is effective under a state's statute of limitations.

Subparagraph 12.2.2 of the AIA General Conditions states: "If, within one year after the date of Substantial Completion … any of the Work is found to be not in accordance with the requirements of the Contract Documents, the Contractor shall correct it promptly…." An owner's right specifically to have defective work corrected by the contractor during the first year after substantial completion is important because most defective work cannot be detached and returned to the manufacturer for repairs.

Specific product warranties are often required in the technical sections of the specifications. Manufacturers, through subcontractors, usually provide warranty certificates prior to substantial completion. The procedure is usually described in Division One.

Record Drawings

Record drawings are sometimes inaccurately referred to as *as-built drawings*. According to Paragraph 3.11 of the AIA General Conditions, contractors are required to maintain copies of all drawings, specifications, addenda, change orders, and other modifications at the site in order to accurately record all changes made during construction. Similarly, they must keep copies of all shop drawings, product data, and samples. These are record drawings.

For example, record drawings might indicate that a door originally intended to be 2'-0" from a corner was installed at 4'-0" from the corner. Architects have occasionally been held liable for what were purported to be as-built drawings when subsequent construction work resulted in damage to concealed utility lines, structural elements, and so on. No drawings can ever depict all as-built conditions.

An owner may want a reproducible set of record drawings when a project is completed. Supplementary Conditions or Division One—General Requirements of the specifications may require contractors to prepare such a set. If the owner requests the architect to prepare such drawings, it is generally considered a change in services.

Final Completion and Final Payment

When the work noted on the *punch list* is completed, the contractor notifies the architect that the project is ready for final inspection and submits a final application for payment. The architect then inspects the project and if, in the architect's opinion, it is complete in accordance with the contract documents, a final certificate for payment is issued to the owner. This payment represents the entire balance due the contractor including retainage.

Before final payment is made, the contractor must submit

- an affidavit that all debts for which the owner or the owner's property could be held responsible have been satisfied, to avoid the filing of liens or claims;
- written consent of the surety to final payment, to avoid surety bonds being voided; and
- other data such as receipts and waivers that an owner may require to demonstrate that there are no outstanding obligations related to the project.

It is solely the responsibility of the owner and the owner's legal counsel to determine the extent and form of required documents. Architects should not advise an owner on, nor judge the legal sufficiency of, legal documents being submitted for final payment.

The AIA General Conditions, subparagraph 9.10.4, states that, upon making final payment, an owner waives all claims against the contractor except those arising from:

.1 liens. Claims, security interests or encumbrances arising out of the Contract and unsettled;

.2 failure of the Work to comply with the requirements of the Contract Documents; or

.3 terms of special warranties required by the Contract Documents.

Upon acceptance of final payment, the contractor waives all claims except those made previously in writing and identified as unsettled at the time final application for payment was made.

Project Close-Out

Close-out procedures are more administrative than contractual. Therefore, these procedures are described in Division One—General Requirements of the specifications, rather than in the AIA General Conditions.

To close out a project, the contractor is usually responsible for

- submission of record drawings and specifications as well as maintenance manuals, warranties, and other record information;
- delivery of tools, spare parts, and extra stock of materials;
- removal of temporary facilities;
- start-up testing of equipment, as well as training of owner's operating/maintenance personnel; and
- final touch-up, repairs, and cleaning.

LESSON 4 QUIZ

1. According to AIA documents, which of the following are NOT considered a part of the construction contract documents?

   I. Specifications
   II. Addenda
   III. Shop drawings
   IV. Owner-Architect Agreement
   V. Supplementary Conditions

   A. IV only
   B. II and III
   C. I and V
   D. III and IV

2. ________ assures an owner that the contractor will execute the work in accordance with the contract.

3. During the course of construction, liability insurance should be maintained by which of the following?

   I. The mortgagee
   II. Trade unions
   III. The surety company
   IV. The owner
   V. The contractor

   A. IV only
   B. I and V
   C. IV and V
   D. II and III

4. Where would provisions for liquidated damages normally be found?

   A. Owner-Contractor Agreement
   B. General Conditions of the Contract for Construction
   C. Additive Alternates to the Contract
   D. Proposal Form and Instructions to Bidders

5. A change during construction in the location of underground lines from that shown on the site utilities plan should be reflected in

   A. record drawings.
   B. architectural drawings.
   C. field sketches.
   D. shop drawings.

6. Architects are responsible for preparing which parts of a Project Manual?

   I. Specifications
   II. General Conditions
   III. Supplementary Conditions
   IV. Owner-Contractor Agreement

   A. I only
   B. I and III
   C. II and III
   D. I, II, III, and IV

7. Which of the following are responsibilities of the owner under the provisions of the AIA General Conditions?

   I. Provide access to the construction site.
   II. Make payments to the contractor.
   III. Make payments to the subcontractors.
   IV. Obtain necessary easements.
   V. Provide necessary copies of construction documents.

   A. II only
   B. II and III
   C. I and IV
   D. II, IV, and V

8. Which of the following statements is TRUE? Contract time

A. is measured in working days.

B. ends at substantial completion.

C. ends at final completion.

D. always starts when the Owner-Contractor Agreement is signed.

9. Under the provisions of the AIA General Conditions, which of the following statements is TRUE?

A. Warranty periods begin at substantial completion.

B. Warranty periods begin at final completion.

C. The general warranty period is one year.

D. Warranties are worthless without bonds to back them up.

10. An architect issues certificates for payment for work performed by the contractor. For what reasons might an architect legitimately nullify all or part of a previously issued certificate for payment?

I. The architect discovers defective work.

II. The owner's lender refuses to release funds.

III. The contractor fails to pay subcontractors.

IV. A pedestrian walking past the project site is injured by a falling piece of lumber and sues the owner.

V. The contractor persistently fails to comply with the contract documents.

A. I, II, III, IV, and V

B. I only

C. I, III, IV, and V

D. III and V

11. Which of the following statements concerning payments to a contractor is FALSE?

A. Payments are based on a schedule of values.

B. Title to materials and equipment passes to the owner upon payment.

C. The owner must pay the contractor within 21 days after the architect issues the certificate for payment.

D. Payments do not normally cover materials in transit.

12. Which of the following parties are responsible for workers' safety on a construction site?

I. The owner

II. The architect

III. The general contractor

IV. The contractor's surety company

A. III only

B. I only

C. I, II, and III

D. I, II, III, and IV

13. An architect, under the provisions of the AIA documents, inspects a project how often during the construction phase?

A. Never, the architect observes the construction.

B. At intervals appropriate to the progress of the construction work.

C. Once, at final completion.

D. Twice, at substantial completion and at final completion.

14. Which of the following is NOT a prerequisite for final payment to the contractor?

A. Final inspection by the architect.

B. Owner's receipt of the certificate of final completion from the architect.

C. Owner's receipt of the contractor's affidavit of payment of debts.

D. Owner's receipt of consent of the contractor's surety company.

15. Which of the following parties must approve shop drawings according to the AIA General Conditions? Check all that apply.

A. Architect

B. Contractor

C. Owner

D. Engineers

E. Subcontractors

LESSON FIVE

COST ESTIMATES

INTRODUCTION

As part of their professional services, architects and their consultants often prepare estimates of construction costs. Although they are expected to apply their professional judgment diligently when predicting costs, their estimates are not guaranteed. The reason is simple: many factors that affect the cost of construction are beyond the control of architects. For example, manufacturers, suppliers, and distributors influence the price of materials and equipment; unions influence labor costs; individual contractors determine how they will prepare their bids; and market forces of supply and demand affect competition.

COST MANAGEMENT

For a detailed description of the architect's role in managing cost, and for the various types of cost estimates, see Lesson 4.

Initial Cost of Materials and Equipment

In order to control construction costs, the architect considers initial cost, installation cost, and long-term cost in specifying products and materials. Poorly defined requirements will result in the selection of inadequate materials and products or those containing unnecessary features. Consequently, the architect must specify products or materials with the necessary properties to achieve the required performance. Building codes establish minimum requirements for certain products and materials; however, their main purpose is to promote safe construction rather than performance.

Most products, materials, and equipment are available with various qualities and features. It is important, for example, to select the proper grade of wood that will provide the necessary strength and desired appearance, the appropriate type of concrete, the desired quality of hardware, the necessary flexibility in an HVAC system, the proper amount and source of light, and the proper type of wall covering, all in accordance with the specific requirements of the particular applications. Another important consideration is that individual products or components may be combined into assemblies and systems. One must consider that the installed cost of an assembly or system may be greater than the sum of the installed costs of the individual elements.

Cost may vary with time and locality. For example, steel may be the most economical structural material in Pittsburgh, but not in Miami. Similarly, plywood or gypsum board shortages may result in price fluctuations from one area to another. It is a mistake to assume that cost information from one project in one location will necessarily apply to a similar project in a different location. Material and labor costs must be projected on the basis of up-to-date information.

Labor Costs

Labor is a significant part, often more than 50 percent, of the total construction cost. Consequently, architects strive to design buildings and prepare drawings and specifications to minimize the amount of on-site labor.

While labor costs have increased, productivity has decreased. Union rules contain numerous requirements that tend to perpetuate traditional and, consequently, more expensive installation methods. Building codes may also restrict or prohibit the implementation of new, labor-saving techniques.

Merit shops, or *open shops*, which are non-union employers, attempt to address the issue of increased labor costs. Some have successfully lowered labor costs while increasing productivity. Most projects are built with either all union labor or no union labor in order to avoid labor problems that may result when non-union construction workers are involved in unionized projects. Some states require that contractors pay a *prevailing wage* to all workers in a trade for public work as a way to equalize union and non-union pay scales. An architect cannot, however, determine whether a project will be constructed by a contractor employing union or non-union labor.

Regardless of the labor situation, drawings and specifications that reduce the amount of on-site labor and encourage the use of factory or shop labor will generally provide a less expensive project. Factory workers work under more controlled and efficient conditions than those for on-site construction workers. All elements fabricated off-site must, of course, be installed on the job. Consequently, on-site labor is always required. Qualified workers must make the installations and connections, which require adequate maneuvering space for workers and equipment.

Construction costs can be reduced through the use of repetitive materials and/or construction details, which enable workers to be more efficient. Construction details that allow work to be performed inside building enclosures will improve construction quality and, at the same time, efficiency. Thus, although labor costs will always be a major factor in the overall cost of construction, they can be controlled.

Long-Term Costs

Operational Costs

Generally, long-term costs are inversely proportional to initial costs; that is, the more money spent initially, the less spent over the long term. For example, high quality mechanical equipment usually operates more efficiently and economically than less expensive equipment. Similarly, the use of solar energy systems may increase initial costs, but result in lower operational costs. It is important to balance initial cost against long-term value.

Some owners may be unwilling to invest in materials and equipment with high initial costs to realize long-term value. A speculative apartment developer, for example, may build a project for quick profit rather than long-term operation. Consequently, the apartment developer may be more interested in the initial construction costs than the long-term operating costs. On the other hand, a government agency investing in a facility projected to last at least 50 years may be more concerned about annual and long-term operating and maintenance costs than initial costs.

The cost of energy required to operate a building is significant. Furthermore, energy sources are not uniformly available. Prices in some industries are regulated by government. Consequently, an architect must select and specify equipment that is efficient, and utilize energy sources projected to have the most stable prices.

Site planning and building design that take advantage of local climatic conditions and incorporate passive solar design and the use of daylighting may also contribute to reducing operational costs. Since long-term costs are significantly greater than initial costs, they deserve an architect's full attention.

Maintenance Costs

Maintenance costs are approximately equal to the sum of all other costs of operating a project. The specification of appropriate materials can affect maintenance costs. Schools, transit facilities, and other buildings that serve large numbers of people must be able to withstand the effects of heavy usage. For example, floor coverings in lobbies of public buildings must be selected for durability, soil and stain resistance, and ease of maintenance. Furnishings and wall coverings must have similar qualities. The selection of an appropriate material for a particular use will result in reduced maintenance costs. Materials used in residential facilities are normally not required to withstand the same wear and tear as those used in commercial facilities, nor are they subjected to commercial cleaning agents.

Labor is also a significant contributor to the total cost of maintenance. Architects' designs can make required maintenance more or less difficult to perform. For example, wall or ceiling mounted toilet partitions provide fewer obstacles to workers cleaning restroom floors than floor mounted partitions. Although they may cost more than floor mounted units, they result in lower maintenance costs during the life of a building.

Provisions for easy access to equipment, and with adequate surrounding work space, will also result in lower maintenance costs. Modular or component systems may minimize replacement or service costs because only faulty units or elements require attention or replacement. In these circumstances, it may be more cost effective to replace a product or system than to repair it.

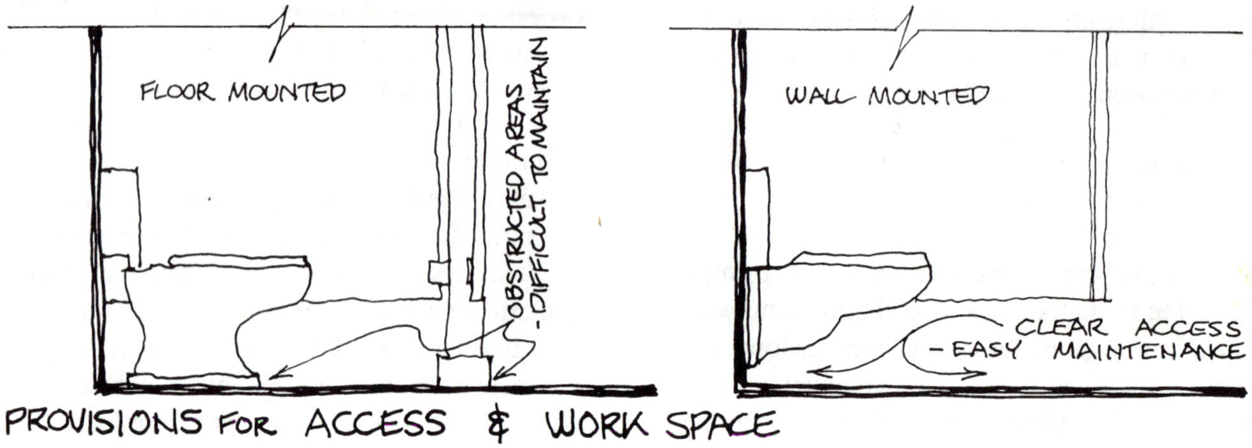

Figure 5.1

Life-Cycle Costs

Life-cycle costs include operational and maintenance costs, taxes, financing, replacement, and renovation. Although financing costs are not within an architect's control, they may be decreased if an architect's services provide for the acceleration of both the design and construction processes. This is the basis for *fast-track* and other delivery approaches that are appropriate when interest rates and financing costs increase.

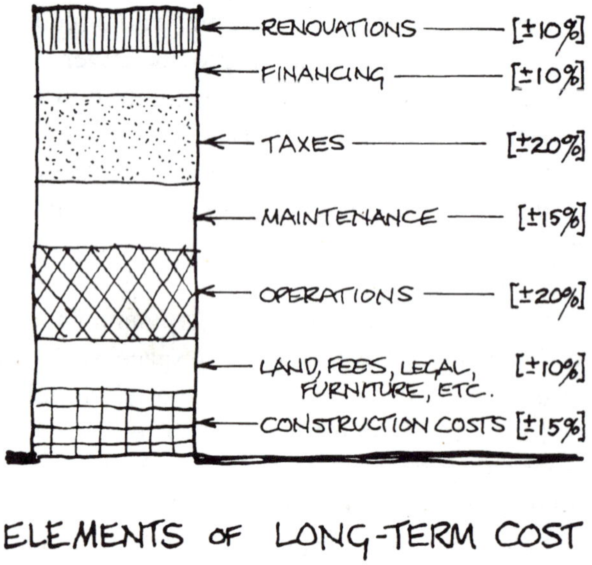

Figure 5.2

Replacement costs can also be predicted and controlled. Generally, high quality products require less maintenance and have a longer life span than lower quality products. They also cost more to purchase. Initial costs and maintenance and replacement costs can be projected and compared over a period of time for various available products.

CONTRACT PROVISIONS

Architect's Responsibilities Under AIA Documents

Architects have a limited responsibility to owners when estimating construction costs of projects. Subparagraph 5.2 of the AIA Owner-Architect Agreement, Document B201, states:

> *Evaluations of the Owner's budget for the Cost of the Work, the preliminary estimate of the Cost of the Work and updated estimates of the Cost of the Work prepared by the Architect, represent the Architect's judgment as a design professional. It is recognized, however, that neither the Architect nor the Owner has control over the cost of labor, materials or equipment; the Contractor's methods of determining bid prices; or competitive bidding, market or negotiating*

> *conditions. Accordingly, the Architect cannot and does not warrant or represent that bids or negotiated prices will not vary from the Owner's budget for the Cost of the Work or from any estimate of the Cost of the Work or evaluation prepared or agreed to by the Architect.*

The architect must study the owner's program and budget for the project to determine if they are reasonably related. For example, a reasonable construction budget for a housing project, in dollars per square foot, may be too low for a hospital. Where a proposed budget is too low, the architect must inform the owner that changes must be made to the program, the budget, or both to bring them into a realistic relationship.

Once the program and budget are reconciled and a schematic design is developed, the architect must, according to subparagraphs 2.2.6 and 5.3 of AIA Document B201, provide the owner with a preliminary estimate of construction cost, based on area, volume, or other unit cost criteria. For example, square foot costs applicable to the particular building type could be used. However, for some buildings where ceiling heights are considerably greater than average, such as auditoriums and gymnasiums, the cost per cubic foot of volume may be more appropriate, since the construction costs are related to volume as well as floor area. Other unit costs, such as the cost per room for hotels or per bed for hospitals, may be more appropriate for estimates for these building types.

Architects attempt to be accurate in making preliminary cost estimates, but accuracy is affected by the available level of detail and other circumstances. They should not create unrealistic project cost expectations and should advise owners about the unpredictability of construction costs.

Subparagraphs 2.2.6, 2.3.2, 2.4.4, and 2.4.5 of AIA Document B201 require the architect to provide updated estimates of construction cost as the design process progresses. Some adjustments may be required to allow for changes requested by the owner, greater detail development in the design documents, or more specific information from structural, mechanical, electrical, and other consultants. Architects may use the subsystem method of cost prediction, rather than area or volume methods, because design development documents contain sufficient detail to prepare such an estimate, which is more accurate than earlier estimates based on schematics. Subparagraph 5.3 of AIA Document B201 states, "[if] the Owner requests detailed cost estimating services, the Architect shall provide such services as an Additional Service under Article 3."

After the construction documents have been prepared, subparagraph 2.4.5 of AIA Document B201 requires the architect to advise the owner of necessary adjustments to earlier estimates. Since it takes a significant amount of time to prepare construction documents, intervening changes in market conditions may cause an owner to revise the requirement of his or her program. For example, an owner may reduce the scope of a proposed office building in response to a weak rental market. Prevailing market conditions may also affect estimated bid prices. If the amount of construction work in the area is considerably different (greater or less) than at the time the previous estimate was prepared, it is likely that bid prices will be different (higher or lower) than previously predicted.

Accuracy Required

Paragraph 5.1 of AIA Document B201 defines the items that are included in an architect's cost estimate. These comprise all elements that the architect designs, specifies, or selects, or for that the architect makes special provisions, as well as the cost of any labor and materials that might be furnished by the owner.

DAUMIER. ROBERT MACAIRE, ARCHITECT

"How now, M. Macaire? This house, which, by your estimate, was to cost me only 70,000 francs, now stands me in 300,000!"

"That isn't my fault. You have a window opening to the south instead of to the north, you want four stories instead of five, and have changed the roof from slate to shingles. I can only be responsible for my original plan. You change it—that's your affair."

Figure 5.3

Certain other costs associated with a project, such as land costs and architect's fees, are not included in this definition and thus are not included in an architect's estimate of construction cost.

Subparagraph 5.2 of AIA Document B201 establishes responsibility for construction cost estimates. It states that evaluations and cost predictions provided by an architect represent the best judgment of a professional familiar with the construction industry, but that these are not a representation or warranty that actual construction costs will not differ from the owner's budget or the cost predictions previously submitted. This statement is essential because architects are not able to control the cost of labor or materials, nor competitive market factors. If an owner is not familiar with these standard clauses, and the actual bid prices exceed the preliminary cost projections, the owner may assume the architect is negligent and not entitled to the fee.

As with all professional services, the architect must exercise reasonable care with the prepared cost projections. If a cost projection is inaccurate because it was negligently prepared, an architect could become liable to the owner for damages.

Fixed Limit of Construction Cost

AIA Document B201 refers to the term owner's *budget*. The architect must treat that budget as if it were such a fixed limit and provide for contingencies for design changes, bidding uncertainties, and cost escalation caused by inflation. The architect must have the authority to decide the quality of materials and equipment to be used. Even the scope of the project may reasonably be adjusted by the architect by reducing the program or including deductive alternates in the bid package to stay within the budget.

Finally, subparagraph 5.4 of AIA Document B201 states that the budget must be adjusted upward to account for changes in the level of prices in the construction industry if bidding or negotiation has not commenced within 90 days after submission of contract documents by the architect to the owner.

If an owner requires that the budget not be exceeded, control of scope and quality must be relinquished to the architect. Architects sometimes put themselves in jeopardy when they agree to a cost limitation, but do not exercise the authority to reduce the scope or quality of a project to keep it within the budget.

Bids in Excess of the Budget

When an architect agrees to design a project to meet the owner's budget, and the lowest bona fide bid or negotiated proposal exceeds that amount, subparagraph 5.6 of AIA Document B201 describes five alternative courses of action the owner may take:

1. Increase the budget and accept the bid amount. This is likely to occur when the difference is small and the funds are available.
2. Rebid or renegotiate the project within a reasonable time. This may occur if market conditions are likely to improve in the near future.
3. Abandon the project and terminate the architect's contract. This may occur if there does not appear to be any reasonable course of action which will bring the project within the budget.
4. Cooperate with the architect to reduce the scope, quality, or both of the project and rebid it in the new form.
5. Implement some other course of action to be determined.

If the fourth alternative is selected, the architect must, without additional charge, modify the drawings and specifications as necessary to comply with the budget. Subparagraph 5.7 of AIA Document B201 states that this is the extent of the architect's responsibility for the Construction Documents Phase. Once the redesign is complete, the architect is entitled to compensation for all services performed whether or not construction is started.

Consultants' Responsibilities Under AIA Documents

Consultants have responsibilities for cost estimates that parallel those of the architect. The AIA Architect-Consultant Agreement, Document C401, describes the responsibilities of engineers providing services on a project.

The definition of construction cost and provisions related to responsibility for cost predictions are almost identical to those in the Owner-Architect Agreement. Thus, the consultant's representations to the architect under AIA Document C401 are no more extensive than those of the architect to the owner.

Consultants must cooperate with the architect to decide what proportion of a project's construction budget must be allocated to their part of the project. For example, an architect and structural engineer may decide that the cost of the structural system is not to exceed 20 percent of the total cost of construction. During the design phase, consultants must submit preliminary estimates of construction cost to the architect for their part of the project. These estimates form a part of those submitted by the architect to the owner for the entire project.

The architect and each consultant should agree on a budget for the consultant's part of the project. This is possible when the work designed and specified by the consultant can clearly be segregated from other work, for example, mechanical work, which is clearly distinct from architectural work.

If the budget for a consultant's part of the project is exceeded by the pertinent part of the low bid, an architect can require that particular consultant, without additional charge, to modify the drawings and specifications in order to bring the cost of the project within the agreed budget. Once that is accomplished, the consultant has no further responsibility with respect to the cost limitation, and is entitled to compensation for services rendered.

If it is impossible to establish a budget for a consultant's portion of a project, such as the structural system, for example, that consultant is responsible for modifying the drawings and specifications to reduce the cost of this part of the project in reasonable proportion to the overall reduction required to bring the total project into the budget. For example, if the structural system represents about 20 percent of the total construction cost, the structural engineer may be responsible for 20 percent of the required reductions. Once this has been achieved, the consultant's responsibility has been fulfilled with regard to the fixed limit of the construction cost and is entitled to be paid for services rendered, without any additional charge for the modifications.

TYPES OF ESTIMATES

Area/Volume Estimates

During the early stages of design, construction costs are usually estimated on the basis of cost per square foot of area or cubic foot of volume. These methods are widely used because area or volume can easily be determined and multiplied by a dollar figure to arrive at an estimated cost.

The method used for calculating area or volume should be consistently applied from project to project. One widely accepted method is described in AIA Document D101, Architectural Area and Volume of Buildings.

Once a building's area or volume has been calculated, an appropriate unit cost must be selected. Generally it is based on actual costs of similar past projects. However, such costs must be adjusted, since costs vary in relation to a project's size, perimeter, height, number of stories, site conditions, type of construction and finish, mechanical equipment, and time and place of construction.

For a firm that has extensive information about construction costs of past projects, determining the appropriate unit cost for a particular building type is relatively easy. For firms that specialize in one building type and that operate in one location, the estimated unit costs are likely to be quite accurate. However, when a firm ventures into an unfamiliar building type or a new geographic market, it must seek help in determining the appropriate unit costs. One source consists of publications such as Means Building Construction Cost Data, Dodge Reports, or Building Design and Construction News. An example of this type of information is shown in Figure 5.4. A local contractor or architect who specializes in the building type under consideration might also provide reliable cost information.

Architects must account for special features in a project and should include contingency amounts because of the lack of available detail in early design stages. An allowance of 10 to 20 percent of the estimated project cost might be an appropriate contingency figure in the schematic design phase.

Subsystems Estimates

As more detailed information becomes available on a project, architects often switch from area/volume methods to subsystems methods of cost estimating. Subsystems methods deal with a project's functional units or assemblies. Subsystems are consistent from project to project

Building Type/System	UNIT	UNIT COSTS IN $ LOW	MEDIUM	HIGH
COLLEGES Classroom & Administration	S.F.	74	95	118
Total project costs	C.F.	5	7	11
Masonry	S.F.	5	6	10
Plumbing	↓	3	5	10
Heating, ventilating, air conditioning	↓	8	11	18
Electrical	↓	6	9	12
Total: Mechanical & Electrical	↓	13	23	36
COLLEGES Science, Engineering, Laboratories	S.F.	101	119	142
Total project costs	C.F.	6	9	10
Equipment	S.F.	4	12	16
Plumbing	↓	5	6	8
Heating, ventilating, air conditioning	↓	6	13	15
Electrical	↓	9	12	16
Total: Mechanical & Electrical	↓	27	37	54
COLLEGES Student Unions	S.F.	72	102	118
Total project costs	C.F.	4	5	7
Plumbing	S.F.	5	7	8
Heating, ventilating, air conditioning	↓	11	13	19
Electrical	↓	6	9	12
Total: Mechanical & Electrical	↓	19	27	30

Figure 5.4

though materials and methods of construction may change.

Subsystems estimates enable comparison between different conceptual solutions during the design phase. An architect deciding between exterior cladding of precast concrete or anodized aluminum, for example, benefits from cost data that allows direct comparison between basic systems. These prices are usually stated in dollars per square foot, allowing the architect to compare the systems on that basis.

Although there are various breakdowns used for the basic subsystems, the following is typical:

Building Costs

1. Foundation
   - Standard
   - Special
2. Substructure
   - Slab on Grade
   - Basement Walls
3. Superstructure
   - Floor Construction
   - Roof Construction
   - Stair Construction

4. Exterior Closure
    - Exterior Walls
    - Exterior Doors and Windows
5. Roofing
6. Interior Construction
    - Partitions
    - Interior Finishes
    - Specialties
7. Conveying Systems
    - Elevators
    - Moving Stairs/Walks
    - Dumbwaiters
    - Pneumatic Tube System
    - Other Conveying Systems
8. Mechanical Systems
    - Plumbing
    - HVAC
    - Fire Protection
    - Medical Gas System
    - Sewage Treatment
    - Solar Energy Mechanical System
9. Electrical Systems
    - Basic Materials and Methods
    - Lighting & Power
    - Special Electrical Systems
    - Communications Systems
    - Electrical Heating Systems
10. General Conditions
11. Equipment
    - Special Equipment
    - Furnishings
    - Special Construction

Site Development Costs

12. Site work
    - Site Preparation
    - Site Improvements
    - Site Utilities
    - Off-site Work

A portion of a detailed breakdown is shown in Table 5.1.

Detailed Estimates

The detailed estimates method of predicting construction costs is sometimes referred to as the *Quantity and Cost Method* or the *Labor and Materials Method*. It requires a detailed calculation of the amount of each type of material and labor necessary to produce the required construction. Costs per unit of material and labor are applied to the calculated quantities to arrive at the total direct cost of the construction work.

Indirect costs must also be added, including

- the contractor's overhead, including insurance, payroll taxes and benefits, general and administrative expenses such as site office salaries, and equipment rental costs;
- General Conditions costs, including project signs, engineering surveys and inspections, tests, drawings and photographs, permits, and repairs and clean-up;
- a contingency amount for cost escalation and unforeseen conditions; and
- contractor's profit calculated as a percentage of total direct construction costs.

Detailed estimates of construction cost are usually a change in services for which the architect must be additionally compensated. Many architects hire cost estimating consultants to prepare detailed estimates to supplement their own skills and experience.

The calculation of quantities of labor and materials is commonly called a *quantity take-off* or *quantity survey*. A properly prepared take-off accurately lists quantities of all materials and notes the various items necessary for the construction of a particular project. Of course, all items must be included and none intentionally omitted from the take-off. Special items should be listed separately from the more common items since these will require extra attention and usually are assigned different unit costs. See Tables 5.2 and 5.3.

SAMPLE OF DETAILED SUBSYSTEMS AND UNITS OF MEASURE

System or Subsystem Name	Unit of Measure
Foundation	Footprint Area
Standard	Footprint Area
Excavation & Backfill	Cubic Yds.
Concrete-Forms	Cubic Yds. Conc.
Masonry	Sq.Ft. Masonry
Special	Footprint Area
Rock Excavation	Cubic Yds. Excav.
Dewatering	Lump Sum
Structural Steel	Tons of Steel
Piles or Caissons	L.F. Piles/Caissons
Underpinning	Cubic Yds. Conc.
Substructure	Sq.Ft.
Slab on Grade	Sq.Ft.
Gran. Fill below Slab	Tons of Fill
Found./Underslab Drain	L.F. Pipe
Concrete-Forms	Cubic Yds. Conc.
Water/Dampproofing	Sq.Ft. Contact Area
Thermal Insulation	Sq.Ft. Contact Area
Basement Walls	Sq.Ft. Wall
Concrete-Forms	Cubic Yds. Conc.
Masonry	Sq.Ft. Masonry
Water/Dampproofing	Sq.Ft. Contact Area
Superstructure	Sq.Ft. Fl. & Rf. Area
Floor Construction	Sq.Ft. Floor Area
Concrete-Forms	Cubic Yds. Conc.
Precast Struc. Components	Sq.Ft. Comps. Area
Structural Steel	Tons of Steel
Rough Carp. Frame Deck	Board Ft. Lumber
Hvy. Timber Prefab Struct.	Board Ft. Lumber
Cementitious Decks	Sq.Ft. Deck
Roof Construction	Sq.Ft. Roof Area
Concrete-Forms	Cubic Yds. Conc.
Precast Struc. Components	Sq.Ft. Comps. Area
Structural Steel	Tons of Steel
Rough Carp. Frame Deck	Board Ft. Lumber
Hvy. Timber Prefab Struct.	Board Ft. Lumber
Cementitious Decks	Sq.Ft. Deck

Table 5.1

EXAMPLE OF DETAILED ESTIMATE (SUMMARY)

QUANTITY TAKE-OFF		_________ ESTIMATE
PROJECT Milano Junior College/Second Increment		JOB NO. JS-104 PG 1
ARCH/ENG/OWNER P. Palladio		PREPARED BY U.F.O.
PROJECT LOCATION Milano, California		CHECKED BY R.F.D.
DESCRIPTION Classroom Building		DATE September 15, 1998
ITEM NO.	DESCRIPTION	TOTAL COST
	SUMMARY OF ESTIMATE	
1a	General Conditions	119,157
2a	Demolition	See Site Estimate
2b	Service Site	See Site Estimate
2c	General Site	See Site Estimate
2d	Off Site Work	See Site Estimate
2e	Underpinning, Shoring	See Site Estimate
2f	Site Utilities	See Site Estimate
2g	Landscaping & Irrigation	See Site Estimate
2h	Piling, Piers, Caissons	None
2i	Excavation & Fill	None
3a	Concrete, Foundations	50,504
3b	Concrete, Structural	669,174
3c	Concrete, Architectural	None
3d	Concrete, Precast	None
3e	Concrete, Slabs on Grade	60,762
3f	Reinforcing	191,756
3g	Cementitious Decks	None
4a	Masonry & Stone	None
5a	Structural Steel	None
5b	Misc. Iron & Arch. Metals	43,402
5c	Metal Siding & Decks	None
*6a	Carpentry, Rough	21,600
*6b	Carpentry, Finish	29,440
6c	Glulam Beams & Trusses	None
6d	Millwork	None
*6e	Rough Hardware	4,651
6f	Stairs & Rails, Wood	None
*7a	Roofing & Rigid Insulation	27,433
7b	Waterproofing	None
*7c	Sheet Metal & Skylights	21,201
*7d	Caulking & Sealants	7,442
7e	Thermal & Sound Insulation	4,735
7f	Arch. Sheet Metal	None
8a	Wood Doors & Frames	11,837
8b	Hollow Metal Work	19,728
8c	Store Front, Sash & Doors	13,415
8d	Glass, Glazing & Sash	18,939

*See next page for details

Table 5.2

EXAMPLE OF DETAILED ESTIMATE (DETAIL SHEET)

QUANTITY TAKE-OFF				____________ ESTIMATE	
PROJECT Milano Junior College/Second Increment				JOB NO. JS-104	PG 4
ARCH/ENG/OWNER P. Palladio				PREPARED BY U.F.O.	
PROJECT LOCATION Milano, California				CHECKED BY R.F.D.	
DESCRIPTION Classroom Building				DATE September 15, 1998	
ITEM NO.	DESCRIPTION	ESTIMATED QUANTITY	UNIT	UNIT COST	TOTAL COST
6a	CARPENTRY, ROUGH				
	Standby	6	MO	3,600	21,600
	Sub Total #6a				21,600
6b	CARPENTRY, FINISH				
	Shelving	810	LF	3.00	2,430
	Install Only:				
	Doors	128	EA	36.00	4,608
	Frames	128	EA	36.00	4,608
	Hardware	128	SETS	48.00	6,144
	Cabinets, Flr.	736	LF	7.00	5,152
	Cabinets, Wall	342	LF	10.00	3,420
	Cabinets, Full Ht.	342	LF	9.00	3,078
	Sub Total #6b				29,440
6e	ROUGH HARDWARE	46,512	SF	.10	4,651
	Sub Total #6e				4,651
7a	ROOFING & RIGID INSULATION	233	SQS	117.74	27,433
	Sub Total #7a				27,433
7c	SHEET METAL & SKYLIGHTS				
	Expansion Jts., Wall	240	LF	24.40	5,856
	Clg.	96	LF	22.40	2,150
	Floor & Roof	144	LF	16.20	2,333
	Genl. SM	23,256	SF	.32	7,442
	Skylights In Cores	342	SF	10.00	3,420
	Sub Total #7c				21,201
7d	CAULKING & SEALANTS		LS		7,442
	Sub Total #7d				7,442

Table 5.3

Subcontract items must be added separately. Once all quantities are available, they can be priced to arrive at the total direct cost of construction. Detailed estimates of construction cost are usually presented in the Construction Specifications Institute (CSI) MasterFormat Division headings.

FACTORS AFFECTING COST

Legal and Administrative Requirements

Legal or administrative factors grouped under the heading of *General Conditions* affect the cost of a project. Legal factors may include provisions that actually or potentially increase a contractor's obligations. For example, if provisions for *liquidated damages* are included in the contract to compensate the owner for any delay in the completion of construction, the contractor must assess that risk and include an allowance for it in his or her bid. If an owner includes a *no damages for delay* clause to preclude the contractor from seeking damages for delays caused by the owner, that risk must also be assessed and priced by the contractor for inclusion in the bid.

Administrative requirements and their costs must also be considered. An owner may require the contractor to purchase various types of insurance, including property insurance, for the project during construction. A project may have unusual or costly requirements for field offices or for submittals of samples and shop drawings. Testing requirements may be extensive, and required project photographs and other documentation may be voluminous.

Complexity of the Project and the Documentation

The size, shape, and complexity of a building affect its cost. For example, large buildings usually cost less per unit of area than small ones. Similarly, compact buildings tend to be less expensive to construct than those with a large or irregular footprint. Unit costs may drop approximately 3 percent for each 10 percent reduction in a building's perimeter. On the other hand, buildings taller than eight or ten stories, or with unusually great floor-to-floor heights, will be more expensive per unit of area than more conventional low-rise buildings.

The complexity or lack of clarity of construction documents may also affect costs. If contractors are confused by the intent of or unsure about aspects of the contract documents, they may include contingency allowances in their bids to allow for unforeseen events. In a competitive bidding climate, however, contractors may tend to minimize this contingency factor in order to remain competitive.

Construction Materials and Methods

The required quality of construction materials and workmanship is a major determinant of construction cost. If a project is a special type of building with stringent requirements such as a *clean* manufacturing facility for electronic components, or if a project requires unusual subsurface preparation or foundations, construction costs will be greater than average. On the other hand, buildings that contain many typical details or that have regular and repetitive floor layouts will be less expensive than average to construct.

Location of the Project

Both labor and material costs vary with the location of a project. For example, the abundance of lumber in the Northwest generally makes wood framing less expensive than in areas where construction lumber must be shipped.

Skilled labor for various construction trades may not be equally available in every location. If construction methods are employed that require specialized skills not locally available, prices tend to rise. The strength of trade unions in an area also influences construction prices: Areas with strong, well established unions often have higher than average labor costs.

The climate also affects construction costs. It can influence an architect's choice of construction materials and may affect both design and construction schedules.

Construction Schedule

A short construction schedule may result in increased construction costs. A contractor may be required to increase work crews and equipment or schedule overtime work to meet deadlines. However, additional workers and equipment can congest the construction site and reduce efficiency.

On the other hand, when a project is extended beyond a normal schedule, the contractor will incur additional overhead costs. The contractor must include in the bid continuing costs for on-site supervision, equipment rentals, sanitary and rubbish services, and similar items.

The weather also affects the construction schedule because work slows in cold or wet weather, and some types of work cannot be performed at all. If, in these circumstances, provisions in the contract require a contractor to pay liquidated damages to an owner if construction is not completed on time, contractors may increase their bid to cover this contingency. However, extensions of time are normally granted for delays caused solely by unexpected inclement weather.

Bidding Environment

The number of projects available for bidding or negotiation and the amount of competition for those projects influence construction costs. In situations where there is a great deal of competition for a small number of projects, contractors may minimize or even ignore uncertainties and contingencies that, under less competitive conditions, would result in higher bids. Architects must assess the level of competition and consider this aspect in preparing construction cost estimates.

OTHER ELEMENTS OF PROJECT COST

The *direct cost of construction* is only one element of the total amount that an owner must expend in the development of a project. However, it is usually the only element covered by an architect's cost estimates. Architects normally do not attempt to predict or directly influence other project costs.

It is important to distinguish between the direct cost of construction and the *project budget*. The project budget may be 30 to 50 percent greater than the direct cost of construction, and may include some or all of the following elements, in addition to the direct cost of construction.

1. Owner's in-house staff costs, legal fees, and fees for outside consultants
2. Land acquisition, including the cost of rezoning, if necessary
3. Demolition of existing structures or other improvements
4. Site work
5. Landscaping
6. Furniture, furnishings, and equipment (FF&E)

7. Special equipment
8. Professional fees for architects, engineers, and special consultants
9. Insurance
10. Financing
11. Taxes during construction
12. Contingencies for unforeseen conditions

Note that items 3 through 7 are sometimes included in the direct cost of construction.

It is important that architects and owners have a common understanding of the amount of money available for construction, and that the owner has sufficient funds available for the other elements of a project's cost.

LESSON 5 QUIZ

1. An architect's preliminary estimate of construction cost would consider all of the following EXCEPT

I. labor and materials supplied by the owner.

II. cooling towers.

III. land acquisition costs.

IV. the architect's fee.

V. metal curtain wall systems.

A. I only

B. I and IV

C. None, everything would be considered

D. III and IV

2. An architect who prepares a preliminary estimate of construction cost according to the subsystems method

A. considers one set of subsystems for concrete buildings and a different set for steel buildings.

B. can easily analyze the cost of alternative exterior wall systems.

C. would most likely use this method during the schematic design phase.

D. would probably be performing an additional service under the AIA Agreement between Owner and Architect.

3. Detailed estimates of construction cost

I. are usually a change in services.

II. are usually basic services.

III. are calculated solely by determining the sum of the required quantity of materials, multiplied by their unit costs.

IV. must include indirect costs such as contractor's overhead and administrative costs, profit, and the cost of General Conditions items.

A. I only

B. I and IV

C. II only

D. I and III

4. If the lowest bid on a project is greater than the budget, the owner has which of the following choices under the provisions of the AIA documents? Check all that apply.

A. Agree to waive the budget

B. Rebid the project when market conditions are more favorable

C. Require the architect to modify the contract documents at no cost to the owner so that the project can be rebid

D. Abandon the project

E. Sue the architect for breach of contract

F. Ask the architect to modify the contract documents and sue if the resubmitted bids still exceed the budget

5. Once an architect has completed a specific type of project, the cost per square foot can be determined and used to predict the cost of a future project of the same type. However, such unit costs must be adjusted to account for which of the following factors?

I. The location of the project.

II. The project's type of construction.

III. The perimeter and configuration of the building.

IV. Whether the owner is a private or public entity.

V. The number of stories of the building.

VI. The source of construction funding.

A. II, III, and IV
B. I, II, and VI
C. III, V, and VI
D. I, II, III, and V

6. In order to meet an owner's budget for a project, the architect has control over which of the following?

A. Quality only
B. Quality and scope
C. Quality and price
D. Quality, scope, and price

7. Architects are able to control which of the following factors affecting construction cost of projects?

A. Cost of labor
B. Cost of materials
C. Contractors' bidding methods
D. None of the above

8. When contractors prepare bid prices, they calculate direct costs, normal overhead, and projected profit. Which of the following factors may also be legitimate considerations?

I. Project funding

II. Potential liquidated damages

III. Competitive conditions

IV. Special insurance requirements

V. The weather

VI. Errors and omissions in the drawings and specifications

A. I and VI
B. II and IV
C. II, III, IV, and V
D. II, IV, V, and VI

9. An architect's preliminary estimate of construction cost is generally

A. based on the National Construction Cost Index.

B. subject to adjustment because of fluctuating market and competitive conditions encountered during bidding.

C. considered to be an additional service under the AIA Owner-Architect Agreement.

D. guaranteed within 5 to 10 percent of the ultimate construction cost.

10. In preparing an initial estimate of construction cost for a building type with which the architect has no experience, an architect may consult with

I. a cost estimating consultant.

II. a contractor experienced in the building type.

III. an architect experienced in the building type.

IV. cost estimating publications.

A. III and IV

B. I, II, and IV

C. I only

D. I, II, III, and IV

11. A contractor's indirect costs include which of the following?

A. General conditions

B. Excavation

C. Architect's fees

D. FF&E

LESSON SIX

TIME AND SCHEDULE

DESIGN SCHEDULING

Establishing a Schedule

In furnishing professional services, an architect must prepare a time schedule that encompasses all phases of production, from initial conceptual planning to the start of construction. The architect must plan the judicious and efficient use of manpower and resources to achieve an economical, functional, and harmonious design, executed within a reasonable period of time, and with an efficient utilization of personnel. The managerial skills required for such planning and scheduling are based on experience and judgment.

To organize the schedule, the architect first separates the design effort into phases, which generally correspond to phases of the AIA standard owner-architect agreements as follows:

1. *Schematic design*, consisting of schematic drawings and other documents that describe the general relationships and space requirements of the project, along with a cost estimate.
2. *Design development*, consisting of preliminary drawings, outline specifications, and other documents that describe the form, size, and materials of a project, and the structural, mechanical, and electrical systems to be utilized. A preliminary cost estimate is also prepared during this phase.
3. *Construction documents*, consisting of working drawings, final specifications, and a final cost estimate.
4. *Bidding or negotiation*, which includes the receipt and evaluation of bids or negotiated

proposals. It may also include preparing addenda to the contract.

5. *Construction administration*, consisting of the services rendered by the architect after bidding or negotiation to assure that the structure is built in accordance with the construction documents. In this phase the architect may issue change orders, approve shop drawings, choose or approve materials and colors, and issue payment approvals.

In complex projects, the five phases described may not be adequate. For example, schematic design may be divided into conceptual design and schematic design. Similarly, the construction documents phase may be organized into several subdivisions, so that work on one subdivision may be completed and bid before the next phase is begun.

The architect must estimate the time required for each phase of the work. The schematic design phase is the most difficult to estimate, since it has the greatest amount of variability. This phase of the work is usually done by a small design team, generally headed by a chief designer, and possibly including an engineer and other specialists. The design concept must be developed out of the skill and experience of the design team working closely with the client and each other. Among the factors affecting the time required for schematic design are:

1. *The size and complexity of the project,* complexity generally being more critical than size.
2. *The quality and completeness of the program information supplied by the client.* If the architect does not have an adequate statement of the client's requirements—a conclusive program—then it will be necessary to prepare one, or to improve what exists. In contrast, an experienced client will often furnish the architect with a thorough and reliable catalog of needs, thereby enabling the architect to begin work immediately. Such a client may provide the architect with information such as project goals, area requirements, functional relationships, zoning information, a site survey, and a budget.
3. *The decision-making ability of the client.* If the client has a decisive representative who has the authority to make decisions, schematic design can proceed at a rapid pace. On the other hand, if decisions require committee approval, or if they cannot be made expeditiously, schematic design time will be prolonged, with consequent loss of momentum. If the client and architect do not have an effective communication system, the process is further delayed.
4. *The nature of the design team.* If the team is well balanced, if they work together harmoniously, if they are skilled and experienced, if they are able to work on the project without interruption, and if they can communicate readily with the client, then schematic design time will be kept to a minimum.

These factors illustrate why it is difficult to plan a time schedule for schematic design. For a simple, conventional project, schematic design can often be completed in one or two months. It is not unusual, however, for the schematic design of a complex project to require 12 months or more.

The design development and construction documents phases of the work are much more predictable than schematic design, assuming schematic design has been thorough and there are no program changes. A team of architects and drafters, headed by a project architect and/or job captain, develops the schematic design into preliminary design drawings, which are then developed into working drawings. If the project is large in scope, staffing must be increased commensurately. The length of time required to produce these drawings may not be directly proportional to the size of the project. A $10,000,000 project, for example, may require only 50 percent more time than a $2,000,000 project. The complexity of a project, rather than its size, determines scheduling and staffing requirements. During the preliminary design and design development phases, close coordination between consultants, client, and designers is vital.

Design development for a typical project takes from two to four months; the construction documents phase may typically require from three to seven months. The bidding or negotiation phase usually requires three to six weeks, regardless of the size of the project.

A less obvious factor that may influence the work schedule is project financing. Whether the client is an individual, a partnership, a corporation, or a public agency, money is required to convert a design into reality. Private clients may borrow money from a bank, while a public agency may have to obtain a bond issue. The client may use the time between work phases to obtain a financial commitment and may, in some cases, postpone authorization to the architect to proceed with a successive phase until financing is secured. This may take weeks or months for a private client and even longer for a public agency.

Client review and approval is customary between phases, and the time required for this will depend on the size and complexity of the

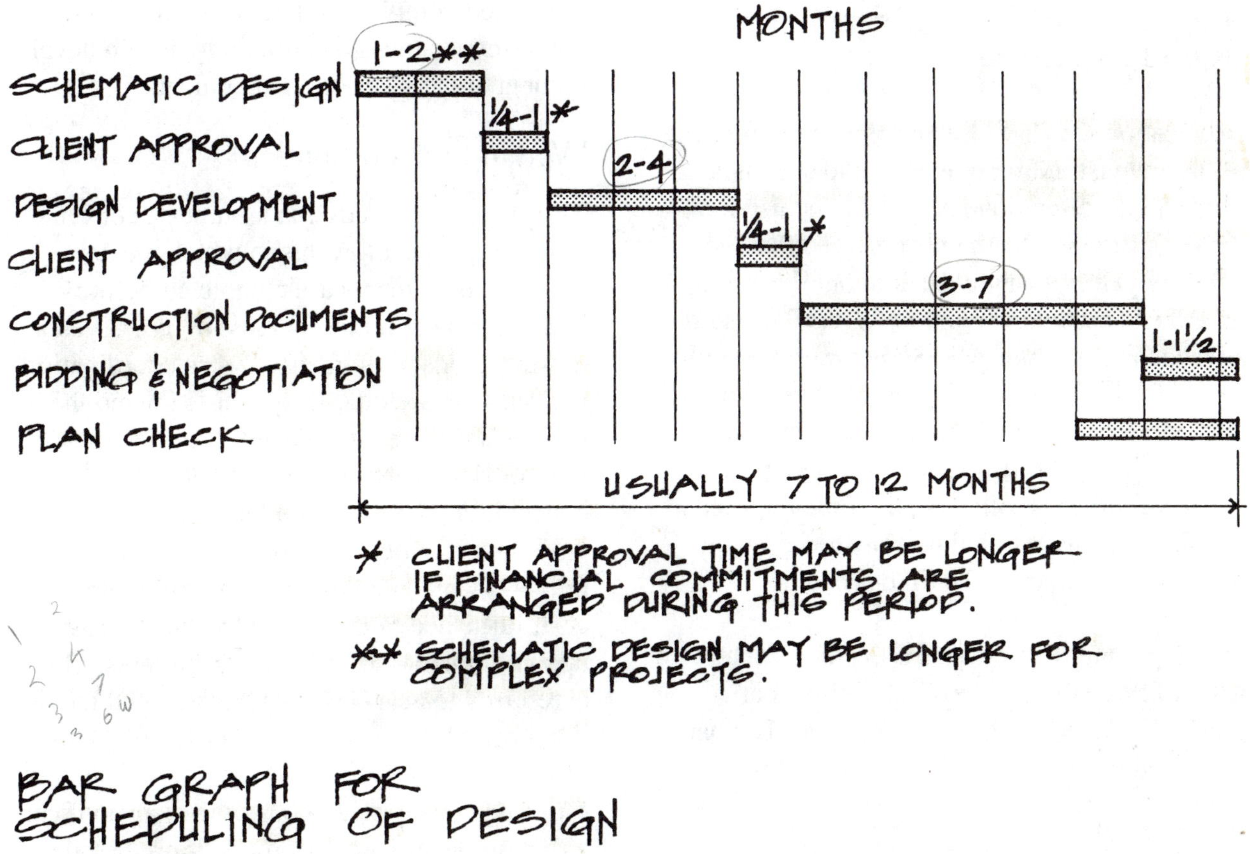

Figure 6.1

project, as well as the ability of the client to make decisions.

Some projects require more than one client approval, which may lengthen the review period. For example, many public school projects require the approval of a state department of education as well as a local school district. Client review and approval usually takes between one week and one month, unless complications arise.

The time required for approval of plans by a building department or other public agency varies considerably, depending on the locality and type of project. For example, a state hospital project, which may require approval by a state agency as well as the local building department, may require up to three months for plan checking. In localities where the checking of plans is less critical, a building permit may be obtained within a week.

Application for a building permit requires the filing of construction drawings and specifications; this is often done near the conclusion of the construction documents phase so that the building permit is obtained at about the same time the construction contract is let. This is not always the case, however. Sometimes the application for the building permit is not made until after the bidding period, while in other cases, the permit is obtained before the bidding phase. Whatever order is followed, the time required for plan approval should be considered by the architect in preparing the time schedule.

In completing the time schedule, the architect assembles all the time estimates into a bar graph, as shown on the previous page. The bar graph indicates ranges of time for each phase. In an actual project, however, a specific period of time would be assigned for each phase.

Contingencies

In organizing the architectural production schedule, the architect must consider the possibility of unexpected problems that may arise. There may be delays with the building department, consultants may need additional time because of unique problems inherent in the project, staffing problems may arise in the architect's office if the work load changes suddenly, or the client may be less decisive than expected. For these and other reasons, it is wise to include a contingency factor in the schedule. If the architect estimates the total required time to be eight months, an additional allowance of at least two to four weeks seems prudent.

The schedule should be flexible and responsive to changing conditions. For example, if the schematic design phase extends beyond its scheduled completion date, it should be possible to reduce the time allotted to design development and construction documents.

Working with a Builder

The preceding discussion assumes a conventional sequence of events in which the construction documents are completed before bidding or negotiation begins. In recent years, however, closer methods of work coordination between the architect-engineer team and the builder have been developed. Many architects now work closely with a contractor from the conceptual phase through the completion of working drawings. A result of this cooperation is often a guarantee of maximum project cost, furnished to the owner by the contractor, upon completion of contract documents. This is referred to as a GMP—a "guaranteed maximum price."

Working closely with a builder has a significant effect on the architect's production schedule. More time must be given to schematic design if the architect is to produce a concept that the contractor considers economical. Design development, likewise, may take more time; however, construction documents will probably take the same time. Since the time during which the drawings are being prepared overlaps actual construction, overall project time is generally shortened. But there are risks in this procedure that the building design may not be fully developed or the components fully resolved.

Regardless of the procedure followed, the working drawings and specifications must be complete, clear, and correct. In some cases where the architect works closely with the contractor who will construct the project, the documents may be less specific, allowing the contractor leeway in procedures, details, and materials. This practice can be risky for both, and hence, should be restricted to common or repetitive projects. With close architect/builder cooperation, the bidding and negotiation phase may be omitted entirely, since these activities become a continuous process.

The total scheduled production time is usually similar to what it would be if the project were done conventionally. The architect's staff hours, however, may be greater because of the time spent coordinating with the contractor and possible redesigning. There are no short cuts; architectural projects require attention to detail, and invariably that takes time.

Extending the Schedule

All creative activity requires time, which should be enough to absorb information and develop ideas, but not so much that momentum and interest lag. For architectural design, an optimum work schedule is one in which the necessary work can be accomplished comfortably without expanding or shortening the schedule.

On a project with an extended schedule, principal team members may retire or take other positions before completion of the work. A recent state college project was delayed four years, between design development and construction documents, because of lack of funding. When the project resumed, the original project architect, mechanical and structural engineers, and key client personnel had made career changes. The resumption of work entailed starting over. The groundwork had to be reestablished, resulting in wasted time and effort.

One of the most significant effects of an extended design schedule is the increased cost due to inflation. In the recent past, inflation ran as high as 1 percent per month. At that rate, a $10,000,000 project which is delayed two months would cost the owner an additional $200,000. The additional cost resulting from the delay of a project may cause it to be terminated or reduced in scope. For example, in the case of the state college project mentioned above, the original project budget could not be increased during the four-year delay, and therefore the scope of the project had to be reduced by about one-third. The facility as finally built was smaller and of lower quality than it would have been without the four-year delay.

Shortening the Schedule

Clients often want their projects completed in as short a time as possible. During periods of inflation, there is additional pressure to shorten the design schedule. The purpose of any schedule, however, is to make optimum use of staff effort and resources. Therefore, to achieve significant reductions in time, one or more of the following methods must be employed:

1. The architectural team works overtime. While this saves time, it is costly and inefficient. A person working a ten-hour day over a long period of time cannot

consistently produce 25 percent more work than someone working an eight-hour day.

2. Hire more people, bring in part-time or freelance staff, or subcontract work to another firm. All of these solutions are possible and will probably save time, but they are also costly and inefficient. New staff people will not be familiar with office procedures or the particular project, and their competence is unknown. Part-time people may be experienced and competent, but they are usually expensive. Subcontracting to another firm is feasible, but this is expensive, and coordination and supervision may be awkward.
3. Reduce the man-hours spent on the project. This generally results in a lower-quality job. Quality work requires adequate time to produce, and if that time is not available, an incomplete set of working drawings and specifications may result. Under these circumstances, one can expect documents that are incomplete, unclear, and likely to contain errors and inconsistencies. That, in turn, implies future problems, delays, and excessive change orders during construction.

Thus, the net effect of a reduced time schedule is likely to be a higher cost for design, a higher cost for construction, and a lower quality project. During periods of high inflation, an owner may be willing to tolerate a degree of increased costs with decreased quality, but this decision should be made only with the client's full appreciation of the consequences.

Methods of shortening both the design and construction schedule, simultaneously, will be described shortly.

CONSTRUCTION SCHEDULING

Establishing a Schedule

By their very nature, all construction projects are complicated, since they involve the work of numerous trades and subcontractors, all of which must be coordinated. Equipment must be utilized efficiently; materials must be ordered, stored, and used in a logical sequence; and accurate time schedules and costs must be recorded.

When a contractor prepares a construction schedule for a project, it is generally based on past experience. But no two projects are ever exactly alike, no two sites are the same, and therefore construction schedule estimates must be tempered with judgment. Contractors must consider a number of factors, including the following:

1. *The construction documents.* If these have been well prepared, relatively few problems or delays may be expected. Conversely, a poor set of working drawings or specifications will lead to disputes among the architect, contractor, and subcontractors. Such disputes consume considerable time and energy.
2. *The architect-engineer.* Some architects and engineers are extremely demanding regarding the interpretation of the contract documents. Others are less demanding and more amenable to changes.
3. *The subcontractors.* The contractor must evaluate their ability to perform the work properly and on time, and to coordinate their work with others.
4. *The contractor's organization.* The skills of the project manager, field superintendent, and the office and field staffs must be considered in relation to the specific project. Some managers and superintendents are more capable of

expediting the work than others. Also, the particular work load of the contractor will influence his ability to divert staff and equipment to and from the project under consideration.

5. *Material dealers.* The contractor must assess their reliability in meeting delivery schedules on time and correctly.
6. *The size and complexity of the project.* Complexity is one of the most critical elements in planning a construction schedule.
7. *Site conditions.* The size and accessibility of a construction site work area are critical factors in schedule planning. So is the condition of the site itself—its drainage, vegetation, subsoil, etc.
8. *The weather.* This is important, especially in the colder areas of the country, where projects may have to be shut down during snowstorms, heavy rainstorms, or periods of extreme cold.
9. *The possibility of labor troubles.*
10. *The possibility of material shortages* or delay in obtaining critical equipment.

The contractor must estimate the time required for each construction operation and the sequence of these operations in order to establish the schedule.

CPM (Critical Path Method)

The first step in developing a "critical path" is the planning phase, in which a diagram is drawn indicating the order in which the various operations comprising the project are to be accomplished. The project is divided into concise tasks called "activities," and these are represented by arrows on the CPM chart.

Each activity has a definite start and finish represented by circles, and referred to as "events" or "nodes." An "event" is defined as that moment when a preceding activity has been completed and the following activity may begin. Important points in the construction process, such as the roofing of a new building, are referred to as "milestone events."

In CPM planning there is no indication of time; the arrows are not drawn to a time scale. The tail of an arrow indicates the start of an activity and the head of an arrow, the finish, and each arrow is associated with a start and finish event. No new activity can be started until activities represented by all the previous arrows have been completed.

The completed CPM diagram is known as a *network diagram.* The network must be continuous, with no gaps or discontinuities.

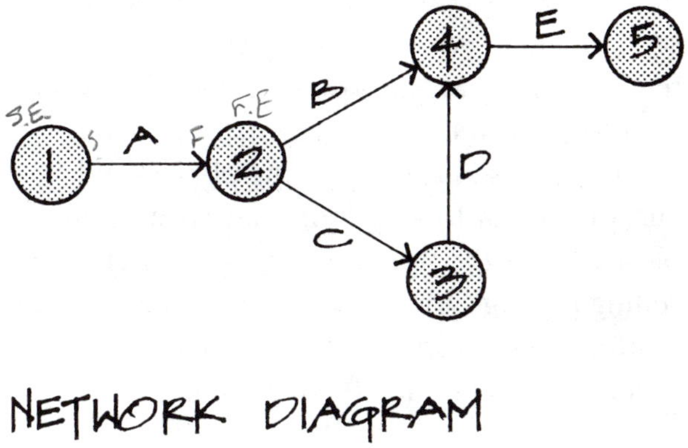

Figure 6.2

In the network diagram shown in Figure 6.2, activity A starts at event 1 and terminates at event 2. Activities B and C cannot start until A is completed. Activities B and C can proceed simultaneously; however, activity D cannot start until C is completed. Activity E, starting with event 4 and finishing with event 5, cannot start until both activities B and D are completed. The construction of a footing supported on drilled cast-in-place concrete piers will now be considered.

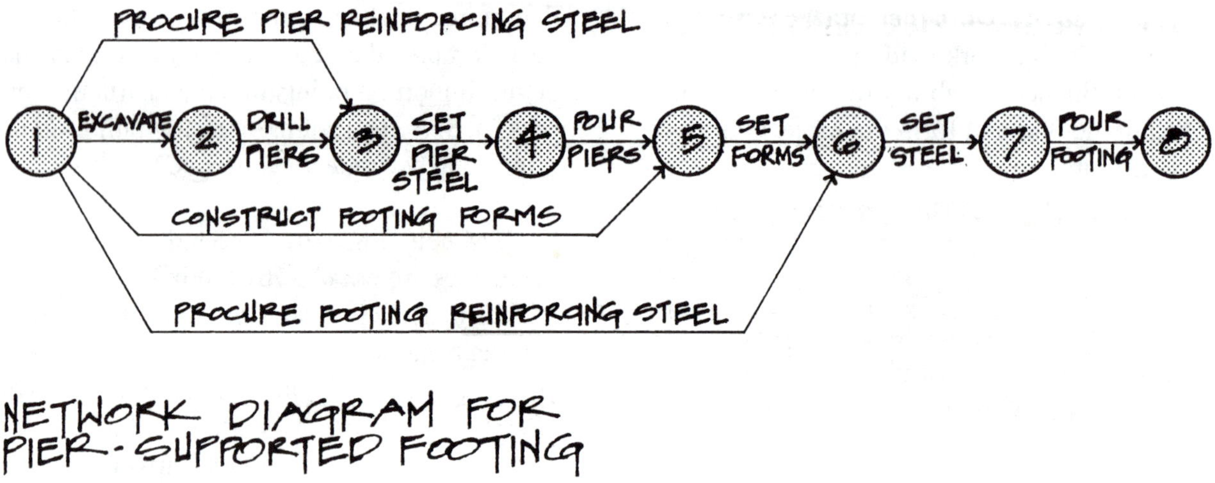

Figure 6.3

Excavation of earth, construction of footing forms, and procurement of reinforcing steel can all proceed independently of each other. Drilling of piers follows excavation. Pier steel cannot be set until after *both* drilling of piers and procurement of pier steel have been completed. Pouring the piers follows setting of pier steel. Footing forms are set after *both* pouring of piers and construction of footing forms have been completed. Setting footing steel proceeds after *both* setting footing forms and procurement of footing reinforcing steel are completed. Finally, pouring footing follows setting footing steel.

The network diagram for the work described in the previous paragraph is shown in Figure 6.3. Note that each activity starts and finishes with an event, shown as a numbered circle, and that the end event always has a higher number than the starting event. Each event number occurs only once in the network.

While the pier-supported footing is a simple project, it serves to illustrate the value of CPM in job planning. The network is a model of the project, and its preparation requires the contractor to analyze the job logically from start to finish. The diagram communicates the job logic far better than any verbal description or bar graph.

Sometimes different portions of a project are planned separately, with separate network diagrams. For example, a project may consist of two buildings with connecting utilities. Events common to both networks are called *interface events*, and are usually shown as in Figure 6.4.

CPM Scheduling

After the project has been divided into concise activities and their logical sequence has been determined and charted in the network diagram, the time required for the project must be determined. Thus far, only the activities and their relationships have been considered; now the element of time is applied to the chart.

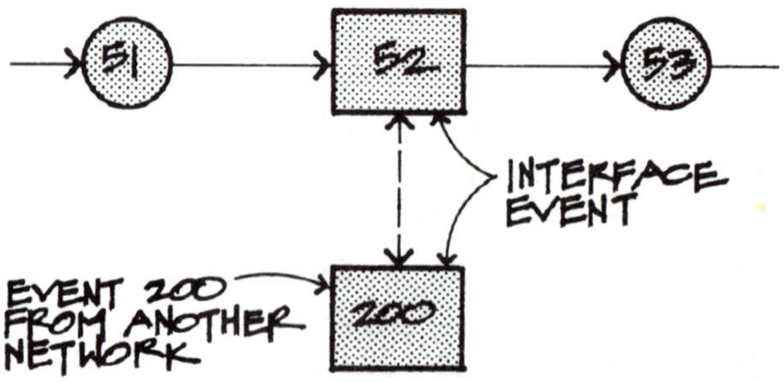

Figure 6.4

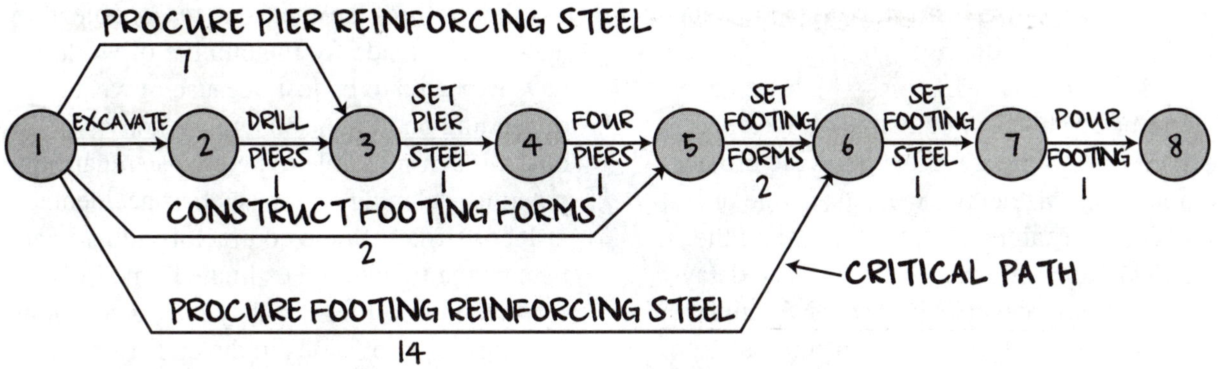

Figure 6.5

The contractor estimates the time required for each activity, based on past experience. A normal working day is taken as the unit of time. The assumption is made that materials and labor will be readily available, and that a normal level of labor and equipment will be utilized. Where subcontractors are involved, the contractor may consult with them regarding the time required to perform their specific activities. The estimated activity times in working days are now noted on the network diagram below each arrow. (See Figure 6.5.)

In preparing an accurate time estimate, the reliability of the subcontractors is critical. A general contractor, therefore, should be familiar with the subcontractors and their work, and consider only those who are pre-qualified or otherwise highly dependable.

Critical Path

The simple project illustrated in the network diagram includes several paths, from start to finish, and each has a varying total time duration. For example, path 1-2-3-4-5-6-7-8 requires a total time of 1 + 1 + 1 + 1 + 2 + 1 + 1 = 8 days. Path 1-5-6-7-8 requires 2 + 2 + 1 + 1 = 6 days. Since *each* path must be traversed to complete the project, the total project time is established by the path with the *longest* total required time. This is known as the critical path, and is generally shown as a heavy line. In the diagram below, the critical path is 1-6-7-8, with a total time of 14 + 1 + 1 = 16 days.

The activities along the critical path are called critical activities—in this case consisting of procuring reinforcing steel, setting footing steel, and pouring footings. If a critical activity is delayed, it will delay the completion of the project. These activities, therefore, must be carefully monitored during construction in order to keep the project on schedule.

Float

All paths in the network diagram, other than the critical path, are called *float paths.* The float is the difference in time duration between the critical path and any other path. Path 1-2-3-4-5-6-7-8, which requires 8 days, has a float value of 8, since it is 8 days shorter than the critical path time of 16 days. Similarly, path 1-5-6-7-8, which requires 6 days, has a float value of 10. The float, then, is a measure of the extra time available for an activity or group of activities.

As long as float time is not exceeded, no delay in project completion time will result. The path 1-2-3-4-5-6-7-8, for example, which we have determined to have a float value of 8, can be delayed up to 8 days without delaying project completion. This delay can occur in one or more activities along the path, providing the total delay does not exceed 8 days. The delay may occur only in activities from 1 through 6, since 6-7 and 7-8 form part of the critical path.

Project Calendar

The contractor, having determined that the finish date of the project is 16 working days after its start, now converts this to calendar days by multiplying by 7/5, since there are five working days in each seven-day week. ($16 \times 7/5 = 22.4$, say 23 calendar days.) Knowing the project starting date, the contractor can calculate the completion date, as well as the start and finish dates of all activities. He now establishes a project calendar, indicating the scheduled starting and completion dates of all the activities within the project. Critical activities are noted in color or boldface, since any delay in the schedule of these activities will delay completion of the project.

If the job schedule has been prepared carefully and realistically, the field work will proceed at an efficient pace. If excessive time has been allowed for certain activities, a more relaxed pace may result, leading to increased labor and overhead costs.

There can be great variation in the duration of construction projects, depending on the factors mentioned previously. However, most building construction projects require from nine to eighteen months.

Contingencies

A realistic schedule should incorporate an allowance for project delays caused by weather or other unforeseen events. A reasonable allowance can be made for the number of working days expected to be lost because of weather, depending on the season and the activity. Obviously, it is impossible to be precise regarding potential delaying factors such as accidents or labor strikes. Some contractors add a fixed percentage to the total estimated time to allow for such possibilities, or they may incorporate contingency provisions in the construction contracts.

CPM Calculations

The example of a pier-supported footing describes a simple project; however, the same logic and scheduling technique is used on large and complex projects. CPM programming can be done at a simple level or a complex one. Computer programs designed for CPM have proven very useful, once the basic activity sequencing and activity times are known. CPM is an extremely helpful planning and management tool, and its use in construction planning and scheduling has become almost universal.

Bar Graphs

Bar graphs have long been used for planning and scheduling construction projects. They indicate the starting and finishing dates of major phases of the work and can be clearly understood by all concerned. Their main disadvantage is that they do not indicate the relationship between the sequence of activities, or the dependency of an activity on the completion of a previous activity. The bar graph therefore is inferior to CPM as a management tool, but superior to CPM as a means of visual communication. Bar graphs, such as the one shown in Figure 6.6, continue to be widely used in construction.

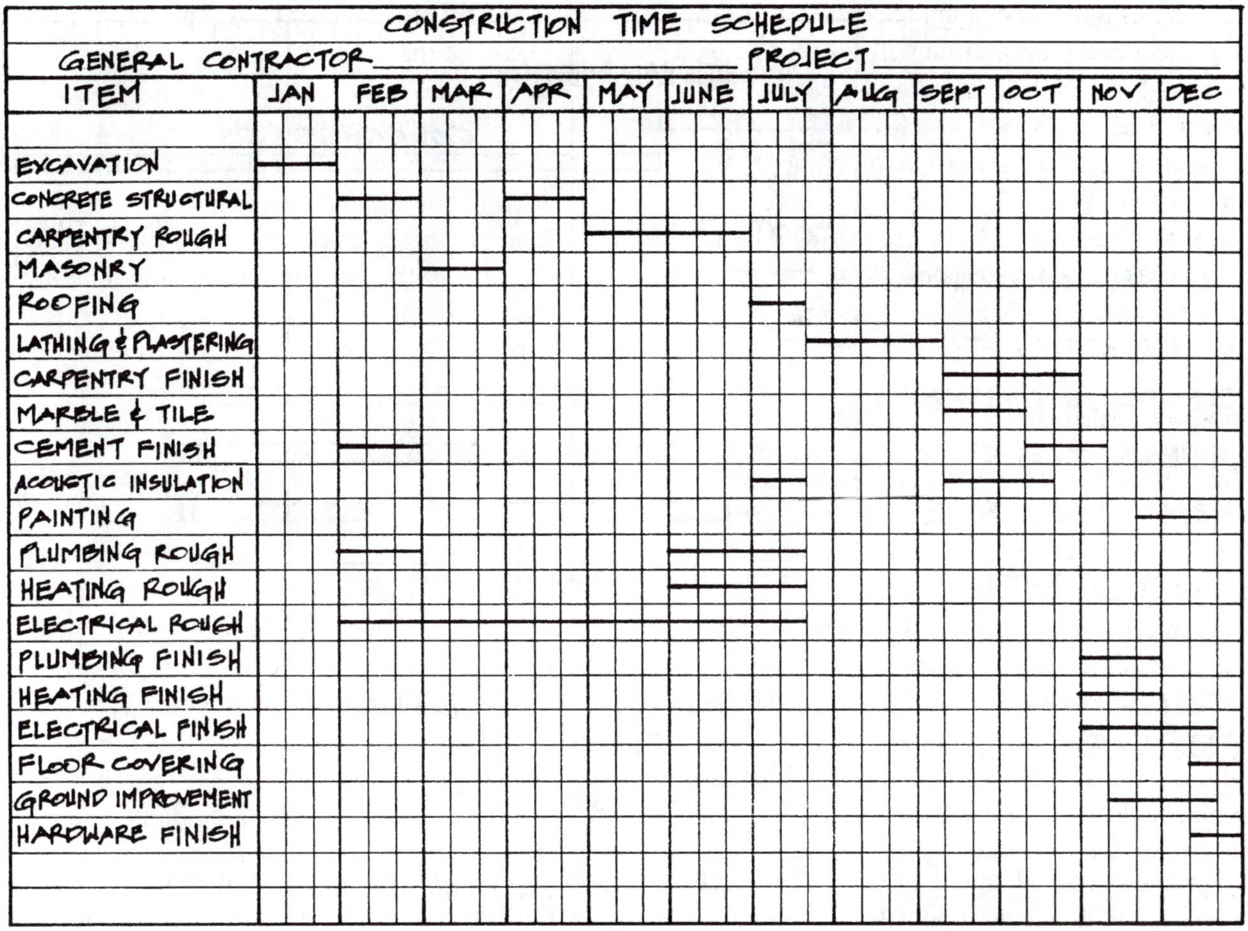

Figure 6.6

Shortening the Schedule

There are a number of reasons why an owner may want the use of his building as quickly as possible. Among these are the demands of business, which is often the case for commercial or industrial facilities. Other reasons may be to minimize the effects of inflation, inclement weather, or the persistent costs of interest on borrowed construction funds.

The CPM method demonstrates that one of the most effective methods to save construction time is to reduce the critical path time. Although the activities on the critical path may amount to only 25 or 30 percent of all the project activities, reducing them reduces the whole construction schedule.

Shortening the durations of the critical activities will very likely increase direct cost, because inefficiency is increased through added overtime work. Increasing the number of workers is also inefficient because supervision and coordination become more difficult. In general, the contractor's direct costs increase as the schedule is compressed into a shorter-than-normal time.

On the other hand, the contractor's overhead *decreases* as the schedule time is shortened. Since the total project cost is the sum of direct

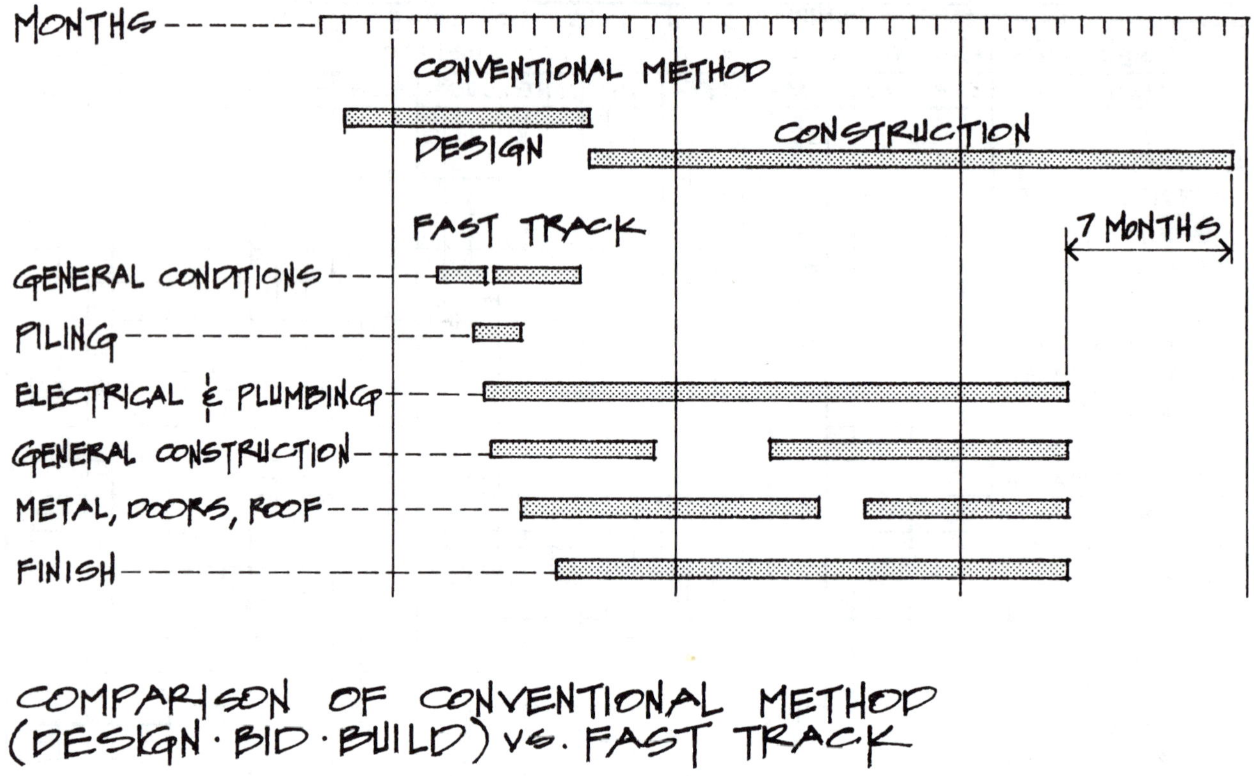

Figure 6.7

costs and overhead, and their effects by shortening the schedule are opposite, a contractor may find it worthwhile to analyze their effects and determine a balance that represents the lowest total project cost. A computer can be highly useful in doing this for a complex job.

Maintaining quality control becomes more difficult as the schedule time is shortened. Errors are more likely to occur because of the increased difficulty of proper supervision. The highest project quality is achieved when the project schedule is normal, that is, neither extended nor shortened.

If it is necessary to shorten the project schedule, the CPM network diagram can be analyzed to determine if the job logic can be modified, or if certain activity durations can be condensed. Individual activity times can be expedited by adding man-hours and equipment, recognizing that this will result in higher direct costs and will place greater demands on supervision.

Fast-Track Scheduling

Shortening design and construction schedules generally results in higher design costs, higher construction costs, and reduced quality. However, by combining the architect/engineer's design schedule with the builder's construction schedule, it is possible to realize an overall saving of time in completing the entire project. This technique is known as "fast-track," "accelerated," or "telescoped" scheduling. In this procedure, the architect first determines the major building elements, such as the column spacing, foundation system, mechanical systems, etc., before the detailed arrangements are worked out. The architect then produces detailed working drawings for a portion of the work on which the contractor may begin construction—site work, utilities, foundations, or possibly framing. Meanwhile, further detailed architectural design continues so that the architect produces his work just slightly ahead of the construction crews.

This approach requires close coordination among architect, engineers, client, and contractors. Since the design concept of building elements is established very early, oversights must be expected, and the correction of errors is generally an integral part of fast-track scheduling. However, major design revisions are all but precluded, except at very great cost.

Fast-track scheduling usually requires staged bidding, in which the project is organized into a number of separate stages or contracts—as many as 20 or 30—that are awarded to different contractors at different times. Thus, it may not be possible to obtain a fixed price for the entire project in advance of construction, as with conventional contracting that employs one general contractor. However, to assure some degree of cost and time control and establish responsibility, a construction manager (CM) may be used to supervise the construction process. Most contractors are able to function either as general contractors or construction managers.

Construction management may also be performed by architectural firms. But large and complex jobs are usually better served by those whose expertise is in the actual construction of buildings.

A comparison of conventional and fast-track scheduling for a $7,000,000 hospital is shown in Figure 6.7, indicating that the construction would be completed seven months earlier if fast-track scheduling were used.

Some architects may find their roles expanded to that of developer, builder, or manager. Whatever the role, it will be essential for the architect to become familiar with new management techniques, since they will have an increasing influence over how future construction work is done.

TIME MANAGEMENT

Fabrication Time

Most products and components of assemblies or systems are specially fabricated for individual construction projects. Whether an item is prefabricated off-site or fabricated or constructed on-site may have an effect on a project's timely completion.

Architects may decide whether to select off-the-shelf, ready-made components, or to design and specify components fabricated in a shop for subsequent installation on-site. Custom designed and fabricated elements may have adverse effects on construction time, thereby favoring manufactured items. Manufactured or prefabricated elements provide many advantages. Design time is shortened, because it is faster to select standard products from a catalog than to custom design new ones. And manufacturers' shop drawings and other submittals are easier to review than those of specialty contractors. Prefabricating elements in a shop reduces the impact of inclement weather on construction time. Work can be performed during winter, rainy days, and even nights if necessary. All of these factors save time.

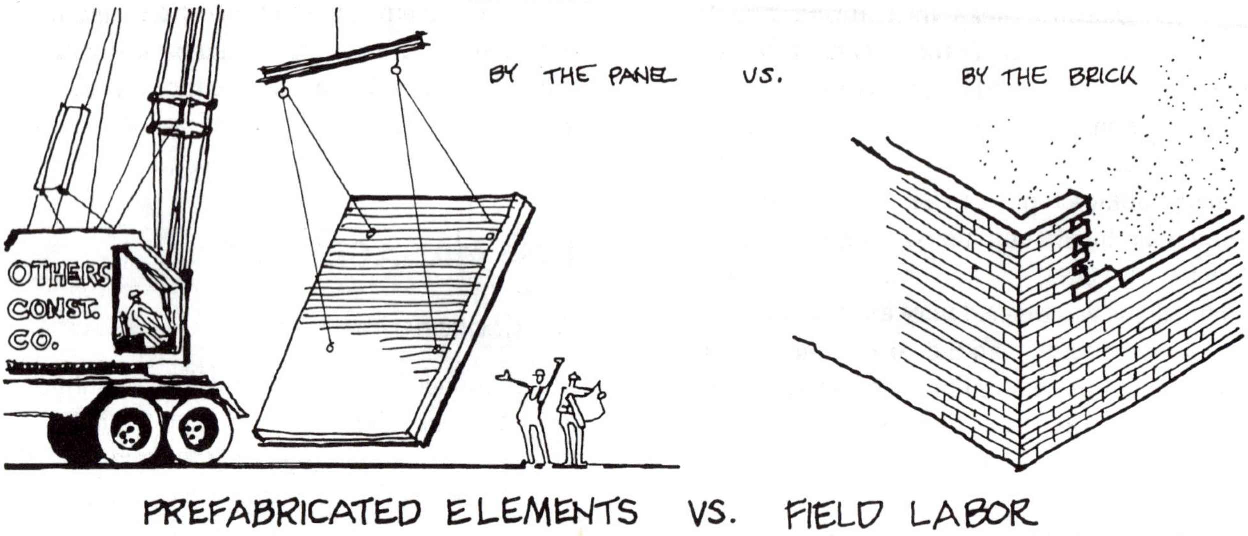

Figure 6.8

Other aspects of construction may be shifted from field to shop, as well. For example, metal-framed panels with brick facing may be mass-produced in a shop and erected on-site, instead of laboriously constructing brick by brick on-site.

Trade union jurisdictions and work rules may also affect construction time. Building construction trade unions may have a vested interest in specific methods and construction processes. Where shop labor is not subject to trade union jurisdiction, it may be possible to bypass certain union rules to shorten construction durations by using prefabricated products and systems.

Erection Time

A project's construction time is affected by, among other things, the extent of prefabrication of its component parts. Erection time may be shortened if a project is composed of mostly discrete building components that have been prefabricated off-site. The various components can be brought on-site, placed, and connected to other elements.

Certain aspects of prefabricated items must be considered, however. Prefabricated elements must, for example, be strong enough to resist lifting and handling. Attachments to other components must be simple, ideally requiring only one construction trade. Adequate clear space must be provided to maneuver prefabricated elements into place. Within these constraints, one may effectively use prefabricated elements to reduce erection time.

The timing of on-site operations must be considered. For example, inclement weather affects the construction time, and may even affect the design process. Winter weather may require that drawings be completed ahead of schedule to allow construction to begin before the onset of cold weather. All of these factors affect the time necessary to erect a building.

Sequencing of Construction Trades

Construction sequence is the order in which the various building trades perform their work, and is within a contractor's control. For example, foundation drains must be placed prior to backfilling. However, the design of a project may contribute to improved sequencing by minimizing the need for "on and off" construction. For example, if a large electrical conduit is placed in the same trench, but above a foundation drain, the construction sequence would be to place the drain, backfill the trench, place the conduit, and complete the backfilling of the trench. Locating the conduit elsewhere allows the contractor to place the drain and conduit independently and do all the backfilling simultaneously, thus eliminating one operation.

The design of a project may limit the ability of various building trades to perform work in a particular sequence. For example, if office partitions are designed to extend to the underside of the floor above, the ceiling installation cannot take place until all partitions are in place. Furthermore, scaffolding required for the ceiling installation must be disassembled and moved from space to space. If the program requirements can be met without the use of full height partitions, or with partitions that extend to the underside of the finished ceiling only, the ceiling contractor can complete larger areas and encounter fewer edge conditions where walls meet ceilings. Furthermore, scaffolds can be moved more efficiently.

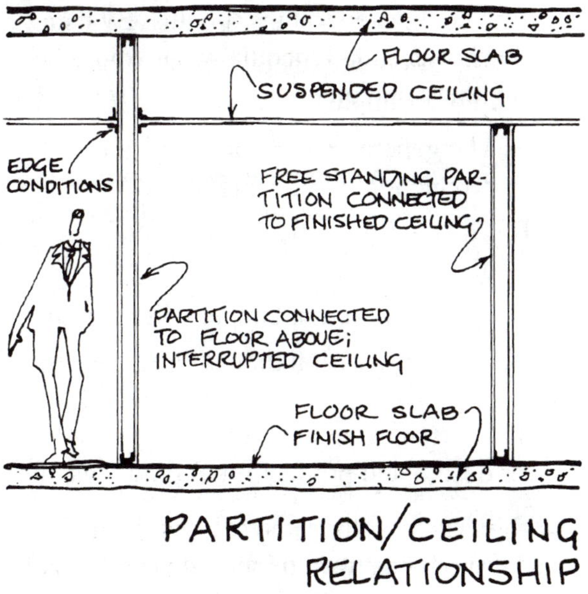

Figure 6.9

Although construction sequencing is the contractor's responsibility, an architect's design can contribute to construction efficiencies by minimizing the need for on-again, off-again labor. Additionally, this will reduce cleanup time, provide for more efficient use of equipment, and avoid potential problems.

Scheduling of Construction Trades

Contractors are responsible for scheduling the various construction trades as well as the sequence of work. Although owners and their architects may establish the total available time for construction as well as interim milestones, they have no responsibility for scheduling construction trades. Architects may, however, establish certain criteria for the contractor's scheduling requirements. Division One, General Requirements, of the specifications may include the following:

1. That all dates be established for ordering and delivery of materials, for submittals (including time for review, revision, and resubmittal, if necessary), and for testing.
2. That scheduling be done according to the Critical Path Method (CPM). CPM schedules are superior to bar chart schedules because they show interrelationships among activities.
3. That the schedule show the time allotted for each activity, as well as the cost, crew size, and equipment requirements for each activity.
4. That subcontractors provide input related to their scope of work.
5. That the schedule be updated monthly by the contractor to reflect the actual progress and current status. If a project is behind schedule, the contractor may be required to propose a plan for regaining lost time.

Comprehensive scheduling requirements which are fairly administered and enforced will ultimately contribute to a project's timely completion.

LESSON 6 QUIZ

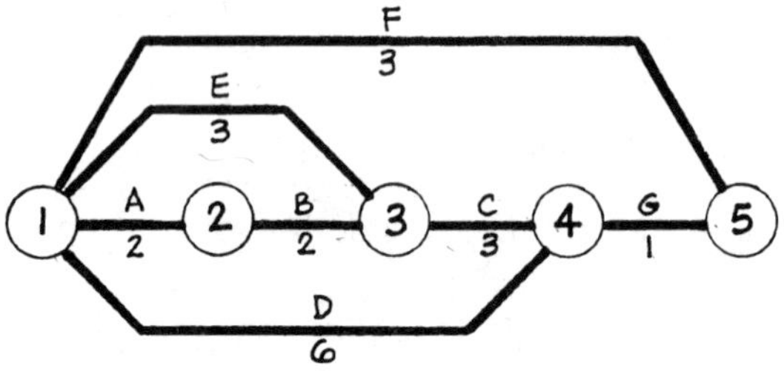

1. With reference to the CPM network diagram above, select the correct statement.

   A. Activity C cannot begin until both B and E are completed.

   B. Activity G cannot begin until F is completed.

   C. Activity F cannot begin until G is completed.

   D. Activity E cannot begin until A is completed.

2. Referring to the same diagram, what is the critical path?

   A. 1-5 **C.** 1-3-4-5

   B. 1-4-5 **D.** 1-2-3-4-5

3. Referring to the same diagram, what is the critical path time?

   A. 3 days **C.** 8 days

   B. 7 days **D.** 20 days

4. Which of the following would normally influence a contractor's construction schedule?

   I. The quality of construction documents

   II. The reliability of material dealers

   III. The total construction cost

   IV. The size of the project

   V. The anticipated weather conditions

   A. I and IV **C.** I, II, IV, and V

   B. II, III, and V **D.** All of the above

5. A type of project scheduling commonly used in situations where minimum construction time is required is called ________.

6. An architect estimates that design and production for a project will take one year. The client, however, requests that the total time be reduced to nine months. By using the shorter time schedule, what would be the likely outcome?

   A. The general quality of design and production will be unaffected.

   B. The quality of the construction documents will be lower.

   C. The construction budget will be higher.

   D. The construction time will be greater.

7. By shortening an architect's time schedule for design and production of drawings, his or her

   A. labor costs would increase.

   B. overhead would increase.

   C. profit would increase.

   D. documents would be unaffected.

8. Arrange the following tasks for a typical project in ascending order of scheduled time, that is, from the task requiring the least time to the most.

I. Schematic design phase

II. Bidding phase

III. Client approval

IV. Design development phase

V. Construction documents phase

A. II, I, III, V, IV

B. I, III, II, V, IV

C. III, I, IV, II, V

D. III, II, I, IV, V

9. An architect's scheduling and staffing requirements for a specific project are dependent on the project's

A. size.

B. cost.

C. complexity.

D. quality.

10. Reducing the critical path time will very likely

A. increase the project cost.

B. extend the construction time.

C. have no effect on the float time.

D. have no effect on project quality.

LESSON SEVEN

DELIVERY METHODS

7

Owner Requirements
Design/Award/Build Delivery Method
Construction Management Delivery Method
Design/Build Delivery Method
Structuring the Architectural Design Team

OWNER REQUIREMENTS

The method of delivering design and construction services is typically based upon an owner's needs and capabilities. A small organization, a small firm, or an individual that wishes to develop a project would typically require full professional design services from an architect and a traditional design/award/build delivery process. An owner who desires to participate in the design process would likely select this traditional method, ensuring that the final project meets all of the owner's criteria. The design/award/build delivery method allows for all design decisions to be made before contracting with a builder.

A large organization or firm that wishes to develop a project may have an in-house staff that has capabilities for project programming, design, engineering, facilities management, construction management, or construction. Such a firm may not require the traditional design/award/build delivery method. An owner also might have certain time frame and/or cost considerations that would require other types of delivery methods. An owner who has a commitment to deliver a project for occupancy in a short time frame may not be able to take the amount of time required of the traditional design/award/build process. Such firms may require other project delivery methods, which typically consist of either the construction management or design/build methods.

DESIGN/AWARD/BUILD DELIVERY METHOD

The design/award/build delivery method typically begins when an owner hires an architect to develop a project program and its subsequent design and construction documents. Bidding of the project to several contractors occurs after all construction documents and specifications have been completed. This allows for the establishment of the lowest reasonable cost for the project. The owner then awards a single prime construction contract to a general contractor to build the project based upon the completed design documents. The architect acts as the owner's agent, representing the owner's interests throughout the design and documentation phases. The architect's services typically include construction administration services.

The architect then acts as an impartial interpreter of the construction documents during construction.

The benefits of the design/award/build delivery process include owner participation in the design of the project and well-established construction costs based upon relatively complete documents. The architect acts in the owner's best interests during design, and the architect acts as an impartial interpreter of the contract documents during construction. This process allows for clear separation of design and construction responsibilities, and allows for simplicity in project scheduling since each phase of the design and construction process is separate.

The design/award/build delivery process, however, requires an extended time period for design and documentation before final costs can be determined and construction can begin. This is a problem if an owner wishes to expedite a project. Also, pricing and constructability experience of the contractor who is to build the project is not available during the design and documentation phases of the project. Lesson 8 discusses the bidding process in further detail.

CONSTRUCTION MANAGEMENT DELIVERY METHOD

The construction management delivery method allows an owner to address constructability and cost issues during design. An owner can also address time issues by utilizing fast-track construction, in which multiple construction contracts are let for different parts of a project as soon as each part of the work is defined enough for a contractor to reasonably commit to a price. In this delivery method, the owner hires or utilizes his or her own construction manager to work with an architect to facilitate the process of design, bidding, and letting of the construction contracts. The construction manager can act as either an advisor to the owner, or as a construction contractor. The construction manager typically has substantial expertise in construction technology, constructability issues, construction scheduling, and construction costs.

A construction manager who acts as an advisor administers the design contracts and works as the owner's representative with the design team. He or she also manages the various construction contracts, but does not have any financial responsibility for the construction of the project. A construction manager may, however, handle some of the typical non-construction activities at the site, such as arranging temporary site facilities, site and construction testing, engineering, building and site layout, and construction site cleaning. Some architecture firms offer construction management services, acting in an advisory role to the owner.

A construction manager who acts as a contractor assumes a vendor relationship with the owner. This person or firm will take on the financial responsibility for the construction of the project, typically utilizing a fixed-price, cost-plus, or guaranteed maximum price cost structure. The construction manager is brought onto the project before design work is complete so that he or she can help resolve constructability and cost issues.

A fixed-price structure allows the manager to establish a guaranteed cost of construction, including his or her own services, before the design is fully documented. The owner is not liable for bid-cost overruns. However, the owner does not obtain any of the savings that might occur from a positive bid climate. A cost-plus structure allows the construction manager to charge the owner the actual construction costs of the project plus a negotiated fee that is agreed to before construction begins. The actual costs are typically determined by the lowest bids received from the manager's subcontractors, plus the cost of any construction work performed by the construction manager's own forces. The guaranteed maximum price structure is a highest-probable-cost limitation for the construction of the project guaranteed by the construction manager. This price is established before design documents are completed, and anticipates the full scope of work and detailing needed to complete the project. Any cost savings from a positive bid climate go to the owner rather than the contractor. However, the contractor becomes responsible for any bid-costs over the guaranteed maximum price.

The advantage of using the construction management delivery method is the ability of the owner to determine the costs of a project before construction documents are complete. The ability to let portions of the work for bid before other portions of the design are complete allows for construction work to commence before all other project drawings are completed. This is a great advantage for an owner who has a short time frame to complete a project due to occupancy requirements or when an owner has to work with high interest rates, which can add substantially to the financing costs of a project. Another benefit of this method is the ability of the construction manager to resolve technological or constructability issues before construction begins, which helps reduce costs due to construction change orders.

The construction management delivery method, however, adds a cost for the construction manager that an owner would not have in the more common design/award/build process. The addition of a construction manager adds complexity to the design and construction team. This can be a benefit if the relationships are managed effectively, but can become problematic if these relationships are not adequately defined and handled. The use of the fast-track construction method also adds to the complexity of the project, requiring the management of multiple bidding periods and multiple prime construction contracts.

DESIGN/BUILD DELIVERY METHOD

The design/build delivery method allows an owner to utilize a single entity that is responsible for both the design and construction of a project. This is the single greatest distinction between this method and both the design/award/build and the construction management methods. A design/build firm can be a single company that has its own architectural and construction staffs, or a company that has its own construction staff that hires an architect to perform design services. A development firm can hire an architect for design services and a contractor for construction services. A design/build firm can also be a joint venture between an architect, construction, and/or a developer.

An owner who wishes to proceed with the design/build process typically issues a request for proposals to selected design/build firms that state the design and performance requirements for the project. The design/build entities submit proposals to the owner that provide a design for the project and the costs for the design development and construction of the project. The selected design/build firm then develops the design, provides construction documents, and builds the project based upon the proposal requirements.

An owner who wishes to have more control over the design of the building can have an architect develop the schematic concept for the project. This can then become part of the request for proposals, which makes the selected design/build firm responsible for the development of the design, the construction documents, and the building of the project.

The advantages to the design/build delivery method include a single source of responsibility for both design and construction of the project, allowing the owner to select from a number of submitted designs. A reliable cost for the project is determined early in the process, and conflicts between the designers and the builders are minimized. This process also facilitates fast-track construction, since the portions of the design work that can be built early can be released for construction before the balance of the design and documentation work is complete.

This delivery method, however, minimizes the ability of the owner to participate in the design of the project. The design/build firm acts solely as a vendor so that the owner does not have an independent agent working for his or her interests. This requires the owner to be adept at managing the design/build contract through construction, or to hire an independent firm to act on his or her behalf. Any design changes would likely require a change order that the owner would have to pay. Since the submitted designs are likely based upon incomplete drawings, disputes may arise regarding the actual scope of work provided in the proposal. Also, a selection that is based solely on the lowest bid may have significant quality issues that would be difficult to address.

STRUCTURING THE ARCHITECTURAL DESIGN TEAM

The architect can act as the sole provider of design services if his or her firm has experienced and qualified in-house staff that can provide the necessary engineering and other specialty services that are required on a project. However, most architects typically form alliances with other firms to provide these services.

In a typical alliance, the architect has the prime contract with the owner and then subcontracts services to other professional firms that act as the architect's consultants for a project. Consultants can include structural, mechanical, electrical, plumbing, civil, or acoustical engineers; landscape design firms; kitchen design consultants; information technology/communications firms; and soil and construction testing services firms.

Architects may also create joint ventures with other firms, creating a single project-based entity with other architecture, engineering, or construction firms that have specific areas of expertise or geographical experience. An architect would typically form a joint-venture with a construction firm as part of a design/build delivery method, and would then act as a vendor rather than as an owner's agent. Acting as a vendor would then require the architect to act on behalf of the joint-venture and its best interests rather than for the owner.

An architect may also act as one of several independent design and engineering firms hired by an owner. In this situation, an owner would typically have some level of project and construction management capabilities to handle and coordinate the different contracts.

LESSON 7 QUIZ

1. A project delivery method that involves an owner hiring someone with constructability and cost expertise to work with the architect during the design phase is called Construction Manager.

2. Which of the following is not an advantage of the typical design/build delivery method?

   A. Facilitates fast-track construction

   B. Provides a reliable project cost early in the process

   C. The owner participates fully in the design process

   D. The design/build firm provides a single source of responsibility for design and construction

3. An architect acts as an owner's agent in which of the following situations?

   I. As a member of a joint venture with a construction company

   II. The design-award-build delivery method

   III. The construction management delivery method

   IV. The design/build delivery method

   A. I and III

   B. II and IV

   C. II and III

   D. I and IV

LESSON EIGHT

THE BIDDING PROCESS

INTRODUCTION

Construction contracts are traditionally awarded to the contractor who submits the lowest bid. If all the bidders on a specific project are equally qualified, and if their bids are based on the same set of bid documents, then price is the only variable. Bidding is therefore a fair and reasonable way to select the contractor for a construction project. When the contractor is selected on this basis, the agreement form used is generally AIA Document A101, Owner-Contractor Agreement (Stipulated Sum).

Alternatively, an owner may negotiate a contract with a general contractor or construction manager selected on the basis of qualifications alone, or qualifications and a fee proposal. In that case, the agreement is usually based on the *cost of the work plus a fee*, with a *guaranteed maximum price*, such as AIA Document A102. Using this method, competitive bidding is limited to subcontractors.

Whether bids for a project are relatively high or low depends on many factors. A competitive climate, in which many bidders compete for few projects, usually results in lower prices. On the other hand, an extremely short bidding period, a short construction period, or unusual or ambiguous requirements can all contribute to higher bid prices from all contractors.

Contractors are normally informed that a project is available for bids by advertisement or by an invitation to bid. The instructions to bidders included in the bid documents cite necessary qualifications for bidders, bid date and location, and other administrative details. There are many legal rules and requirements applicable to bidding. Architects must pay particular attention to procedures for dealing with:

1. Informalities and irregularities
2. Selection of alternates
3. Mistakes in bids
4. Determination of the successful bidder

This lesson discusses the bidding process and its potential problems.

BIDDING ENVIRONMENT

Supply and Demand

The economic law of supply and demand applies to construction as much as to any other element in a free market. When demand exceeds available supply, construction products and services are relatively expensive. Conversely, when supply exceeds demand, prices usually decrease.

In most localities, there are relatively few qualified contractors available for any given size and type of project. If many owners want to build at the same time and the supply of contractors is limited, demand for contractors' services is high, and bid prices therefore increase. Each contractor is aware that many other projects are available, and he or she need not submit a very competitive bid.

Architects must therefore be knowledgeable about the level and type of activity in the local construction market. They may suggest either accelerating or delaying a bid date, to avoid a bidding environment in which many similar projects are available to contractors. Owners usually benefit from lower prices if their projects are bid when other similar work is scarce.

Labor Rules and Construction Methods

Most contractors have relatively small full-time staffs. When necessary, they hire workers from local trade union hiring halls or, if a project has nonunion labor, from the available labor pool in the area. At times, certain skilled workers may be fully engaged on projects already under construction. An architect considering a steel structure, for example, may find the available supply of iron workers is insufficient to accommodate a new project at the bid date. If such workers have to be recruited and brought into the area, prices will increase. In that case, the architect may decide to change the structure to concrete, if concrete workers are available.

Specific construction methods are not equally popular everywhere. Attitudes in this regard are shaped by local traditions, the availability of certain materials, or local contractors' experience. It is generally unwise for an architect to design for, or specify the use of, materials and methods that are unusual in a specific locality. If contractors are inexperienced with the specified materials or methods, bid prices may be unusually high.

TIME ALLOWED FOR BIDDING

Contractors need adequate time to study bid documents and prepare bids. Otherwise, they must include contingencies, which may lead to unnecessarily high bids.

The optimum length of time depends upon many factors, such as a project's size and complexity and the number of similar projects being bid at the same time. In general, two weeks is the minimum time necessary to prepare a bid. Three or four weeks, or more, may be justified in some circumstances. There is usually very little improvement in the quality and accuracy of bids prepared during a lengthy bidding period, as compared to a brief one. In practice, general contractors are dependent on subcontractors for many components of a bid. Subcontractors usually do not submit their prices until the last minute, to avoid giving their competition advance information and being undercut as a result.

An owner can improve the quality of bids by providing an adequate number of sets of contract documents to bidders. It is important that subcontractors and suppliers who are bidding have access to documents. Providing bid documents to local *plan rooms*, and listing projects with services such as F.W. Dodge facilitates access and can speed bid preparation.

TIME ALLOWED FOR COMPLETION OF CONSTRUCTION

Bidding documents must clearly state an owner's requirements concerning the length of the construction period, so that contractors can determine the amount of labor and equipment necessary to complete the required work in a timely manner.

A predetermined date of completion is a factor over which an architect often has little or no control. For example, a school board may require that a new school be ready for occupancy at the beginning of the school year, or a developer may require that an office building be completed in time to satisfy occupancy commitments made to prospective tenants. An owner may establish the length of a construction period based on financing requirements and other business factors.

When an unrealistically short construction period is demanded by an owner, the required number of workers and the amount of equipment required to meet such a schedule tends to congest the construction site. Efficiency therefore decreases and progress is impeded. When contractors are forced to work overtime or provide additional shifts for long periods of time, similar problems occur. Doubling the number of workers and working hours is inefficient and does not result in twice as much work.

If a construction schedule is unrealistic, a contractor's problems may be further compounded when the proposed contract contains provisions for *liquidated damages*. For example, a contract may stipulate a construction period of 300 days with $5,000 damages per day payable to the owner for each additional day required to complete the project. If a contractor estimates that it will take 320 days to complete the work, he or she will probably add $100,000 ($5,000 × 20 days) to the bid price to allow for payment of the potential liquidated damages. Therefore, the owner may pay for contingencies that never occur.

Contractors have an economic interest in constructing projects in the shortest possible time. However, an unrealistically short construction period may result in higher bid prices, with no apparent benefit to the owner.

Finally, the bidding documents should make clear the means and extent to which the contract will provide for extending the allotted construction period to allow for delays caused by *acts of God*, such as a tornado, earthquake, or other events beyond the control of owner and contractor.

EFFECTS OF UNUSUAL REQUIREMENTS

Role of Supplementary Conditions

The purpose of supplementary conditions is to modify the general conditions of the contract because of unusual or special circumstances or to accommodate specific project requirements. Contractors familiar with the AIA General Conditions, Document A201, expect the supplementary conditions to cite the special requirements of a project.

A contractor may learn, for example, that he or she will be required to purchase property insurance for the project, rather than the owner, who normally provides such insurance. Under these circumstances, the cost of such insurance must be included in the bid price. Modifying provisions of the general conditions may result in greater risk to the contractor, as well as increased cost. For example, a contractor may have to assume the risk that actual subsurface conditions will differ from those anticipated and shown in the contract documents.

The supplementary conditions may restrict the contractor to certain working hours, or to a certain portion of the site. These factors will influence the contractor's bid. The owner may be exempt from sales taxes or regulatory agencies' fees, and these unique conditions will likewise influence the contractor's bid. Bidders must therefore carefully study the supplementary conditions.

Role of Pre-Bid Conference

A pre-bid conference provides a practical way to inform contractors about unique or special aspects of a project. An architect can describe important factors that contribute to a project's success, but which may not be evident from the construction documents. For example, the reasons for certain design decisions are not generally apparent from simply studying the construction documents. By itself, a detail or system may not appear to be the optimum solution; however, when considered in the overall design context, it may clearly be the best choice. Only the architect understands the rationale behind all elements of the contract documents. Information conveyed to contractors at a pre-bid conference may preclude substitution of other solutions or details.

Contractors usually have a relatively short period of time in which to familiarize themselves with documents that may have taken months to prepare, so it is unlikely that they will immediately grasp all the nuances of the design. Properly conducted pre-bid conferences can avoid potential misunderstandings by providing the architect an opportunity to explain the design intent and to discuss anticipated results with contractors.

At pre-bid conferences, the architect and owner can reiterate the instruction to bidders, outline the bid documents and highlight unusual provisions, conduct site tours, and record questions from bidders. Answers to questions must be formalized in an *addendum* to become part of the bid documents.

EFFECTS OF AMBIGUOUS DOCUMENTS AND ARBITRARY REQUIREMENTS

If construction documents are ambiguous and there are two or more reasonable interpretations, a bidder will usually choose the less expensive one. For example, if neither paint nor stain is specified for interior wood trim, a contractor would probably be justified in omitting the cost in his or her bid, despite the fact that no other finish is specified and interior wood trim is usually painted. If the contractor was subsequently asked to paint or stain the trim, a change order to reflect the additional cost would be required.

Some architects have special provisions in the contract documents to cover situations where more than one interpretation is possible. For example, they may state that contractors are to base their price on the most difficult or most expensive alternative wherever more than one interpretation is possible. Other documents may require contractors to bring all questionable aspects of the contract documents to the architect's attention prior to the bid date or else waive future claims for extra costs.

Arbitrary requirements may also affect bid prices. If the specifications require unnecessarily complicated procedures or cumbersome or restrictive methods of construction, or if the phraseology is obscure, bidders will increase their bids to meet these requirements. The contract sum will consequently reflect these higher prices. If during the course of construction, the contractor and architect agree on simplified procedures, methods, or language, the owner will receive little or no credit for these revisions.

ADVERTISEMENT AND INVITATION FOR BIDS

Public agencies are usually required by law to solicit competitive bids from qualified contractors. To do so, they publish legal notices in one or more newspapers.

Local laws at the project locations may specify requirements for an advertisement for bids. The following items are often included:

1. The name and location of the project.
2. The name and address of the owner and architect.
3. The date and time that bids are due, as well as the location.
4. The nature and scope of the project, as well as whether a single general contract or multiple trade contracts are required.
5. A description of the project. Normally the description includes the size of the project, principal materials and construction systems, and other pertinent technical data.
6. Whether or not bids will be opened and read publicly. On many publicly funded projects, bids are required to be read aloud or a tabulation of bids provided.
7. Where and how bidding documents may be obtained, including the type and amount of bid deposit, if any, required to obtain them.
8. The names and locations of places where bidding documents may be inspected by interested bidders.
9. Details about proper procedures for submitting bids.
10. The type and amount of bid security, if any, that is required.

TYPICAL ADVERTISEMENT FOR BIDS

Bids are invited from qualified contractors licensed in Virginia on the following basis:

PROJECT NAME AND LOCATION

Grandcondo Management Building
11 Suffolk Street
Arlington VA 22206

NAME AND ADDRESS OF OWNER

Grandcondo Unit Owners Association
15 Suffolk Street
Arlington VA 22206

NAME OF ARCHITECT

Charles McCarthy, AIA
3004 Vine Street
Alexandria VA 22304

SUMMARY DESCRIPTION OF THE PROJECT

The Project is to house the management and maintenance staffs of the Grandcondo. There will be approximately 2,700 square feet under roof, divided approximately 1/3 to office and 2/3 to maintenance. Construction will be concrete footings, brick and block bearing walls, wood rafters, and asphalt shingles. The office will be finished with gypsum drywall over wood framing. The maintenance area will be semi-finished, with exposed, painted block and some gypsum drywall over wood framing. There are toilet facilities and a unit kitchen. There is an unroofed, outdoor yard area that will be walled in. There is considerable free area around the Project location.

AVAILABILITY OF DOCUMENTS

Documents will be available after 8:00 AM on September 20, 2010.

ISSUING OFFICE

GUOA
Attn: Sheila Dubois
15 Suffolk Street
Arlington VA 22206
703-555-5555

DEPOSIT

$25.00 per set of Documents.

Make check payable to Grandcondo Unit Owners Association.

Deposit will be refunded (less cost to replace damages) for sets returned in good condition at the end of the bid period.

SUBBIDDERS

Bidding documents will not be issued to Subbidders.

COPIES OF BID REQUIRED

One.

BID SECURITY

None required.

(continued)

Table 8.1

TYPICAL ADVERTISEMENT FOR BIDS

NAME OF PARTY TO RECEIVE BIDS

Jackie Hays, President
Grandcondo Unit Owners Association
15 Suffolk Street
Arlington VA 22206

TIME AND DATE FOR RECEIPT OF BIDS

October 12, 2010 at 12 Noon.

BIDS TO BE HELD OPEN

Bids may not be withdrawn for 90 days from October 12, 2010.

BID OPENING

Bids will be opened and read publicly on October 12, 2010, at 1:00 PM at the Issuing Office.

QUALIFICATIONS OF CONTRACTORS

Bidders must be licensed in Virginia. They must also submit, upon request, a certified statement, with appropriate references, indicating that they have completed at least three projects in the last three years, each with a contract value of at least $250,000.

BONDS

A Performance Bond/Labor and Material Payment Bond will each be required in an amount equal to 100 percent of the Contract Sum. The form of the Bonds will be AIA Document A312, Performance Bond and Payment Bond, 2010 Edition. Bonds will be written to the benefit of the Grandcondo Unit Owners Association.

AGREEMENT

AIA Document A107, Standard Form of Agreement Between Owner and Contractor for a Project of Limited Scope, 2007 Edition.

CONTRACT TIME

Contract Time from Notice to Proceed until Substantial Completion will be 130 calendar days.

Table 8.1 (continued)

11. Whether or not the owner has the right to waive informalities in the bidding process or to accept a bid other than the lowest. On publicly funded projects, owners usually are not permitted to accept bids that do not conform precisely to requirements or to accept any but the lowest bid.
12. The date the advertisement was first published and the name and address of the party responsible for publishing it.

The text of an advertisement for bids is usually included in the project manual as a matter of information and record. See Table 8.1.

When projects are privately financed, invitations to bid are often used instead of public advertisements. The architect recommends contractors who are known to be skillful, responsible, qualified, and likely to be interested in bidding the project, based on its size and type. If the owner concurs, those contractors are sent an invitation containing information substantially the same as that found in a typical bid advertisement.

Invitations should be sent to as many contractors as necessary to provide adequate price competition. Usually, six bids are adequate, but the number may be more or less depending on the nature of the project and the level of activity in the local construction industry. The invitation to bid should also be included in the project manual.

INSTRUCTIONS TO BIDDERS

General

The Instructions to Bidders, AIA Document A701, is designed for use with the General Conditions, AIA Document A201. Its purpose is to establish the basic requirements for bidding.

The Instructions to Bidders defines important terms, either directly or by reference to definitions in the AIA General Conditions, in order to avoid ambiguity. For example, the bidding documents comprise the following:

1. The advertisement or invitation for bids.
2. The instructions to bidders and any supplementary instructions.
3. The bid form.
4. Sample bidding and contract forms.
5. The proposed contract documents, including addenda, if they are issued prior to receipt of bids.

The AIA Instructions to Bidders also provides that each bidder, by submitting a bid, implicitly states that he or she understands the bidding documents, has visited the project site, and has become familiar with the local conditions under which the work is to be performed. Local conditions in this context include ordinances, labor rules, and physical conditions. This statement by bidders is useful in avoiding claims for additional costs for work required by conditions not anticipated. Each bidder also states that the bid is based, without exception, on the information in the bidding documents. Consequently, bidders may not base their bid price on substitute materials even if they consider them to be equal to those specified.

Bidding documents are normally issued only to prime contractors in complete sets to prevent subcontractors from submitting bids based on partial documents related only to their part of the project. For example, an electrical subcontractor using only the electrical drawings and specifications might not be aware of important requirements, such as connections to equipment contained in the mechanical drawings or in Division One of the specifications, and might not include the cost of such requirements in his or her bid.

Role and Timing of Addenda and Substitutions

The AIA Instructions to Bidders requires bidders to request, at least seven days prior to the bid date, that the architect clarify any unclear or ambiguous conditions existing in the drawings and/or specifications. For example, if the contract documents are inconsistent and show two different types of carpeting for the same floor area, a contractor might inadvertently base his or her bid on the more expensive one. If the architect's intent is to use the less expensive carpet, the bids will be unnecessarily high.

A contractor may propose an equally good, but less expensive, *substitution* for a material, system, or product. Bidders who propose substitutions must submit technical documentation to the architect for review and evaluation at least 10 days before the bid date. If a substitution is acceptable to the architect and owner, the architect will issue an *addendum* to all bidders at least four days prior to the bid date. Only substitutions approved by addenda may be used as a basis for bid prices. Bidders must acknowledge in their bids that they have received and bid on all addenda issued. Bidders should only rely on information appearing in addenda, never on information provided orally.

By providing a procedure for addenda and substitutions, the AIA Instructions to Bidders provides a means for the architect to control costs and reduce errors and potential future claims. See Figure 8.1.

Bidding Requirements

Qualifications of Bidders

Either before or after the bid date, contractors may be required to prove that they are qualified to perform the work required by the bidding documents. Factors to be considered may include the scale and type of projects completed, a financial statement, physical plant (machinery and equipment), and personnel (especially supervisory staff). The quality of a bidder's work can be evaluated by reputation, references, and visits to recently completed projects.

Some public agencies maintain lists of contractors whom they have investigated and *pre-qualified*. Under certain circumstances, only pre-qualified contractors may bid on the agency's projects. Similarly, if a contractor is bondable, it indicates that a surety company has investigated many of the factors mentioned above and has pre-qualified that contractor to be financially capable.

Form and Completeness of Bids

Normally, bidders must submit their bid on the form included in the bidding documents. This procedure allows the owner to compare bid proposals on the same terms and conditions. Bids must be either typed or written in ink, with the price shown in both numerals and words. If there is any discrepancy between them, the words prevail over the numbers. If the bid documents include alternates, each must be bid. If the project is divided into several separately identified parts, of which some are awarded while others are not, a bidder may withdraw his or her bids if he or she is not awarded a contract for all the parts. Finally, every bid proposal must be signed by a person legally authorized to bind the contractor. If the contractor is a corporation, appropriate officers must sign, and the corporate seal should be applied. A partner should sign on behalf of a partnership. See Figure 8.2.

On occasion, bidders may attach a list of qualifications or exclusions to their bids. Because the AIA Instructions to Bidders requires bidders to furnish bids based solely on the bid documents, the submittal of qualifications or exclusions violates the bidding requirements. The owner can either reject the bid, rebid the project, or award the contract and adjust the contract sum by change order if the qualifications or exclusions affect cost. The latter option is not permitted on publicly funded projects.

Timeliness of Bids

The AIA Instructions to Bidders provides rules for timely submission of bids. They are designed to preclude bidders from obtaining a competitive advantage by submitting a bid after the established deadline. Bids must be in writing and transmitted by mail, courier, fax, or electronic mail to the stipulated person and place prior to the established deadline. Bids received orally or over the telephone should not be considered.

There have been cases where bids were submitted one minute past the deadline and consequently rejected. When this circumstance is challenged in court, the ruling is usually based on whether the late bidder obtained any advantage. If bids had not yet been opened at the time of the late submittal, and if the rejected bid was only slightly late, a court would probably consider it to be a legitimate bid.

TYPICAL ADDENDUM

ADDENDUM #1 *March 6, 1998*

TO DRAWINGS AND PROJECT MANUAL FOR
GRANDCONDO MANAGEMENT BUILDING
DATED FEBRUARY 8, 1998
CHARLES McCARTHY, AIA, ARCHITECT

This Addendum covers changes to the Drawings and Project Manual and will become a part of the Contract Documents. Bidder shall include all effects which these items may have on its proposal.

Sheet A1 - Site/Roof Plan

Item 1.

Delete base course shown under the portland cement concrete paving in 2/A1, 3/A1 and 4/A1.

Item 2.

The sanitary outlet from the building has been relocated on the same line, but to the south of the building.

Sheet A2 - Floor Plans

Item 3.

In Room 107-Restroom, Room 110-Lockers, and Room 111-Washroom, delete ceramic tile as the wall finish. Instead, use finish #1 - Drywall Painted. For both the walls and the ceilings in the above rooms, use:

IPS-12:	*1st Coat*	-	*Latex Primer (TT-P-650)*
	2nd Coat	-	*Polyester epoxy (TT-C-545)*
	3rd Coat	-	*Polyester epoxy (TT-C-545)*

Not less than 4.0 mils dry film thickness

This change is also reflected on Sheet A5, Elevations 11, 13, 15, 16, and 17.

Item 4.

Delete the vinyl base in the following rooms in the maintenance area: 109, 112, 113, 114, 115, 116, and 119.

Figure 8.1

TYPICAL BID FORM

FROM:

Name of Bidder ______________________________

Address ______________________________

____

Telephone ______________________________

Area Code *Telephone Number*

TO:

ATTENTION:

A2.01 *BASE BID: The undersigned, having inspected the construction site and familiarized him/herself with all conditions likely to be encountered affecting the cost and schedule of work, and having examined all of the Contract Documents, hereby proposes to furnish all labor, materials, tools, equipment, and services required to perform all of the work in strict accordance with the Contract Documents as prepared by*

for the

for the Base Bid Sum of:

____

(Use Words)

______________________________ *Dollars ($* ______________ *)*

(Use Figures)

and if this proposal is accepted, will execute a formal contract equal in form to that bound in the Contract Documents, to this effect.

A2.02 *ADDENDA: The undersigned further acknowledges receipt of addenda as listed below and represents that any additions or modifications to, or deletions from the work called for in these Addenda, are included in the Base Bid Sum, and Unit Prices if affected thereby.*

ADDENDUM NO.	*DATED*	*ADDENDUM NO.*	*DATED*
________	________	________	________
________	________	________	________
________	________	________	________

(NOTE: If no Addenda have been received, write in "NONE.")

Figure 8.2

TYPICAL BID FORM

A. *This proposal shall remain valid, for acceptance by for a period of not less than thirty (30) days from the bid due date.*

B. *All requirements concerning licensing and all other local, state, and national laws have been or will be complied with and that no legal requirements will be violated in the execution of the work if the proposal is accepted.*

C. *No person or persons or company other than the firm listed below or otherwise indicated hereinafter have any interest whatsoever in the proposal or the contract that may be entered into as a result thereof. This proposal is submitted in good faith, without collusion or fraud.*

D. *The person or persons signing this proposal is/are fully authorized to sign on behalf of the conditions and provisions thereof.*

SUBMITTED THIS ______________________ *DAY OF* ______________

NAME OF FIRM ______________________________________

STREET ADDRESS ____________________________________

CITY ______________________ *STATE* _________ *ZIP CODE* __________

OPERATING AS: (Complete, and strike out words that do not apply)

A CORPORATION UNDER THE LAWS OF THE STATE OF ______________

A PARTNERSHIP (GIVE FULL NAMES OF ALL PARTNERS)

____

____

ENCLOSED: One (1) construction schedule

Figure 8.2 (continued)

Bids for private work can be opened publicly or privately. Bids for public work must be opened publicly. If opened publicly, amounts for bids and alternates are read aloud and a written summary is distributed to bidders upon request.

The AIA Instructions to Bidders states that the owner has the right to reject any or all bids. The low bidder is not automatically awarded the contract. Bids must be evaluated, and a contract may be awarded to a contractor other than the low bidder. This may occur even on publicly funded work, since instructions to bidders usually state that the award will be to the *lowest responsible* or *lowest and best* bidder. However, under such circumstances, the owner and architect must be able to substantiate that the low bidder is, in fact, not responsible.

Validity of Bids

A valid bid must conform to the requirements of the advertisement or invitation to bid and the instructions to bidders. If a bid proposes more, less, or work other than that required in the documents, it may be considered unresponsive. Objectionable deviations may be concerned with the form of the bid, the work itself, time for performance, bonding requirements, or terms of the proposed contract. Many owners have funding limitations. If all the bids are for an amount greater than the available funds, they may all be rejected by the owner even if they are otherwise proper.

Negotiated Contracts

For *cost of the work plus a fee* contracts, the general contractor (GC) or construction manager (CM) selected by the owner solicits bids from subcontractors when the bid documents are ready. On *fast-track* projects, certain trades requiring long lead times, such as structural steel, curtain wall, or elevators, may be bid in advance of other trades. After the GC or CM receives and evaluates the bids, they are submitted to the owner for approval. For *stipulated sum* contracts, the owner has limited opportunity to approve subcontractors and no opportunity to approve the subcontractors' bid amounts, before entering into an agreement with the GC.

Waiver of Informalities and Irregularities

Private Work

Under the provisions of the AIA Instructions to Bidders, an owner has the right to waive any irregularity or informality of any bid proposal and to accept the bid that he or she believes to be in his or her best interests. This right should only be exercised with regard for the integrity of the bidding process. It should not be used to promote favoritism, nor as a subterfuge to allow a bidder to gain additional time or information in order to manipulate subcontractors and prices for a successful bid.

The crucial considerations are the significance of the deviations from the provisions of the instructions to bidders and the extent of the potential damages to other bidders resulting from the waiver of irregularities.

Public Work

Agencies that award contracts from publicly funded projects have a duty to procure the best quality construction at the lowest price, consistent with budget limitations and other constraints imposed by law. Everything they do is subject to public scrutiny and criticism, and the agencies must avoid even the appearance of impropriety. As a result, public officials are very meticulous about the bid form and the bidding process. They are unlikely to waive informalities or irregularities in the bidding process, which could be construed as favoritism in violation of the public trust. Under these circumstances, it is not uncommon for an unsuccessful bidder to file a legal action, preventing a bid award, and, consequently, delaying the project.

Alternates

Types

Alternates are changes to specific parts of the drawings or specifications that allow an owner to tailor bid proposals to fit the available funds, or to allow a choice between alternative materials or equipment. Alternates may be either additive (an addition to the base bid) or deductive (a deduction from the base bid). Additive alternates are preferable, because credit to the contract sum for deleted work is usually less than the additional cost for the installation of the same work.

Selection Procedures

The AIA Instructions to Bidders gives an owner the right to accept any alternates, in any order or combination. The low bidder is determined by the sum of the base bid and the cost of additive alternates, and the credit of deductive alternates. An owner may choose alternates calculated to buy the most (or best) construction for the available funds, or may select a combination of alternates that appear to be in his or her best interests. An owner should not select alternates merely to manipulate the bidding process to give one bidder an advantage at the expense of the others.

To avoid such manipulation, public agencies may prescribe that alternates be accepted or rejected in order of priority as they appear in the bid documents. For example, a public agency could not accept the base bid plus alternates 4 and 5 without first accepting alternates 1, 2, and 3.

Unit Prices

Where the quantity of a particular type of work is unknown, or varies from that shown in the contract documents, the cost of the work may be determined on the basis of *unit prices*. The units involved, such as cubic yards of concrete, must be described accurately, and base quantities must be reasonably easy to determine. Often, the contract documents specify the quantity of materials on which to calculate the base bid. Bid forms should provide space to list all unit prices that a contractor is required to quote. Since prices for added work may be different from those for deleted work, all unit prices should be quoted as either *additions to* or *deletions from* the contract. Unit costs cannot be used to determine the low bid, but rather to adjust costs once the contract is awarded.

Withdrawal or Correction of Bids

By submitting a bid, each bidder agrees not to modify or withdraw the bid for a specified period of time after the bid date—usually 30 to 60 days. If a bid is submitted before the due date, however, it may be withdrawn and resubmitted within the original bidding period and before the bid opening.

A bidder may wish to withdraw the bid if, for example, he or she finds soon after the opening that the bid contains substantial mathematical errors. Furthermore, he or she may determine that a large item of work has been inadvertently omitted, or a subcontractor's bid has been misstated. The law recognizes the basic unfairness of binding a contractor to a bid that contains an honest mistake. On the other hand, permitting a bid withdrawal may open the door to manipulation of the bidding process and possible favoritism. Regardless, clerical mistakes, or those caused by the omission of a substantial portion of the work, are usually adequate justification to withdraw a bid.

Even on publicly funded work, the withdrawal of a bid may be allowed. However, changes and corrections after submittal and bid opening are not allowed. The need to avoid any appearance of favoritism and to serve the public interest in obtaining the lowest price override these considerations. For both private and public work, contractors are responsible for their own errors in judgment. For example, if a contractor estimates that certain work will require one crew three days to complete, but in actuality requires three crews two weeks to complete, the owner bears no legal or financial responsibility. The contractor must either absorb the additional costs or refuse to enter into the contract, with consequential forfeiture of his or her bid security as well as potential legal action.

Determination of Successful Bidder

Paragraph 5.3 of the AIA Instructions to Bidders considers the determination of the successful bidder. It states that the owner intends to award a contract to the lowest bidder if all bidding requirements are met and if sufficient construction funds are available to award such a contract. However, it also states that the owner reserves the right to waive informalities and irregularities in bids and to award a construction contract to someone other than the low bidder.

Contractors are entitled to rely on the integrity of the bidding process. The affirmation of intent to award a contract to the low bidder is important. If a contractor suspects that an owner might award the construction contract based on criteria other than price, there is no incentive to bid competitively.

At times, an owner may make minor changes in the contract documents before the construction contract is signed. If so, such changes and their effect on the contract sum can be negotiated with the low bidder. If major changes are desired, as for example where all bids are well in excess of the project budget, an owner may reject all bids and rebid the project based on substantially revised contract documents. The original low bidder may not necessarily be the successful bidder on the second bid based on the revised documents.

The Instructions to Bidders also outlines requirements for data to be submitted after the bid date and prior to the award of a construction contract. For example, if they have not been pre-qualified, contractors being considered may be asked to submit a statement of their qualifications.

After the successful bidder has been selected for award of the construction contract, he or she must submit information on suppliers and subcontractors that are proposed for major portions of the work. The owner and architect then have an opportunity to review the general contractor's list and to make reasonable objections to any of the proposed suppliers or subcontractors. Under these circumstances, the contractor may withdraw the name of a supplier or subcontractor and propose a substitute with the bid price being adjusted accordingly. If a substitute supplier or subcontractor is proposed by the general contractor, the owner may either accept the substitution (with modified price) or disqualify the bidder. Once approved, proposed suppliers and subcontractors may not be substituted after the contract award.

Finally, the AIA Instructions to Bidders covers an owner's right to require a Performance Bond and Payment Bond, as provided for in the advertisement or invitation for bids. The cost of this bond is included in the bid amount. It also describes the form to be used for the contract for construction.

Bid Security

Form

The advertisement or invitation to bid should state whether a bid bond or a bid security is required, and if so, in what form. Usual forms of bid security include bid bonds, cashier's checks, or certified checks. The advertisement or invitation should also state the period of time for which the securities will be held by the owner.

Adequacy

The required amount of bid security may be stated either as a fixed sum of money or as a percentage of the bid price. Normally the amount, using either method, is between 5 and 10 percent of the estimated contract price.

Although all bidders are required to provide a bid security, once the bids are opened owners usually retain only that of the three or four lowest bidders until a contractor actually signs the construction contract. It is unlikely that other bidders would be awarded the contract, and, consequently, it is unnecessary to hold their security. An owner has the right to sign a contract for the lowest bid amount. If the low bidder refuses to sign, he or she must forfeit the bid security with the understanding that the amount of the bid security will fully or partially cover the difference in cost between the low bid and the second low bid to eliminate or minimize the owner's financial loss.

LESSON 8 QUIZ

1. Which of the following statements describes the primary purpose of a pre-bid conference?

   A. Subcontractors deliver their sub-bids to the general contractor.

   B. The general contractor meets with the owner to discuss project funding and budget.

   C. The architect describes special aspects of the project's design or construction.

   D. The architect issues all addenda to describe changes made since the bidding documents were issued.

2. On private construction projects, the owner may use which of the following methods to request proposals for the construction of an addition to the building?

   I. Publish an advertisement for bids.

   II. Send an invitation for bids to pre-qualified contractors.

   III. Negotiate with a construction manager.

   IV. Call the contractor who built the original structure.

   A. I only

   B. II only

   C. I, II, and IV

   D. I, II, III, and IV

3. Which of the following documents are part of the bidding documents? Check all that apply.

   A. Drawings

   B. Instructions to bidders

   C. Shop drawings

   D. Addenda

4. Under normal competitive bidding using AIA documents, an owner may waive which of the following irregularities or informalities?

   I. Late bids

   II. Bids submitted on non-standard bid forms

   III. Incomplete bids

   IV. Omission of a required performance bond

   A. I and II

   B. II and III

   C. I, II, and III

   D. I, III, and IV

5. Which of the following factors have NO effect on the ideal length of the bidding period?

   A. Size of the project

   B. Complexity of the project

   C. Number of bidders

   D. The distribution of documents to local plan rooms

6. If a bid security is required as part of a contractor's proposal, which of the following is the most appropriate form?

   A. Performance bond

   B. Binding covenant

   C. Certified check

   D. Indemnity agreement

7. Once a general contractor is determined to be the low bidder on a project, he or she is required to submit a list of the subcontractors and major suppliers proposed for the project. Which of the following statements about that list is INCORRECT?

A. It is for the owner's and architect's information only.

B. The owner may make reasonable objection.

C. The architect may reasonably reject any of the listed subcontractors.

D. If any parties are substituted at the owner's or architect's request and additional costs result, the owner must pay the price increase or release the bidder without penalty.

8. According to the AIA Instructions to Bidders, bidding documents are issued to

A. general contractors only.

B. general contractors and all sub-bidders.

C. general contractors and selected sub-bidders.

D. general contractors, sub-bidders, and all major material suppliers.

9. Which of the following is NOT included when totaling bids to determine the low bidder?

A. Deductive alternates

B. Unit prices

C. Base bid

D. Additive alternates

10. Where a bidder is required to offer proposals including amounts to be added to or subtracted from the base bid for the addition or deletion of specific items of work, these proposals are known as

A. change in services.

B. change orders.

C. alternates.

D. addenda.

LESSON NINE

CONSTRUCTION CONTRACT ADMINISTRATION

9

GENERAL CONCEPTS

Introduction

Construction contract administration encompasses a wide range of services performed by an architect after completion of the bidding or negotiation phase, and extending up to and sometimes beyond completion of the construction of the project. Use of the terms *supervision* and *inspection* to describe the architect's services during the construction phase should be avoided because of potential professional liability problems. Supervision should be left to the contractor and inspection to building department inspectors and testing agencies. It should be recognized that there is no single word to describe all the services customarily performed by an architect during the construction phase.

The single best source of information about an architect's customary construction phase services—construction contract administration—is found in the AIA Owner-Architect Agreement (Document B201). Article 2.6 describes in detail what is expected of an architect during the construction phase. This description of services is paralleled in the AIA General Conditions of the Contract for Construction (Document A201). Throughout this course, references are made to these two important AIA Documents.

For this course, for the licensing examination, and throughout an architect's professional career, familiarity with the provisions in these two standard AIA documents is a fundamental requirement. Therefore, candidates should initially scan both documents, especially the index of the AIA General Conditions, so that the material in this course can be studied in context. Much of the material in this course will explain the provisions in the AIA documents.

Principles of Agency Law

An *agent* is defined as someone who is authorized to act on behalf of another party, the *principal*. Both persons and corporations can act as agents for other persons or corporations. The authority of an agent is usually established by an agreement between the parties, but it can also be determined by the agent's conduct and implied conditions.

During the construction phase, an architect normally performs construction contract administration as the owner's agent. In order to avoid liability problems, the agreement between the architect and the owner should spell out the scope of the architect's *agency* and establish that the architect is authorized to act on behalf of the owner only to the extent set forth in the contract. In the absence of a written contract, the scope of the architect's agency will be implied or determined by general principles of law. When the architect is acting as an agent for the owner, it is important that the owner's identity be disclosed, because an agent can be held personally liable if he or she acts on behalf of an undisclosed principal. This can have direct implications in regard to who is responsible for paying for the construction work, among other things.

Standard Contract Requirements

The scope of the architect's role as the owner's agent is established by Subparagraph 2.6.1.1 in AIA Document B201:

> *The Architect shall provide administration of the Contract between the Owner and the Contractor as set forth below and in AIA Document A201-2007, General Conditions of the Contract for Construction.*

Note that this provision about the scope of the architect's construction contract administration services is in the Owner-Architect agreement, but it makes a direct reference to the General Conditions, which is part of the Owner-Contractor agreement. The reason for this is both simple and important: the Owner-Architect agreement establishes the scope of the architect's authority to act as the owner's agent, and the contractor is informed about this authority by parallel provisions in the General Conditions of the contract for construction. The contractor would not normally see the Owner-Architect agreement because he or she is not a party to it. It would make little sense for the architect to be authorized to act as the owner's agent during construction if the contractor were not informed about the architect's role and thus unaware that the architect could act on behalf of the owner in this manner.

Article 2.6 of B201 describes the contract administration services an architect agrees to provide. Compensation for the architect's services is provided for in Article 6 of B201. Exceeding the scope of these services opens the architect to potential liability and financial loss. Certain services, such as redesign caused by changes in the owner's program, may be provided for additional compensation (see B201, Paragraphs 3.3 and 3.3.1). Other services, such as offering legal or insurance advice, should not be provided at all.

The architect must know and understand the contract terms that establish the scope of his or her agency. Because he or she is the owner's representative, the architect must be careful not to act in a way that either causes problems for the owner or creates a professional liability exposure. For example, the contract normally gives the architect the authority to reject work that does not conform to the requirements of the contract documents (B201, Subparagraph 2.6.2.2, and A201, Subparagraph 4.2.6), but the architect usually would not have authority to stop the work without independent written authorization from the owner. Serious consequences, such as claims for delay or the potential claims by workers injured during the course of construction, can arise from the architect's right to stop the work. This has led to the elimination of the contract language that formerly gave the architect the right to stop the work. *Only the owner can order the work to be stopped.*

In addition, the construction contract makes the contractor responsible for construction means, methods, techniques, sequences, procedures, and for safety precautions in carrying out the work. Therefore, the architect, as the owner's agent or otherwise, should not involve himself or herself in *how* the contractor performs this work—only whether the work performed meets the requirements of the contract documents. The architect should also avoid direct contact with subcontractors and suppliers, except in accordance with the provisions of the contract documents. For example, A201, Subparagraph 9.6.3, permits the architect to furnish subcontractors with information on the percentages of completion or the amounts of payment applied for by the contractor. If the architect deals with subcontractors outside of permissible contract bounds, he or she can become liable for interfering with the contractor's contractual relationships with subcontractors.

When an owner requires extensive representation by the architect at the site, the architect may assign one or more of his or her staff as Project Representative(s). The duties, responsibilities, and limitations of the authority of the Project Representative are described in B207. These services are compensated as a Change in Services.

The architect customarily engages consultants, such as structural, mechanical, and electrical engineers, to provide services as part of the Owner-Architect Agreement. A properly prepared Architect-Consultant Agreement (C401) binds the consultant to the architect according to the same terms as the architect is bound to the owner.

Professional Liability Coverage

Professional liability insurance protects an architect against claims which may arise out of his or her negligent acts, errors, or omissions during the performance of professional services. It is sometimes called *errors and omissions* insurance or *malpractice* insurance. Although this insurance is very costly, most established firms carry it. Sophisticated clients, such as corporations and some government agencies, often require, as a condition of the Owner-Architect Agreement, that the architect have professional liability insurance.

As with any insurance policy, the professional liability policy contains exclusions to limit certain types of claims for which the insurance does not apply. Some exclusions are quite broad, such as the exclusion for claims arising out of any express warranties or guarantees that the architect may have agreed to. Other exclusions apply to specific aspects of rendering services, such as claims arising out of cost estimates being exceeded.

Professional liability insurance covers the architect's liability for professional *negligence*. Negligence is defined as a failure to meet the ordinary *standard of care* expected of an architect under the same or similar circumstances as those associated with actual allegations of negligence in a specific case. Before an architect can be found liable for negligence, a plaintiff (the person bringing the claim) must allege and prove that there was a legal duty owed by the architect, the architect breached that duty (the architect did something he or she should not have done, or failed to do something he or she should have done), and the breach of the duty was the proximate cause of (was somehow directly related to) actual injury or damage suffered by the person bringing the claim. All of these factors must be present before an architect can be found liable for negligence. Because an architect's professional liability is based on the law of negligence, professional liability insurance policies do not provide coverage for intentionally wrongful acts.

During a trial, the professional standard of care is determined by the testimony of expert witnesses—usually other architects—who are knowledgeable about what is ordinarily expected of an architect under the circumstances. The testimony of expert witnesses plays a significant role in determining whether liability is imposed on an architect for damages or injuries that occur during the course of a project.

GENERAL LEGAL PRINCIPLES

Liens

A *mechanic's lien* is a legal claim against someone's property. Liens are authorized by statutes in every state to protect people who expend labor or provide materials to improve someone else's property. Because the labor or materials cannot be repossessed once they are incorporated into the property, the right to a lien protects the worker's right to payment. If the recipient of the labor or materials fails to pay for them, his or her property can be attached by a lien to ensure that there will be collateral for the debt. Thus, the property itself becomes the security for the money owed if it is not otherwise paid, and if the claimant is forced to go to court to collect. After a mechanic's lien is filed, the claimant still must file a lawsuit to prove his or her right to payment and to get a court order to foreclose on the property.

Mechanics' liens are carefully defined by statute, and all statutory requirements must be complied with or the lien may be invalidated. Lien statutes customarily require that liens be filed within 30 to 90 days, depending on the jurisdiction, after the last work is performed or materials are installed. Some jurisdictions require subcontractors to file a notice of lien before they can actually file a lien. In all jurisdictions, general contractors have lien rights. Whether or not lien rights extend to sub-subcontractors or material suppliers depends on the lien laws of the jurisdiction in which the project is located. Architects' lien rights vary from state to state. In a few states, architects have absolutely no right to a lien for any architectural services. In others, the project must proceed into construction before an architect has lien rights. And, in other states, an architect can assert a lien even if only design services have been performed, if he or she is not paid. As noted, the lien laws vary considerably from state to state, and they must be checked carefully before a lien is filed. Since the ARE is a national exam, questions on lien laws are general in nature.

In order to protect the owner from liens if a contractor fails to pay his or her subcontractors or material suppliers, the standard AIA contract documents set up several layers of protection. If liens are actually filed, or if the architect learns that liens may be filed, the owner must get his or her attorney involved. After all, it is the owner's property against which the lien claimants are asserting their claims. However, the architect must be familiar with lien-related concepts, because both the owner and his or her attorney often will seek technical advice from the architect.

(1) The first level of protection for the owner to guard against liens involves the *retainage*. By withholding (retaining) a small percentage of the money owed to the contractor, as the construction progresses, the owner will have funds to pay for labor or materials if the contractor fails to make payment. Retainage also can be used to pay for the correction of work performed improperly. Obviously, once the work has been performed properly and all bills have been paid, the retainage must be paid to the contractor at the end of the construction phase. However, the retainage should never be reduced or released without the written permission of the surety company that issued the performance bond for the contractor. This will be covered later in detail in the discussion on Bonds and Insurance. Retainage is provided for in the AIA Owner-Contractor Agreement, Document A101, Article 5. In addition, Subparagraph 9.5.1 in the AIA General Conditions (A201) states in part:

> *The Architect may withhold a Certificate for Payment in whole or in part, to the extent reasonably necessary to protect the Owner... because of:*
>
> .2 *third party claims filed or reasonable evidence indicating probable filing of such claims;*
>
> .3 *failure of the Contractor to make payments properly to Subcontractors or for labor, materials or equipment;...*

Therefore, if the architect has information that the contractor is not paying his or her bills, he or she can require adjustments in the contractor's Application for Payment to withhold funds to protect the owner.

(2) Another level of protection for the owner is to require the contractor to provide a Labor and Material Payment Bond. This is a bond that guarantees payment for labor and materials if the contractor fails to pay for them. This will be covered in detail later in this lesson.

(3) A further level of protection is to require the contractor to submit an affidavit and releases of liens before the contractor is entitled to receive final payment, and in some cases progress payments, from the owner. This is provided for in the AIA General Conditions (A201) in Subparagraph 9.10.2 which states in part:

> *Neither final payment nor any remaining retained percentage shall become due until the Contractor submits to the Architect (1) an affidavit that payrolls, bills for materials and equipment, and other indebtedness connected with the Work for which the Owner or the Owner's property might be responsible or encumbered (less amounts withheld by Owner) have been paid or otherwise satisfied,...and (5) if required by the Owner, other data establishing payment or satisfaction of obligations, such as receipts, releases and waivers of liens, claims, security interests or encumbrances arising out of the Contract, to the extent and in such form as may be designated by the Owner. If a Subcontractor refuses to furnish a release or waiver required by the Owner, the Contractor may furnish a bond satisfactory to the Owner to indemnify the Owner against such lien. If such lien remains unsatisfied after payments are made, the Contractor shall refund to the Owner all moneys that the Owner may be compelled to pay in discharging such lien, including all costs and reasonable attorneys' fees.*

In order to facilitate receipt of the contractor's affidavit and release of liens, the contractor can submit properly completed copies of AIA Document G706, Contractor's Affidavit of Payment of Debts and Claims and AIA Document G706A, Contractor's Affidavit of Release of Liens. These documents include a notarized statement by the contractor that all subcontractors, material suppliers, and others have been paid. Any exceptions must be noted in the form. In addition, the form requires the contractor to attach separate releases or waivers of liens from subcontractors and materials or equipment suppliers to the extent required by the owner. There is no AIA form for the release of lien required from subcontractors, but most contractors have developed their own forms. In general, they require the subcontractor or supplier to state that he or she has been paid for all labor and materials for the project and that he or she releases all rights to assert a lien against the property. If the contractor is unable to secure a release of lien from all subcontractors or material or equipment suppliers, he or she can provide a lien bond to protect the owner from loss in the event a lien is filed.

All of the above procedures—retainage, bonds, and releases of lien—are designed to protect the owner from having to pay twice for labor or materials. If the owner pays the contractor, but the contractor fails to pay his or her subcontractors or suppliers, the latter parties can assert a lien against the owner's property. If they are successful in proving their claim of nonpayment, the owner must either pay them directly, or else risk losing his or her property. Because mechanic's liens have serious consequences, the law and standard AIA contract documents provide important protections that must be understood by the architect as part of his or her construction phase services.

Arbitration

The architect's specific role in the arbitration process will be covered later in this lesson. In general, arbitration is widely used in the construction industry as a means of resolving disputes. Examples of arbitration clauses can be found in standard AIA documents dating back to the late 19th century. Today, standard construction industry documents typically require mandatory, binding arbitration of any unresolved disputes that arise between the parties to the contract. The arbitration proceedings usually are conducted under the auspices of the American Arbitration Association (AAA), but the parties can agree to non-AAA arbitration. The AAA has developed a set of Construction Industry Arbitration Rules to govern the arbitration process. (See Figure 9.1 for a sample Demand for Arbitration form.)

In all but a few states, contract clauses requiring arbitration of *future* disputes are valid and binding. In a few states, arbitration clauses are not binding in regard to future disputes, but even in those states, the parties can agree to binding arbitration *after* a dispute arises. The courts generally take a very liberal attitude toward arbitration and normally will require matters to be arbitrated if there is any basis for doing so, such as an agreement between the parties to arbitrate their disputes. No one can be compelled to arbitrate if he or she has not agreed to arbitration, but once there is an agreement to arbitrate, it is very difficult to avoid arbitration.

The standard AIA contract documents contain a broad arbitration clause calling for arbitration of any claim or dispute. These clauses can be found in Paragraph 4.3 of the AIA Owner-Architect Agreement (B102) and in Paragraph 15.4 of the AIA General Conditions (A201). Note that AIA Document B201 incorporates AIA General Conditions (A201) by reference.

In arbitration, the arbitrator hears the arguments and evidence presented by each party, and he or she then makes a decision based on the facts and the law as he or she understands them. The arbitrator is not required to give any reasons for his or her decision, and the decision is final and binding and cannot be appealed except on very narrow grounds such as fraud or collusion between the arbitrator and a party. Once the

TYPICAL ADVERTISEMENT FOR BIDS

American Arbitration Association

CONSTRUCTION INDUSTRY ARBITRATION RULES

To institute proceedings, please send three copies of this demand and the arbitration agreement, with the administrative fee as provided in the rules, to the AAA. Send the original demand to the respondent.

DEMAND FOR ARBITRATION

DATE: Sept. 30, 1998

TO: Name Empire Trust Company, Inc.
(of the Party on Whom the Demand Is Made)

Address 255 Royal Way

City and State Kingville, ME ZIP Code 04226

Telephone (207) 555-3282 Fax (207) 555-3283

Name of Representative ______ (if Known)

Representative's Address ______

Name of Firm (if Applicable) ______

City and State ______ ZIP Code ______

Telephone ( ) ______ Fax ______

The named claimant, a party to an arbitration agreement contained in a written contract, dated May 20, 1998 and providing for arbitration under the Construction Industry Arbitration Rules of the American Arbitration Association, hereby demands arbitration thereunder.

THE NATURE OF THE DISPUTE: Under the contract dated May 20, 1998, Smith Brothers Construction Co., Inc. demands payment of $90,000, such sum representing final payment due on construction contract with Empire Trust Company, Inc. Empire Trust Company contests this demand, claiming that construction of its building is not complete by virtue of several unresolved problems.

THE CLAIM OR RELIEF SOUGHT (the Amount, if Any): $90,000 plus interest at the rate of 12 percent per annum.

Please indicate the industry category of each party.

CLAIMANT: ☐ Owner ☐ Architect ☐ Landscape Architect ☐ Engineer ☒ Contractor
☐ Subcontractor (Specify. ______) ☐ Interior Designer ☐ Other ______

RESPONDENT: ☒ Owner ☐ Architect ☐ Landscape Architect ☐ Engineer ☐ Contractor
☐ Subcontractor (Specify. ______) ☐ Interior Designer ☐ Other ______

HEARING LOCALE REQUESTED: Kingville, ME
(City and State)

You are hereby notified that copies of our arbitration agreement and this demand are being filed with the American Arbitration Association at its Bangor, ME office, with a request that it commence administration of the arbitration. Under the rules, you may file an answering statement within ten days after notice from the administrator.

Signed Karl Klink Title Attorney
(May Be Signed by a Representative)

Name of Claimant Smith Brothers Construction Co., Inc.

Address (to Be Used in Connection with This Case) 6506 Broadway

City and State Bangor, ME ZIP Code 04401

Telephone (207) 555-3663 Fax (207) 555-3664

Name of Representative Karl Klink

Name of Firm (if Applicable) Klink, Blink, and Fink, Attorneys

Representative's Address 65 Moop St.

City and State Bangor, ME ZIP Code 04401

Telephone (207) 555-8274 Fax (207) 555-8275

☐ **MEDIATION is a nonbinding process. The mediator assists the parties in working out a solution that is acceptable to them. If you wish for the AAA to contact the other parties to ascertain whether they wish to mediate this matter, please check this box (there is no additional administrative fee for this service).**

Form C12–6/93

Figure 9.1

arbitration has been held, the case normally is over, regardless of the outcome. The standard arbitration clause states that the award shall be final, and judgment may be entered upon it in any court having jurisdiction.

Even though the standard AIA documents call for mandatory arbitration, there is no general legal requirement that the parties must agree to arbitration. It is not uncommon in non-AIA contracts for the arbitration clause to be eliminated. The parties can always agree not to arbitrate even after signing a contract containing an arbitration clause. If there is no agreement to arbitrate, the dispute or claim must be dealt with in court, as with any other legal matter.

There is no way to determine beforehand whether arbitration or litigation is preferable. The AIA and other construction industry organizations have decided that arbitration is preferred, and they have therefore included arbitration clauses in their standard contract documents. For each supposed advantage of arbitration there may be arguments to the contrary. Nonetheless, arbitration is considered advantageous because it is usually quicker than a trial, it can be less costly, the matter can be heard by an arbitrator who has expertise in the construction industry, and it is private. This last factor is a particular benefit to architects who normally are concerned about their professional reputations and want to avoid court proceedings that are a matter of public record.

BONDS AND INSURANCE

General

The term *risk management* acknowledges that risk is ever present during the design and construction phases of a project. Since no one wants unnecessary exposure to risk, there are several forms of protection available. They include: programs to prevent loss (such as TQM, or Total Quality Management), bonds to guarantee performance, insurance to pool risk, liens to obtain payment, and retainage to ensure completion and correction of deficiencies. Examples of risk and available protection include:

- A shored embankment gives way and a crane plunges into the excavation, causing injury and property damage.

  The contractor is protected by liability insurance and workers' compensation insurance.

- A subcontractor goes bankrupt from accumulated losses on previous projects, causing workers and material suppliers to go unpaid.

  The owner is protected by the contractor's labor and material payment bond.

- Material shipped from overseas is mishandled, causing breakage and shortage of critical building components on the construction site.

  The contractor is protected by his or her liability insurance, and the owner is protected by his or her "builder's risk" insurance if he or she has paid for the material.

- An architect specifies toilet fixtures which do not comply with the Americans with Disabilities Act, causing the owner to replace all the fixtures to avoid potential lawsuits.

  The architect is protected by professional liability insurance if he or she is found to be negligent. However, given the high deductibles in most such policies, this expense would most likely be paid out of pocket.

- A contractor is the low bidder on a project but refuses to enter into an agreement, causing the owner to accept a higher bid and to delay the start of construction.

  The owner is protected by the contractor's bid bond or bid deposit.

Although bonds and insurance are often discussed together, and both are provided by the insurance industry, there are fundamental differences between the two. A *bond* is essentially a guarantee—the bonding company's role being similar to that of a co-signer on a bank loan. The *surety*, the party that issues the bond, guarantees that the *principal*, the contractor, will perform certain acts to be undertaken for the benefit of the *obligee*, the owner. There is no expectation of loss with a bond, since the owner, contractor, and surety all believe the contractor is capable of properly performing the work. If they did not, presumably, the owner would not hire the contractor, the contractor would not agree to undertake the work, and the surety would not agree to issue the bond. Despite this confidence, bonds are usually required because of the significant financial risk associated with most construction projects.

Insurance, on the other hand, anticipates the possibility of loss and is written with the expectation that events will occur to cause a loss. However, it is not known which individual in a group of insureds will suffer the loss. By pooling relatively small amounts of money—the premiums—a large sum can be aggregated to pay or compensate those members of the group of insureds who suffer a loss. As with automobile or home insurance, the relatively small premiums paid by all car or home owners enable the insurance company to pay out large sums to the relatively few individuals who are involved in car accidents or whose homes burn down or are burglarized. Now we will discuss specific forms of construction bonds and insurance.

Bid Bonds

A bid bond is a bond furnished by a bidder, as part of his or her bid submission, to guarantee two things:

1. That the bidder will, in fact, enter into a contract with the owner at the price and on the terms stated in his or her bid; and
2. That the bidder will provide a performance and labor and material payment bond to guarantee that the work will be properly carried out and paid for.

If the bidder fails to do these two things, the surety or bonding company is liable for any extra costs, up to the penal amount of the bond, incurred by the owner in good faith in order to enter into a contract with another contractor.

Occasionally, the bid documents may permit bidders to submit a certified check in some stated amount, possibly 5 percent of the bid, or a set dollar amount in lieu of a bid bond. The principle remains the same: if the selected bidder fails to enter into the contract, the owner can use the bid security to pay for any increased costs or other damages suffered by having to contract with another party.

Performance Payment Bonds

The performance bond and payment bond guarantee precisely what their titles indicate—performance of the work and payment for labor and materials. These bonds are discussed together because they are often written on a combined form. (See AIA Document A312, Performance Bond and Payment Bond.)

A312 is a performance bond and payment bond form which combines two separate bonds into one form. It is not a single combined performance and payment bond, however. Certain federal and state laws, called *Miller Acts*, require the use of bonds on public works in lieu of allowing the contractor mechanic's lien rights. The A312-2010 version was drafted to address concerns from the surety industry about judicial interpretations from various state courts.

The amount of the performance bond normally is 100 percent of the contract amount. The issuance of a 100 percent bond indicates that the contractor's surety company believes the contractor can carry out the work. If the contractor requests a lower bond amount, it could suggest that the surety company may not be willing to write a bond for the contractor in excess of a certain limit because the surety does not believe the contractor has the capability to perform that volume of work, either on its own or in combination with other projects under contract.

Sometimes, an owner may decide to eliminate the requirement for a performance bond because he or she believes the contractor has sufficient strength to carry out the work, and the owner is not concerned that the contractor will default on the contract or go out of business. This can result in a saving of between one and one and a half percent of the contract price. Nonetheless, the decision to eliminate the requirement for a performance bond should be made by the owner only after consultation with his or her attorney and other advisors. The architect might be asked his or her opinion about the contractor's stability and for a recommendation in this regard. However, any recommendation should advise the owner about the risk associated with not having a bond, no matter how stable and capable the contractor may appear to be.

In the event of a default by the contractor, the performance bond requires the surety company to either (1) complete the contract in accordance with its terms and conditions, or (2) obtain bids to enable another contractor, under contract to either the surety or the owner, to complete the contract in accordance with its terms and conditions. The surety's financial liability extends to the penal amount on the face of the bond. Under the standard AIA bond form, any suit to enforce the bond must be brought within two years from the date on which final payment under the contract is due.

Not all contractors and their sureties use the standard AIA bond forms, so the owner's attorney and insurance advisors should review the actual bond forms submitted by the contractor to determine what notices may be required, about extensions of time or change orders, for example, and the procedures in the event of a default. Architects' professional liability insurance excludes coverage for claims arising out of giving or failing to give advice about insurance and bonds. Architects can avoid exposing themselves to this uninsured liability by not attempting to interpret bond terminology for their clients.

The labor and material payment bond, as indicated above, protects the owner against claims by subcontractors and suppliers who are not paid by the contractor. This bond gives those parties the direct right to sue on the bond in order to collect payment from the surety. The owner has no liability for any costs or expenses associated with any such suit. The AIA Performance and Payment Bond (A312) extends protection to parties having a direct contract with the contractor, to parties having a contract with a subcontractor of the contractor, and to water, gas, power, light, heat, oil, gasoline, telephone service, and equipment rental directly applicable to the contract. There is no requirement that a claimant must file a lien before attempting to collect payment from the surety company. In fact, suing on the bond is a clear alternative to filing a lien in an attempt to collect from the owner.

With both the performance bond and the labor and material payment bond, the retainage, if any, held by the owner can be used by the surety to reduce its financial losses. In the event of a default or failure of payment by the contractor, the owner must still pay the contract amount. The surety only pays for any excess costs caused by the contractor's default, up to the penal amount of the bond. Because the retainage is viewed as the surety's money in the event of a default, *the retainage should never be reduced or released without the surety's written permission*. Even if things are going smoothly on a project, and a reduction or release of retainage is in order, the contractor could be in trouble on other projects. In that case, the surety probably would want to monitor all of the contractor's finances. To facilitate getting the surety's permission for a reduction or release of retainage, AIA has developed two forms: Consent of Surety to Reduction in or Partial Release of Retainage, Document G707A, and Consent of Surety Company to Final Payment, Document G707. These two forms alert the surety that the retainage will be released upon the surety's consent and that the surety's obligations remain intact even though this money will be paid to the contractor. If the surety has any objections to this, it can refuse to execute the documents. In that case, the retainage should not be reduced or released. The consent of the surety to final payment is a contract condition contained in Subparagraph 9.10.2 of the AIA General Conditions (A201).

Contractor's Insurance

Construction contracts normally require the contractor to carry various types of insurance to protect against risks or liabilities associated with the construction of a project. This insurance is important because it can protect the owner and architect, as well as the contractor, from claims or financial loss in the event of injury or damage during the course of the work. When insurance is in effect, claimants normally look to the insurance for a prompt settlement of their claim. Without insurance, however, lawsuits are often filed in an attempt to collect damages from any party with assets who might have been responsible for causing the problem.

The AIA General Conditions require the contractor to carry several different types of insurance to protect against loss from various categories of potential claims. It is the owner's responsibility to stipulate which types of insurance coverage and the limits of coverage the contractor must carry for the project. Often, because of the architect's experience from previous projects, the owner will ask the architect for advice about insurance coverage and limits. The architect should *never* give insurance advice, because he or she is not an insurance agent or broker and his or her professional liability insurance does *not* cover claims arising out of the giving of insurance advice. Instead, the architect should offer to assist the owner's insurance agent in determining which and how much insurance is appropriate for the project. The architect can tell the owner what his or her experience on prior similar projects has been, but the architect should make it very clear that he or she is not making any recommendations about insurance for the specific project, and that the final decision about insurance must be made by the owner on the basis of professional insurance advice.

To assist the architect in dealing with the owner on matters of insurance, as well as to document the owner's instructions about insurance coverages and limits for the project, AIA has developed Document G612, Owner's Instructions to the Architect. This form is in a checklist format that can be used by the owner and his or her insurance agent to decide which bonds and insurance will be required, as well as the limits of coverage. After the form is completed by the owner, it is sent to the architect to be inserted into the bidding and contract documents for the project.

Another useful AIA document in establishing insurance requirements is the Guide for Supplementary Conditions, Document A503. The insurance requirements in the AIA General Conditions are general insurance requirements. These provisions must *always* be supplemented in the Supplementary Conditions to establish specific project insurance requirements and limits of coverage. The AIA General Conditions can never be used without appropriate Supplementary Conditions in regard to insurance requirements. The AIA Guide for Supplementary Conditions (A503) provides guidance and useful sample language in this regard.

To provide a detailed analysis of bonds and insurance, AIA publishes *Construction Bonds and Insurance*, by Bernard B. Rothschild, FAIA. This publication is a detailed analysis of bond and insurance requirements contained in the AIA standard construction documents, and it includes sample contract language, a glossary of insurance terminology, and sample insurance policies.

By law, all employers must carry workers' compensation insurance to protect employees in case of job-related injuries. Therefore, the contractor is required, by Subparagraph 11.1.1.1 of the AIA General Conditions, to carry this coverage for his or her construction workers. The theory behind workers' compensation insurance is that workers suffering job-related injuries should be able to get a prompt and fair insurance settlement, instead of having to bring a lawsuit to recover damages for their injuries. The employer is required by law to pay the premiums for workers' compensation insurance. These premiums are included in the contractor's overhead costs and are therefore reflected in his or her bid prices.

By statute, the employer who is required to pay the workers' compensation insurance premium is given immunity from a separate lawsuit by the worker who collects the insurance benefit after an injury. The worker, however, is not precluded from suing any other party who might have been responsible for the injury. Thus, an architect may be sued directly by a construction worker after he or she is injured and collects workers' compensation insurance benefits. Because the worker cannot sue the contractor, the injured worker can look for a third party to sue (i.e., the architect) if he or she does not feel the insurance has provided adequate compensation.

These *third party suits* are based on the idea that the architect somehow had a legal duty to properly direct and manage the construction work. If the plaintiff can establish that the architect had this duty, and that the architect breached the duty, the architect can be held liable for the injuries. For this reason, the word *supervision* should *never* be used to describe the architect's construction phase services, or as a synonym for construction contract administration. Several court decisions have held that *supervision* means a right to manage, direct, and control. When supervision is used in the context of a construction project, it is intended to describe the contractor's function—the contractor, and not the architect, directs, manages, and controls the construction work. If the architect has a contractual obligation to *supervise*, he or she can be held liable for any injuries or damage that occur because he or she failed to properly direct, manage, and control the construction work. Thus, to avoid liability to construction workers and to limit their recovery for job-related injuries to the workers' compensation insurance, an architect should not undertake, by contract or otherwise, any duties that might be interpreted as part of the contractor's scope of work.

Other liability insurance required of the contractor covers claims by his or her employees that are not covered by workers' compensation insurance, claims by non-employees because of bodily injury, sickness, or disease, and claims for damages insured by the usual personal injury liability coverages (i.e., libel, slander, etc.). In addition, the contractor is required to carry insurance for claims for damages other than to the work itself (an adjacent building, for example) and for claims arising out of the ownership, maintenance, or use of a motor vehicle. The contractor's workers' compensation insurance, employer's liability insurance, comprehensive general liability insurance, personal liability insurance, and automobile liability insurance policies normally protect against the claims described above.

An important adjunct to the contractor's insurance coverages is required by Subparagraph 11.1.1.7 of the AIA General Conditions (A201). This provision requires the contractor's insurance to include contractual liability insurance applicable to the contractor's obligations to indemnify and hold the owner and architect harmless from certain types of claims, as set forth in Paragraph 3.18 of the AIA General Conditions. Indemnification is an agreement between two parties not to hold one of the parties liable for future legal action or fines. Contractual liability insurance provides coverage for a liability assumed by contract. A contractual provision calling for indemnification is a contractually assumed liability. Paragraph 3.18 requires the contractor to indemnify the owner and architect if they are sued by a construction worker who is injured during the course of the work. By requiring contractual liability insurance, usually written by an endorsement to the contractor's comprehensive general liability policy, the owner and architect can be assured that there is a party with financial responsibility—the insurance company—standing behind the contractor's obligation to indemnify them in the event of this type of claim.

Owner's Insurance

The owner has an insurable interest in the work as it progresses. Under the provisions of the AIA General Conditions and many non-AIA construction contracts, title to the work passes to the owner each month as the owner makes the progress payments to the contractor. Therefore, it is considered appropriate for the owner to carry the insurance on the property itself. On occasion, the contractor, rather than the owner, will be required to carry the property insurance. This occurs when the owner is relatively unsophisticated and does not want to have the responsibility for the property insurance or when the contractor can arrange this coverage more conveniently than the owner can. In any event, it should be recognized that the AIA General Conditions place the responsibility for carrying the property insurance with the owner. If the contractor is to carry the property insurance, the AIA Guide for Supplementary Conditions (A503) contains suggested language to achieve this objective.

The property insurance is sometimes referred to as the *builder's risk* or the *builder's risk-all risk* insurance. Because the insurance is often carried by the owner, the term *builder's risk* is a misnomer. In addition, the coverage is defined by the policy terms, conditions, and exclusions, so it does not literally cover *all* risks. And, this coverage can be written on a specified perils basis rather than on an *all risk* basis. The function of the property (or *builder's risk*) insurance is to protect the property itself against such risks as fire, theft, vandalism, and malicious mischief. Typically, the coverage is written for the full insurable value of the work on either a completed value or a reporting form. The completed value is coverage for the full value of the work, with a single premium paid. The reporting form requires a monthly adjustment as the value of the work increases during construction. If it is to be for a lesser amount, the owner is obligated to notify the contractor prior to commencement of the work so that the contractor can arrange his or her own coverage to further protect himself or herself and the subcontractors. The cost of arranging this latter coverage can be charged to the owner. The AIA General Conditions also requires the owner to maintain boiler and machinery insurance, if required by the contract documents or by law.

If a loss occurs that is covered by the property insurance, the owner is required to act as a trustee for the insurance proceeds received in the event of a covered loss. If the contractor is carrying the property insurance, the contractor would act as the trustee in the event of a loss. The owner is required to pay the contractor, and the contractor in turn to pay the subcontractors, a just share of the insurance proceeds received in settlement of claims. Funds received in settlement of claims should be placed in a separate account, pending distribution to the parties.

Normally, if a loss occurs, the owner or his or her attorney will become involved, but they may look to the architect for technical assistance. The architect should limit his or her involvement to providing professional advice. He or she should not attempt to determine the rights of the respective parties, act as a custodian for the insurance proceeds, or make any determination about how the proceeds are to distributed. These decisions must be made by the owner on the basis of advice from his or her attorney.

The architect should be aware of the concept of subrogation. *Subrogation* is a procedure by which an insurance company, after it pays a loss to its insured, can attempt to recover this amount from some other party who may have actually caused the loss. An insurance company cannot recover from its own insured, whether or not the insured was at fault. However, if someone else was at fault, the concept of subrogation enables the insurance company to "step into the shoes" of its insured in an attempt to recover its loss.

The standard AIA General Conditions contains a *waiver of subrogation* clause (Subparagraph 11.3.7) that precludes the parties from seeking to recover any money from each other for any loss covered by the property insurance. Because the parties are waiving their rights to recover from each other, the property insurance company would not have any right to do so either, since the insurance company has no greater rights than its insured. Most insurance policies prohibit the insured from waiving any rights *after* a loss occurs, but it is not customary for the policy to prohibit this prior to the occurrence of a loss. Presumably, if the insurance company is aware of the contract terms calling for the waiver, it can determine the appropriate premium, knowing that it will not be able to offset any amount paid out in the event of a loss.

Another aspect of the property insurance involves occupancy of the work by the owner prior to substantial completion. If it becomes necessary for the owner to occupy the work prematurely, the contractor and the company providing the property insurance must agree to this. The insurance company's consent must be evidenced by an endorsement to the insurance policy.

The above discussion covers the standard property insurance coverages for a typical project. Neither the owner nor the contractor is limited to the standard coverages. If special project requirements or personal concerns of either party necessitate additional coverages, the parties can consult their insurance agents and arrange whatever insurance is appropriate to protect against the risk of loss.

Certificate of Insurance

The Certificate of Insurance is a memorandum that outlines the types and limits of the insurance coverages carried by the contractor for the project. Subparagraph 11.1.3 of the AIA General Conditions (A201) requires the contractor to provide certificates of insurance acceptable to the owner. The certificates must contain a provision stating that the owner will be given written notice at least 30 days before the underlying insurance policies can be cancelled. The notice requirement gives the owner and the contractor a reasonable opportunity to arrange replacement coverage or effect a termination of the contract if replacement insurance cannot be arranged.

AIA has developed Document G715, Supplemental Attachment for ACORD Certificate of Insurance 25-S, to facilitate the communication of this information. The certificate of insurance must be filed with the owner before the work commences. If the certificate is sent by the contractor to the architect, the architect should promptly forward it to the owner with instructions that it be reviewed by the owner's insurance agent to determine whether the contractor's insurance coverages comply with construction contract requirements.

The Certificate of Insurance is normally prepared and signed by the contractor's insurance agent, and there can be no guarantee that he or she studied the contract requirements carefully before completing the form. There is always the possibility that the contractor's insurance agent simply listed on the certificate of insurance the coverages normally carried by the contractor, rather than the coverages required by the contract documents. The owner's insurance agent, and not the architect, should analyze the certificate of insurance for compliance with contract requirements. If it does not appear that the contractor's insurance is in compliance, the architect should be notified immediately so that he or she can inform the contractor not to commence work until the proper insurance is arranged. If the required insurance is not in effect prior to the commencement of the work, there is a risk that an uninsured loss could occur.

There is no requirement in the AIA General Conditions that the owner must file a Certificate of Insurance with the contractor. Thus, there is no AIA form to serve as a certificate for the insurance carried by the owner for the project. However, Subparagraph 11.3.6 of the AIA General Conditions requires the owner to *file with the Contractor a copy of each policy that includes insurance coverages required* of the owner before an exposure to loss occurs. In other words, for those insurance coverages carried by the owner, the contract requires the owner to give a copy of each policy to the contractor. Here, too, the architect's role is limited to transmitting the documents. The architect should not make any substantive judgments or recommendations about the adequacy of the owner's insurance. If the contractor has any concerns in this regard, he or she should direct them to the owner, via the architect.

SUBCONTRACTORS AND MATERIAL SUPPLIERS

Subcontractors and material suppliers are parties who have contracts with the general contractor to provide labor and/or materials required in connection with the work. There is no contractual relationship between the owner or the architect and subcontractors and material suppliers. Contractors generally are very sensitive about their contractual relationships with subcontractors and material suppliers and object to any interference by the architect with these relationships. Except under narrowly prescribed circumstances, the architect should not communicate directly with subcontractors and material suppliers. All such communication should be directed to and through the contractor. Likewise, all communications from subcontractors and material suppliers also should come through the contractor to the architect. This is particularly appropriate in regard to shop drawing and sample submittals, which should not be accepted by the architect if they have not been checked and approved by the contractor first. This will be covered in detail in Lesson 10.

Article 5 of the AIA General Conditions requires the contractor, as soon as practical after the contract is awarded, to give the owner and the architect written notice about the subcontractors that the contractor proposes to use for major portions of the work. Although the owner does not have a right of prior approval for subcontractors, the contract gives the owner and the architect the right to raise reasonable and timely objections to any subcontractors or material suppliers. If the owner or the architect has a reasonable objection, the contractor is precluded from contracting with any such party. In that case, the contractor must subcontract with someone else, and the owner will be liable for any increased costs. Paragraph 5.3 of the AIA General Conditions (A201) requires that all subcontracts be in writing so that the subcontractor is bound to the contractor, to the extent of the work to be performed, to the same extent as the contractor is bound to the owner. If the contract is based on documents other than the standard AIA General Conditions, the architect should review the contract carefully to determine the scope of his or her authority in regard to approving or rejecting subcontractors and material suppliers.

As noted above, the architect should refrain from communicating directly with subcontractors and material suppliers. It may seem expeditious for an architect to speak directly with a subcontractor or field superintendent in order to give instructions, discuss changes, or evaluate the quality of construction. If the architect does this, however, the architect risks a claim by the contractor that the architect's communication has increased costs, caused delay, or otherwise interfered with the contractor's ability to properly supervise and manage the construction process. All such communications should be directed by the architect to the contractor, with some minor exceptions which will be discussed elsewhere.

ARBITRATION

General arbitration concepts have already been covered in detail, but there are also several related aspects of dispute resolution during construction contract administration that directly involve the architect. Under standard contract documents, the architect serves as the initial decision maker when disputes arise between the owner and the contractor, and the architect can be called as a witness in any ensuing arbitration proceeding.

Architect's Role as Initial Decision Maker

The standard AIA Owner-Architect Agreement (B201) requires the architect to interpret the drawings and specifications and make decisions on matters in question between the owner and contractor. Subparagraphs 2.6.2.3 and 2.6.2.4 state:

> The Architect shall interpret and decide matters concerning performance under, and requirements of, the Contract Documents on written request of either the Owner or Contractor. The Architect's response to such requests shall be made in writing within any time limits agreed upon or otherwise with reasonable promptness.
>
> *Interpretations and decisions of the Architect shall be consistent with the intent of and reasonably inferable from the Contract Documents and shall be in writing or in the form of drawings. When making such interpretations and decisions, the Architect shall endeavor to secure faithful performance by both Owner and Contractor, shall not show partiality to either, and shall not be liable for results of interpretations or decisions rendered in good faith.*

These provisions are paralleled in the AIA General Conditions in Subparagraphs 4.2.11 and 4.2.12, which bind the owner and the contractor to initially refer claims, disputes, and other matters in question to the architect for a decision. Because the architect is familiar with the contract documents and does not have an economic interest in the construction contract, the law gives the architect limited immunity when the architect serves in the role of an initial decision maker after such disputes or claims arise. The architect cannot be held liable for the consequences of any decisions made in good faith, regardless of whether the decisions favor the owner or the contractor. The architect may not show partiality to either party when serving in this role.

The architect's decisions and interpretations must be consistent with the intent of, and be reasonably inferable from, the contract documents. If the matter in question involves aesthetic effect, the architect's decision is final, if consistent with the intent of the contract documents. Other decisions are subject to arbitration or mediation upon written demand by either the owner or the contractor. See Article 15 of the AIA General Conditions. After a matter is referred to the architect for a determination, the architect must act within specified time limits. Neither party can demand arbitration until the architect has rendered a written decision or until 10 days after the parties have presented their evidence to the architect, if the architect has not rendered a decision by that time. These limitations do not apply if the position of architect is vacant, if the claim relates to a mechanic's lien, or under various other circumstances. The time within which the parties must demand arbitration of a dispute depends on the nature of the architect's decision. If the decision is in writing and states that it is *final but subject to mediation and arbitration* (AIA General Conditions, Subparagraph 15.4). Subparagraph 15.2.6.1 states that either party within 30 days from the date of an initial decision, demand in writing that the other party file for mediation within 60 days of the initial decision.

The architect's role as the initial decision maker is a logical extension of the architect's role as the party who prepared the drawings and specifications. Being familiar with the documents, the architect normally can make a prompt and knowledgeable determination, based on the intent of the contract documents, when either the owner or the contractor has a question or when a dispute arises between them about contract requirements. Except in regard to aesthetic effect, the architect's decisions are always subject to appeal by a demand for arbitration. If both the owner and the contractor are satisfied with the architect's decision, the matter is resolved, and no one will incur the expense and inconvenience of having to refer the matter to an outside party for resolution. If either party is dissatisfied with the architect's decision, he or she has the option of demanding arbitration. If the architect were not permitted to make initial determinations, every question or dispute would have to be referred to an outside party whenever the owner or the contractor disagreed about the intent of the contract documents.

Architect as a Witness

In the AIA Owner-Architect Agreement (B201), Subparagraph 3.3.1.8 describes the architect's services in connection with any dispute resolution proceeding or legal proceeding involving the project. This service is performed as a Change in Services, with the architect being compensated for his or her time to prepare and serve as a witness or perform other services.

The architect could be called as a witness due to knowledge of facts connected with the project, or the architect could be called as an expert witness because of technical expertise as a design professional. As an expert witness, the architect's function is to aid the court or arbitration panel by giving testimony about technical matters that may be beyond the scope or knowledge of the judge, jury, or arbitrator who will be deciding the case. Often, this involves testimony about the *professional standard of care* during the performance of services or *customary practices* in the construction industry in regard to the issues in question.

It is important to recognize that, when testifying as an expert witness, the architect must be factual, objective, and knowledgeable about construction industry practices related to the issues under deliberation. The architect must base the testimony on an adequate independent investigation of the facts before forming an opinion. Otherwise, the architect risks being embarrassed on cross-examination and having his or her credibility as a witness and a professional seriously undermined.

LESSON 9 QUIZ

1. AIA Document B201, the Owner-Architect Agreement, gives the architect the right to do which of the following?

   A. Determine construction techniques and procedures

   B. Reject work not conforming the requirements of the contract documents

   C. Stop work without written authorization from the owner

   D. Have unlimited direct contact with subcontractors and material suppliers

2. The architect's legal status in representing the owner's interests during construction is defined in

   A. the Owner-Architect Agreement.

   B. the Owner-Contractor Agreement.

   C. the General Conditions.

   D. A and C

3. Professional liability coverage insures the architect against

   A. claims arising out of any express warranties or guarantees to which the architect agreed.

   B. claims arising because of cost estimates being exceeded.

   C. professional negligence.

   D. intentionally wrongful acts.

4. Which of the following statements concerning mechanic's lien laws is INCORRECT?

   A. General contractors have lien rights.

   B. Subcontractors have lien rights.

   C. Material suppliers have lien rights.

   D. Mechanic's lien laws are uniform from state to state.

5. Which of the following actions may an owner take if a construction contract calls for a retainage? Check all that apply.

   A. Use the money to pay for labor or materials if the contractor fails to make these payments

   B. Use the money to pay for the correction of work performed improperly

   C. Hold back money from the contractor at the owner's discretion

   D. Reduce or release the retainage entirely without the written permission of the surety company that issued the performance bond for the contractor

6. AIA Document A201, the General Conditions, requires that, in order to receive final payment from the owner, the contractor must

   A. submit to the architect an affidavit that all payrolls, bills, etc., connected with the work have been paid.

   B. obtain consent of surety to final payment.

   C. if required by the owner, submit other data establishing payment by furnishing receipts, releases, and waivers of lien arising out of the contract.

   D. do all of the above.

7. A formal decision-making process where the decision is final and binding is called Arbitration.

8. A bond involves which of the following parties?

I. The surety—the party that issues the bond

II. The principal—the contractor

III. The obligee—the owner

IV. The obligator—the architect

A. I, II, III, and IV

B. I, III, and IV

C. I, II, and III

D. II and IV

9. When a bid bond is furnished by the low bidder, who subsequently fails to enter into a contract with the owner at the price and on the terms stated in the bid,

A. the surety is liable for the additional cost incurred by the owner to enter into another contract.

B. the owner must sue the low bidder for the additional costs.

C. the second lowest bidder pays the surety the difference in cost between his or her bid and the lowest bid.

D. the owner and the contractor must enter into arbitration to determine the extent of the penalty under the bond.

10. The amount of the performance bond is normally what percentage of the contract sum?

A. 75%

B. 50%

C. 66 2/3%

D. 100%

11. Construction contracts normally require which of the following parties to carry liability insurance?

I. The contractor

II. The owner

III. The architect

IV. The mortgagee

V. The bonding company

A. I only

B. I and II

C. V only

D. I, II, III, IV, and V

12. Under which circumstances should the architect *supervise* construction?

A. Whenever possible

B. Never

C. Only if the contractor agrees to otherwise *manage, direct, and control* the project

D. Only if no one else is available for this task

13. The owner must obtain the approval of which of the following parties if he or she wishes to occupy the work prior to substantial completion?

I. The contractor

II. The architect

III. The insurance company

IV. The surety

A. I only

B. I and III

C. I, II, and III

D. III and IV

14. The owner normally carries builder's risk (or builder's risk-all risk) insurance because

A. it is required in the Owner-Contractor Agreement.

B. it is required in the Owner-Architect Agreement.

C. no builder can be trusted entirely to perform all of the required work.

D. title to construction work passes to the owner in increments as payments are made for each portion of the work.

15. On a construction project, there is a disagreement as to whether a specified paint has been properly mixed to match the color sample submitted by the contractor. Who is responsible for interpreting the specifications and making a determination regarding the correct color?

A. The architect

B. The contractor

C. The owner

D. An arbitrator

10

SUBMITTALS, SITE VISITS, TESTING, AND INSPECTION

MATERIAL AND EQUIPMENT SUBMITTALS

Shop Drawings, Product Data, and Samples

Introduction

Shop drawings and samples are submitted by the contractor to the architect during the construction phase of the project to show how the contractor intends to perform the work called for by the contract documents. Shop drawings are needed because the architect's drawings and specifications are never absolutely "complete." Even if it were possible for an architect to prepare complete documents, it would not be in the owner's best interest to do so. Various manufacturers can provide acceptable alternative products to fulfill the performance called for by the architect's design and contract documents. However, comparable products may differ in detail, depending on the manufacturer's or fabricator's particular processes and procedures. By permitting the contractor to select from one of several sources for specific items of material, systems, or equipment, the owner benefits from the process of competitive bidding without sacrificing performance.

The purpose of shop drawing, product data, and sample submittals is to enable the architect to see whether the specific items selected by the contractor meet the architect's design intent as expressed in the contract documents. This lesson will cover the standard contract requirements related to shop drawings, product data, and samples, recommended procedures for processing them, and some common problems encountered when either the architect or the contractor deviates from the contract requirements.

Basic Concepts

Shop drawings are defined in the AIA General Conditions, Subparagraph 3.12.1, as follows:

> *Shop Drawings are drawings, diagrams, schedules and other data specifically prepared for the Work by the Contractor or a Subcontractor, Sub-subcontractor, manufacturer, supplier or distributor to illustrate some portion of the Work.*

Product data are defined in the AIA General Conditions, Subparagraph 3.12.2, as follows:

> *Product Data are illustrations, standard schedules, performance charts, instructions, brochures, diagrams and other information furnished by the Contractor to illustrate materials or equipment for some portion of the Work.*

Samples are defined in the AIA General Conditions, Subparagraph 3.12.3, as follows:

> *Samples are physical examples that illustrate materials, equipment or workmanship and establish standards by which the Work will be judged.*

It should be recognized that shop drawings are *not* part of the contract documents. See A201, Subparagraph 3.12.4. The contract documents are prepared by the architect; the shop drawings are prepared by the contractor. Contract documents and shop drawings are complementary. However, shop drawings cannot be relied upon to remedy poorly prepared drawings and specifications. The drawings and specifications show the intended results to be achieved by the contractor; the shop drawings show how the contractor will achieve those results. Product data and samples serve a similar purpose. Samples also establish the standards by which subsequent construction will be judged. After a sample has been approved, the contractor is required to fabricate and install the work to match the approved sample.

Contractor's Responsibilities

As noted above, the contractor is responsible for preparation of the shop drawings. Normally, the shop drawings are original documents, not merely reproductions of the architect's drawings. Preparation is done by the contractor, subcontractors, material suppliers, or fabricators. Paragraph 3.12 of the AIA General Conditions (Document A201) describes the requirements for submittals. In particular, note the following two subparagraphs:

> *The Contractor shall review for compliance with the Contract Documents, approve and submit to the Architect Shop Drawings, Product Data, Samples and similar submittals required by the Contract Documents in accordance with the submittal schedule approved by the Architect or, in the absence of an approved submittal*

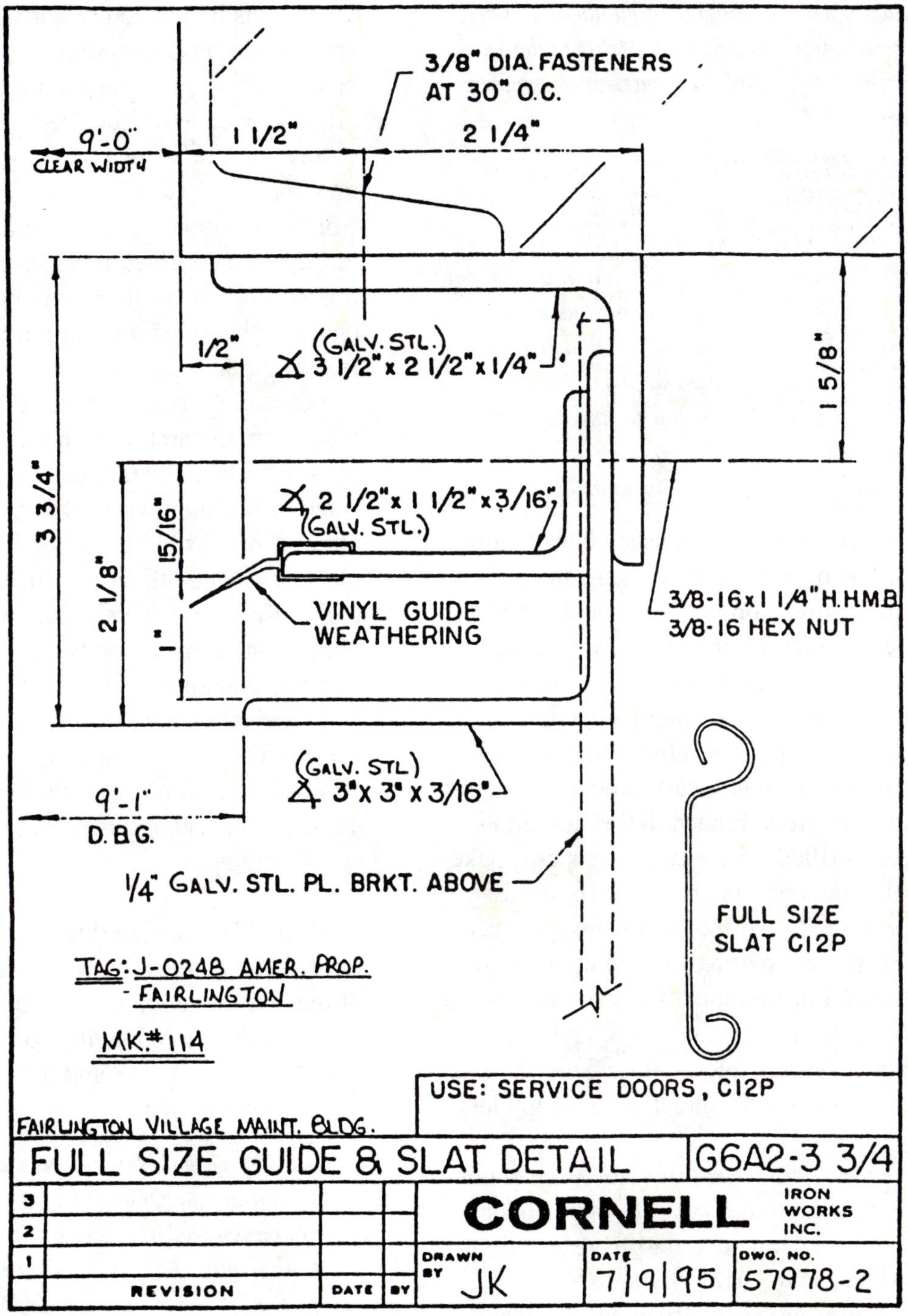

TYPICAL SHOP DRAWING

Figure 10.2

> *schedule, with reasonable promptness and in such sequence as to cause no delay in the Work or in the activities of the Owner or of separate contractors.*
>
> *By submitting Shop Drawings, Product Data, Samples and similar submittals, the Contractor represents to the Owner and Architect that the Contractor has (1) reviewed and approved them, (2) determined and verified materials, field measurements and field construction criteria related thereto, or will do so and (3) checked and coordinated the information contained within such submittals with the requirements of the Work and of the Contract Documents.*

The above two subparagraphs state the contractor's responsibilities for processing and approving shop drawings, product data, and samples. First, the contractor must review and submit the required shop drawings, product data, and samples in an orderly manner. Second, the contractor must approve submittals to signify that all materials, field measurements, and field construction criteria related to the submittals have been verified. These two provisions make it clear that the contractor has a responsibility to check and approve shop drawings, product data, and samples before forwarding them to the architect. Furthermore, when submittals are returned by the architect to the contractor, the contractor is responsible for distribution of the documents to the subcontractors and suppliers.

In many cases, work performed by one subcontractor is related to the work of other subcontractors. The general contractor is responsible for coordinating the work of the various subcontractors and suppliers including coordination of their submittals. For example, the specifications may require the mechanical and electrical contractors to prepare a composite shop drawing showing how ductwork, piping, and major electrical conduits are coordinated with structural members and the reflected ceiling plan.

Deviations by the contractor from the General Conditions are common. Shop drawings, product data, and samples may, for example, be forwarded to the architect without adequate, or any, review by the contractor. Sometimes they may be forwarded to the architect directly from subcontractors and suppliers, rather than through the contractor. An architect, in order not to delay the project, may be tempted to check and approve the submittals without insisting that the contractor meet his or her responsibility to review the submittals first. In that event, the architect could be liable for damages if errors are later detected in the fabricated or installed work. Because the architect attempted to expedite the work, the contractor may have been relieved of his or her contractual obligations. The contractor can claim those obligations were waived when the architect checked and approved the submittals without prior review and approval by the contractor. The architect must insist that all submittals for review be transmitted by the contractor only after the contractor has previously checked and approved them.

The AIA General Conditions also covers deviations that may exist between the requirements of the contract documents and information contained in the shop drawings or other submittals. Subparagraphs 3.12.8 and 3.12.9 specifically address deviations:

> *…the Work shall be in accordance with approved submittals except that the Contractor shall not be relieved of responsibility for deviations from requirements of the Contract Documents by the Architect's approval of Shop Drawings, Product Data, Samples or similar submittals unless the Contractor has specifically informed the Architect in writing of such deviation at the time of submittal and (1) the Architect has given written approval to the specific deviation as a minor change in the Work, or (2) a Change Order or Construction Change Directive has*

> *been issued authorizing the deviation. The Contractor shall not be relieved of responsibility for errors or omissions in Shop Drawings, Product Data, Samples or similar submittals by the Architect's approval thereof.*
>
> *The Contractor shall direct specific attention, in writing or on resubmitted Shop Drawings, Product Data, Samples or similar submittals, to revisions other than those requested by the Architect on previous submittals.*

Thus, the contractor must call the architect's attention to deviations, and the architect must give specific written approval before the contractor is permitted to deviate from the requirements of the contract documents. An approved shop drawing does *not* take precedence over the contract documents.

By knowing and enforcing the requirements of Paragraph 3.12 in the AIA General Conditions, the architect can properly process shop drawings. By adhering to the contract requirements, the architect can reduce exposure to liability if the contractor "slips something by" in the shop drawings or other submittals. If the contractor has deviated from the submittal requirements of the General Conditions, the architect may reject either the submittals or the subsequent nonconforming work.

To avoid delays and additional costs, the procedures for submitting shop drawings, product data, and samples should be reviewed in detail at the pre-construction meeting. If the architect makes it clear at that point that he or she will strictly adhere to contract requirements for submittals, it is likely that submittal problems will be significantly reduced or eliminated during the construction phase.

Architect's Responsibilities

The architect's responsibilities for reviewing and approving submittals of shop drawings, product data, and samples are covered in both the AIA Owner-Architect Agreement (Document B201) and the AIA General Conditions. Subparagraph 2.6.4.1 in B201 states:

> *The Architect shall review the Contractor's submittal schedule and shall not unreasonably delay or withhold approval.*
>
> *The Architect's action in reviewing submittals shall be taken in accordance with the approved submittal schedule or, in the absence of an approved submittal schedule, with reasonable promptness while allowing sufficient time in the Architect's professional judgment to permit adequate review.*

Subparagraph 2.6.4.2 states:

> *In accordance with the Architect-approved submittal schedule, the Architect shall review and approve or take other appropriate action upon the Contractor's submittals such as Shop Drawings, Product Data and Samples, but only for the limited purpose of checking for conformance with information given and the design concept expressed in the Contract Documents. Review of such submittals is not for the purpose of determining the accuracy and completeness of other information such as dimensions, quantities, and installation or performance of equipment or systems, which are the Contractor's responsibility.*
>
> *The Architect's review shall not constitute approval of safety precautions or, unless otherwise specifically stated by the Architect, of construction means, methods, techniques, sequences or procedures.*
>
> *The Architect's approval of a specific item shall not indicate approval of an assembly of which the item is a component.*

Subparagraph 2.6.4.3 states:

> *If the Contract Documents specifically require the Contractor to provide professional*

design services or certifications by a design professional related to systems, materials or equipment, the Architect shall specify the appropriate performance and design criteria that such services must satisfy. The Architect shall review Shop Drawings and other submittals related to the Work designed or certified by the design professional retained by the Contractor that bear such professional's seal and signature when submitted to the Architect. The Architect shall be entitled to rely upon the adequacy, accuracy and completeness of the services, certifications and approvals performed or provided by such design professionals.

The architect reviews and approves submittals only for the limited purpose of checking for conformance with the design concept and with the information given in the Contract Documents. This limitation is very important. Consequently, the shop drawing approval stamp recommended by the AIA contains the qualifying language.

The architect's approval is not a "double check" of the contractor. The contractor remains responsible for checking field measurements and other construction criteria.

The architect's involvement in the approval process is to determine if the specific products selected by the contractor are consistent with the architect's design intent as expressed in the contract documents. The contractor is permitted to select a product from one of a number of manufacturers whose products meet the architect's specifications. If the contractor misinterpreted the contract requirements, additional cost and/or construction delays could result. By giving the architect an opportunity to review shop drawings, misinterpretations can be rectified before the work is fabricated or installed.

The architect is also responsible for preparing proper specifications for shop drawing, product data, and sample submittals. The architect should limit submittals to those items that truly need the architect's review. For example, submittals are essential for natural materials, such as wood or stone, which have characteristics unique to a particular forest or quarry. Submittals are also essential for *design-build* work, such as curtain wall systems, where design criteria are specified, but the details to achieve them are unique to a particular manufacturer.

If the architect requires too few submittals, the architect will not be able to determine if the contractor has properly interpreted contract requirements. If the architect requires unnecessary submittals, administrative expense and the opportunity for error will increase. If the contractor submits shop drawings for items for which shop drawings are not required by the specifications, the architect should return them to the contractor without review.

When a submittal is for work designed by a consultant to the architect, such as structural steel, the consultant is required to review and take action on the submittal. However, the contractor must initially transmit the submittal to the architect. The architect reviews the submittal and transmits it to the consultant. Once the consultant has stamped the submittal with action taken, it is then stamped by the architect and returned to the contractor.

Procedures

The following procedures for processing shop drawings, product data, and samples are derived from specific AIA contract document requirements and other recommended AIA practices. While the General Conditions and the Owner-Architect Agreement state the contractual duties of the architect and contractor, the Specifications Division 1, General Requirements, covers project-specific procedural requirements. Section 01300 in the MasterFormat headings created by the Construction Specifications Institute (CSI) is entitled "Submittals." Individual architectural offices often develop their own procedures, based on the type of project and their particular practice. In general, however, most firms adhere to the basic procedures described below.

- **Recordkeeping**

  The architect must maintain detailed records to log the dates when shop drawing, product data, and sample submittals are received from the contractor, and when these are returned. The records should also indicate the action taken by the architect (approved, approved as corrected, revise and resubmit, or not approved). This log will avoid potential claims by the contractor or others that submittals were either not approved or not returned within a reasonable time.

  AIA Document G712 Shop Drawing and Sample Record provides a convenient way to document the receipt and return of submittals.

  The architect should retain copies of each submittal with the project records. This is particularly important when a shop drawing is resubmitted several times. Multiple submittals often suggest a serious problem. The record of each submittal conveys the extent and nature of the problem and the deviations from the contract documents cited by the architect. If only the final, or approved, submittal is retained, it may not be possible to explain why several submittals were required. Such evidence may be necessary if the contractor files a delay claim alleging that shop drawings were not approved within a reasonable time.

  The architect should check with the owner before disposing of shop drawings after the project's completion even if there is no likelihood of a claim. Many clients rely on the architect to maintain project records. If additions, remodeling, or repair are required years later, the owner may ask the architect to provide copies of documents required to perform such work. Shop drawings are typically saved in electronic format, often PDF, and then archived either on a separate server or data storage media. If the owner indicates that he or she does not need the shop drawings and there is no other reason to keep them, the architect can then dispose of them. Often, the statute of limitations serves as a guide to how long records should be kept.

- **Timeliness**

  The AIA General Conditions and the Owner-Architect Agreement require both the architect and the contractor to process submittals within a "reasonable" time. Other contracts not using AIA documents sometimes establish a precise time for processing of submittals, such as 10 working days or two weeks.

  The AIA approach provides both the architect and the contractor flexibility in processing shop drawings within the context of project requirements. The architect or the contractor may have special requirements for submittals that take priority. These priorities need to be established at the start of construction. For example, if the contractor knows that certain items require a long lead time for ordering, the contractor can request the architect to process the related shop drawings as quickly as possible. Conversely, if the contractor submits shop drawings for fixtures that will not be needed for several months, the architect can defer acting on them in order to give priority to the shop drawings for items needed sooner.

  The contractor can be requested to provide a schedule showing when submittals will be sent to the architect and when they must be returned to the contractor. The architect can then discuss with the contractor potential adjustments to the schedule to reflect the requirements of the architect and consultants to process these submittals. The schedule must allow sufficient time to transmit submittals. Additional time is necessary when the submittal requires review by consultants.

- **Changes and Revisions**

  Shop drawings should not be a means for the architect to change the design. If design changes are made that affect cost and/or time, a change order must be issued. This procedure is covered in detail in Lesson 2. However, it is acceptable to make minor revisions, which can be indicated on the shop drawings. If the contractor claims that changes to the shop drawings increase project cost or time, the contractor is entitled to request a change order before proceeding with the work.

  As previously noted, the contractor is required to direct the architect's attention to revisions in shop drawings, particularly on resubmittals. If the contractor fails to call revisions to the architect's attention, the contractor will not be relieved of the responsibility to meet contract requirements, even though the architect has approved the submittal. The architect cannot be expected to review any aspect of the shop drawings for which he or she has not been made aware.

Standard Formats

Many architectural firms and major clients have standard formats for processing submittals. Two major publishers of standard specifications, including those related to submittals, are the Construction Specifications Institute (CSI) and the Production Systems Division (PSD) of the AIA. Many firms view the CSI and PSD publications as industry standards and use them as reference documents for the development of their own specifications library.

Construction Specifications Institute (CSI)

The CSI is a professional organization dedicated to the development and improvement of construction communication documents, including specifications. CSI has established the Construction Sciences Research Foundation to develop, among other things, computerized construction communication systems. To date, major CSI/CSRF efforts have included COM-SPEC and SPECTEXT to make computer programs for specifications and hard copy master specifications available to architects and others who prepare specifications.

Professional Systems Division (PSD)

PSD was established by the American Institute of Architects to develop computerized master systems for architects and engineers. The first such system is MasterSpec®, a text-based master specification system, available by subscription, that includes comprehensive, current text for specifying many construction products, systems, and materials. Because MasterSpec® is comprehensive, the architect must edit the master specification text to make it fit specific project requirements. The edited specification text can be either retyped or processed by computer. A more recent specification master produced by PSD is SPECSystem, which is a knowledge-based system. The purpose of MasterSpec® and SPECSystem is to save architectural firms the time and money required to develop their own specification systems.

SITE VISITS

Contract Requirements

During the construction phase, the architect is required to visit the site as part of the construction contract administration basic services.

Subparagraph 2.6.2.1 of the standard AIA Owner-Architect Agreement (Document B201) states:

> *The Architect shall visit the site at intervals appropriate to the stage of construction, or as otherwise required in Section 3.3.3, to become generally familiar with the progress and quality of the portion of the Work completed, and to determine, in general, if the Work observed is being performed in a manner indicating that the Work, when fully completed, will be in accordance with the Contract Documents. However, the Architect shall not be required to make exhaustive or continuous on-site inspections to check the quality or quantity of the Work. On the basis of the site visits the Architect shall keep the Owner reasonably informed about the progress and quality of the portion of the Work completed, and report to the Owner (1) known deviations from the Contract Documents and from the most recent construction schedule submitted by the Contractor, and (2) defects and deficiencies observed in the Work.*

The phrase *at intervals appropriate to the stage of construction* is intended to give the architect some flexibility in determining when site visits are necessary. For some projects, the owner may require the architect to visit the site in accordance with a pre-established schedule. If so, the schedule should be incorporated into the Owner-Architect Agreement, and the architect's fee should allow for the time required for these services.

Previous versions of the AIA documents used the term *periodic* to describe the architect's visits to the site, which implied that visits were required to be at regular intervals. Clearly, the need for the architect to observe the work at the site cannot be predetermined with such rigidity, and the current version affords the architect more flexibility.

The standard AIA Owner-Architect Agreement (Document B201) contains the following provision:

> *The Owner shall provide the Architect access to the Project site prior to commencement of the Work and shall obligate the Contractor to provide the Architect access to the Work wherever it is in preparation or progress.*

A provision in the General Conditions (Document A201) states:

> *The Contractor shall provide the Owner and Architect access to the Work in preparation and progress wherever located.*

Both provisions obligate the contractor to provide access for the architect to observe the work, whether it is at the site, in fabrication, or stored off-site.

Architect's Role

During the construction phase, the architect's principal function is to advise the owner whether the contractor is performing the work in accordance with the contract documents. The architect is the judge of the contractor's performance, and the architect's judgment is relied upon by the owner. In this regard, Subparagraph 4.2.2 of the AIA General Conditions (Document A201) is almost identical to Subparagraph 2.6.2.1 of the Owner-Architect Agreement (Document B201).

Meetings and Reports

Throughout the construction phase, the architect participates in meetings, some of which the architect conducts, while attending others as an observer. Regardless of the architect's role, documentation of each meeting is an important aspect of project administration. Proper documentation reduces the possibility of misunderstandings and can provide the necessary records should any claims be made.

Pre-Bid Meeting

A pre-bid meeting may be held shortly after the bidding documents are distributed to contractors, but well before the bids are due and opened. A pre-bid meeting is often appropriate for large or complex projects, in order to give bidders an opportunity to ask questions about the bidding documents or other matters. Often, major subcontractors and suppliers will attend the pre-bid meeting, along with the general contractors.

Statements made at a pre-bid meeting should not be considered to modify bidding requirements or the requirements of the contract documents. Bidding documents often contain provisions that prohibit reliance on oral representations made by the architect or any other party. A recording or transcript of statements made at the pre-bid meeting can be made in order to document what was said and by whom.

If changes to the bidding documents are required because of clarifications made during a pre-bid meeting, the architect should issue an addendum. Subparagraph 1.1.1 in the AIA General Conditions (Document A201) states that addenda become part of the contract documents. Minutes prepared after the pre-bid meeting should be distributed to all parties who were in attendance, with a request that comments or corrections be made within a short time. Any changes to the minutes should be circulated to those receiving the original version.

Pre-Construction Meeting

After bids are received and the construction contract has been prepared, a pre-construction meeting is scheduled with the selected general contractor to review the contract documents and to establish administrative procedures. Usually, one of the major items of discussion is the schedule for submissions of shop drawings, product data, and samples. Other major topics may include the submission of the contractor's schedule of values and applications for payment.

Payments to the contractor are described in detail in Lesson 11. The architect may utilize this meeting to explain the contract document requirements for proper checking, review, and approval of shop drawings by the contractor, before they are forwarded to the architect for review.

If a project involves multiple prime contractors, scheduling and coordination procedures become even more important. Holding a pre-construction meeting allows all prime participants in the project to state their requirements and define what they will expect from each other. By having a clear and mutual understanding among all parties, the likelihood of confusion and delays is minimized. Minutes of the pre-construction meeting should be prepared and distributed to all attendees for corrections and comments, as described above for the pre-bid meeting.

During Construction

The need for formal meetings during construction depends on how smoothly the project is proceeding. If there are few problems, scheduled meetings may not be necessary. The architect's normal site visits may be sufficient to answer questions and resolve problems. However, if the project is delayed or if problems occur because of changes or poorly prepared contract documents, regular or special meetings may become necessary.

To facilitate communication, the architect and contractor may agree to hold weekly meetings at a set time to review the progress of the project since the previous meeting and to review upcoming work. Each party's responsibilities can be identified to arrive at a clear understanding. It is essential to prepare and distribute job meeting minutes and other documentation about items that have been discussed at these meetings. Either the architect, the contractor, or some other party may assume the responsibility for the preparation and distribution of the minutes. It is important that someone is assigned this task and that it is carried out in a responsible manner.

Some owners and architects prefer to hold regularly scheduled construction site meetings, such as weekly, whether or not serious problems exist. If the architect knows in advance that it is required to attend such weekly meetings, compensation for construction phase services must reflect the time required for this activity.

In addition to meetings, the architect is required to visit the site at *intervals appropriate to the stage of construction* as described in AIA Owner-Architect Agreement (Document B201) Subparagraph 2.6.2.1. These site visits are necessary so that the architect can become generally familiar with the progress and quality of the work to perform the construction phase administrative services. In order to inform the owner and contractor of the architect's observations at the site, it is essential that a field report be prepared for each visit. AIA Document G711 Architect's Field Report is intended for this purpose. The report should be worded carefully to avoid language that might be used in a professional liability claim against the architect. Observations should be factual, directions to the contractor should be consistent with the contract documents, and comments about construction means, methods, safety, or other responsibilities of the contractor should be avoided. The architect's consultants must also adhere to this policy.

Safety

Safety at the construction site is always a concern. Construction can be a hazardous activity, and there are serious implications for everyone if construction workers, or members of the public, are injured or killed as a result of construction accidents. In addition to the personal trauma caused by a serious accident, insurance rates can go up, fines and penalties can be imposed by federal and state safety agencies, and liability lawsuits may be filed against anyone who might have been responsible for the accident or for not having taken precautions to prevent it.

With the enactment in 1970 of the Occupational Safety and Health Act (OSHA), much attention has been focused on construction site safety. It is important to recognize that OSHA applies to *employers* who are legally responsible for the workplace safety of their employees. At a construction site, the contractor and subcontractors have the responsibility for OSHA compliance. The architect's responsibility for OSHA compliance extends only to the architect's own employees who may visit the site. The responsibility for enforcing compliance with OSHA requirements rests with the federal or state agency charged with OSHA enforcement in the particular jurisdiction. The architect has no legal responsibility for enforcing OSHA, and it is recommended that specific references to OSHA compliance in the Supplementary Conditions or elsewhere in the contract documents be omitted. Otherwise, it could be argued that, even though the architect does not have a legal duty to enforce OSHA standards, the architect does have a responsibility for ensuring that the contractor meets the requirements of the contract documents. If the contract documents require OSHA compliance, the architect could then be held to have a duty to enforce those contract requirements. The enforcement of OSHA standards should remain a government responsibility.

Architect's Role

The standard AIA contract documents clearly state that the architect is *not* responsible for construction site safety. Subparagraphs 4.2.2 and 4.2.3 of the General Conditions (Document A201) state the following:

> *...the Architect will not have control over, charge of, or responsibility for, the construction means, methods, techniques, sequences or procedures, or for the safety precautions and programs in connection with the Work, since these are solely the Contractor's rights and responsibilities...*
>
> *On the basis of the site visits, the Architect will keep the Owner reasonably informed about the progress and quality of the portion of the Work completed, and report to the Owner (1) known deviations from the Contract Documents and from the most recent construction schedule submitted by the Contractor, and (2) defects and deficiencies observed in the Work. The Architect will not be responsible for the Contractor's failure to perform the Work in accordance with the requirements of the Contract Documents. The Architect will not have control over or charge of and will not be responsible for acts or omissions of the Contractor, Subcontractors, or their agents or employees, or any other persons or entities performing portions of the Work.*

Subparagraphs 2.6.1.2, 2.6.2.1, and 2.6.2.2 of the Owner-Architect Agreement (Document B201) reiterate these provisions.

However, as a minimum, the architect's responsibility in regard to safety is to inform the owner if the architect observes or is notified of unsafe conditions or practices at the construction site. The architect cannot simply ignore obviously unsafe conditions. If someone is injured or killed, the architect can be named as a defendant in a subsequent lawsuit. In a recent court decision, a state supreme court held a design professional liable for failing to take action when he observed an obviously unsafe condition at a site. The court held that despite contract language excusing the professional from liability for site safety, an architect or engineer cannot simply ignore unsafe conditions. It is the professional's duty to alert someone who can take steps to remedy the problem. This does not suggest that an architect must act as a safety inspector or make a special effort to look for unsafe conditions. The contract documents state that the architect has no such duty. However, as a professional licensed to protect the public health, safety, and welfare, an architect cannot ignore obviously unsafe conditions on the construction site. The architect must warn the contractor or other responsible parties about the hazards, so that they can be corrected.

Contractor's Responsibilities

Unlike the architect, the contractor has clearly defined responsibilities for construction site safety. Article 10 in the AIA General Conditions (Document A201) is entitled *Protection of Persons and Property.* Subparagraph 10.1 states:

> *The Contractor shall be responsible for initiating, maintaining and supervising all safety precautions and programs in connection with the performance of the Contract.*

Article 10 describes the basic safety requirements the contractor must meet. The contractor's responsibilities extend to all employees on the work, all other persons who might be affected by the work, the work itself, including material and equipment stored at the site or elsewhere, and property, structures, roadways, etc., that are adjacent to the site. The contractor is required to comply with all safety laws and regulations and to give all notices required by them. Danger signs and other warning devices must be erected. The contractor must notify utility companies when construction operations may affect gas, electric, or other services. When hazardous materials, such as explosives, are used, Subparagraph 10.2.4 calls for *utmost care* and supervision by properly qualified personnel. Provisions are also included that obligate the contractor to promptly remedy all damage or loss caused by acts or omissions of the contractor, or by anyone for whom the contractor is responsible.

The contractor is required to designate a responsible member of the organization at the site whose duty it is to prevent accidents. Usually, the person is the contractor's superintendent, unless the contractor designates someone else and informs the owner and the architect in writing.

The contract gives the contractor discretion to act in any emergency affecting the safety of persons or property to prevent threatened damage, injury, or loss. If additional costs are incurred, or delays result from such emergency actions, the contractor is entitled to make a claim. If the claim is valid, a change order must be issued by the owner to increase the contract sum and/or time.

Responsible contractors recognize their duties to maintain proper safety programs at construction sites, and they make safety a top priority among their employees. Good employees are too valuable to lose through job-related injuries or death. Furthermore, the contractor does not want to incur increased workers' compensation insurance rates and OSHA citations for violating safety laws. The Associated General Contractors of America (AGC) has developed comprehensive construction safety recommendations, which many contractors utilize. Safety is one area where there should be no disagreements between the contractor and architect. If the architect observes an unsafe condition, the contractor should welcome being informed about it so that remedial action can be taken. If the contractor does not respect safe conditions, and appreciate being advised about unsafe conditions, accidents are likely to occur.

Evaluating Work Against Contract Documents

As noted previously, one of the architect's primary services as the owner's agent during construction is to determine whether the work being performed by the contractor is in accordance with the contract documents. The architect must be familiar with the contents of the contract documents, including the drawings, specifications, and all other contract documents, such as the Owner-Contractor Agreement and General and Supplementary Conditions. When problems and questions arise in the field, answers can be usually found in the contract documents. Conversely, problems often are magnified when answers are based on inadequate research, or if instructions are given that vary from contract requirements.

When evaluating the contractor's performance, the architect should refrain from subjective opinions or general comments, such as, "The contractor's work is poor." If the work does not conform to the specifications, the architect's report should cite the requirements of the specifications, and objectively report how the work fails to meet those requirements.

The contractor's obligation is to perform the work in accordance with the contract documents. The architect represents the owner's interests in determining when the work is, or is not, in accordance with the contract documents. Subparagraph 2.6.2.3 of the Owner-Architect Agreement (Document B201) states, in part:

> *The Architect shall interpret and decide matters concerning performance under, and requirements of, the Contract Documents...*

The AIA General Conditions (Document A201) Subparagraphs 4.2.11 and 4.2.12 also give the architect the initial responsibility for judging whether the contractor and the owner are fulfilling their respective contractual responsibilities. The architect must respond to requests by the owner or contractor for interpretations *with reasonable promptness*.

In addition to interpreting compliance with the contract documents, the architect has a primary role in settling claims and disputes. A claim involving additional cost, time delays, concealed or unknown conditions, or other contractual conflicts, must be made by the claimant within 21 days of the occurrence. The architect must respond to the claim within 10 days of receipt. If the claim is not resolved, the claimant may submit additional supporting data, modify the claim, or reiterate the initial claim. If the claim remains unresolved, and if the above procedures have been followed, then the architect's decision is subject to binding arbitration. Lesson 9 discusses arbitration in detail.

Rejecting Work

The architect has the right to reject work that does not comply with the requirements of the contract documents. The pertinent contract provisions are found in the AIA Owner-Architect Agreement (Document B201), Subparagraph 2.6.2.2:

> *The Architect has the authority to reject Work that does not conform to the Contract Documents. Whenever the Architect considers it necessary or advisable, the Architect shall have authority to require inspection or testing of the Work in accordance with the provisions of the Contract Documents, whether or not such Work is fabricated, installed or completed....*

The related provision in Subparagraph 4.2.6 of the AIA General Conditions includes the following statement:

> *...neither this authority of the Architect nor a decision made in good faith either to exercise or not to exercise such authority shall give rise to a duty or responsibility of the Architect to the Contractor, Subcontractors, material and equipment suppliers, their agents or employees, or other persons or entities performing portions of the Work.*

This sentence makes it clear that the architect's right to reject work is limited only to protecting the owner's interest in obtaining the work required by the contract documents.

The architect's right to reject work authorizes the architect to require the contractor to remove and replace work that does not conform to the requirements of the contract documents. The right to reject work does not include stopping the work.

Stopping the Work

The AIA contract documents do not give the architect the right to stop the work. That authority is specifically reserved for the owner.

Paragraph 2.3 of the AIA General Conditions (Document A201) states that the *owner* has the right to order the contractor to "stop the Work" if the contractor fails to correct work that is not in accordance with the requirements of the contract documents or persistently fails to carry out the work in accordance with the contract documents. This stop work order must be in writing and signed by the owner or by an agent of the owner who has the owner's written authorization to order the contractor to stop the work. The general duties of the architect as the owner's agent during construction do not include this authority. The contract clearly states that the authority to stop the work, if given to an agent, must be so stated in writing.

Because of the serious consequences of an order to stop the work, the owner should consult with his or her legal counsel before taking this step. If the problem causing the owner to issue a stop work order is not corrected, the next step probably would be termination of the contract, which is described in the AIA General Conditions (Document A201) Paragraph 14.2. Termination is a last resort because of the substantial legal and financial consequences involved.

Subparagraph 2.6.2.2 of the AIA Owner-Architect Agreement (Document B201) gives the architect the right to *reject* improper work, in order to protect the owner. The right to reject work can be an effective deterrent without creating the liability exposure associated with the right to *stop* the work. In fact, at least one court has specifically rejected an injured worker's claim in a lawsuit that the architect's right to reject work implied the right to stop work. In that case, the court held that the right to stop work cannot be implied, but must be specifically stated in the contract.

TESTING AND INSPECTION

Requirements of the Contract Documents

In order to determine whether the contractor's work conforms to the requirements of the contract documents, the specifications may require that certain elements of the work be tested. Public laws and regulations may also require special testing and inspection. The architect has independent authority to order certain work to be tested if such testing is believed to be necessary or advisable for the proper implementation of the intent of the contract documents.

Local code officials inspect the work for compliance with building, fire, mechanical, electrical, and other codes. Theoretically, once a building permit has been issued, the contract documents should include all work required to comply with the codes. If an inspector requires work necessary to comply with the code that is not shown in the contract documents, the contractor may be entitled to a change order.

The AIA General Conditions (Document A201) differentiates between routine testing and inspection required by the contract documents and additional testing and inspection directed by the owner. Subparagraph 13.5.1 describes routine testing that is not usually observed by the architect. Subparagraph 13.5.2 describes additional testing.

Subparagraph 13.5.2 requires the contractor to give the architect timely notice that work is ready for testing or inspection to enable the architect to observe these activities. The architect's observation of inspections and tests usually takes place, if practical, at the sources of supply. The architect must respond promptly when notified by the contractor when and where testing will take place.

If the architect, the owner, or a public authority determines that work requires additional testing or inspection, the architect must obtain written permission from the owner before ordering the contractor to proceed with the test or inspection. The owner's prior approval is required because the owner must pay for the test or inspection if the results show that the work meets the requirements of the contract documents. If the work is found to be deficient, however, the contractor must pay for the testing or inspection, including the compensation for the architect's related additional services. See Subparagraph 13.5.3 in the AIA General Conditions (Document A201) for the applicable contract provisions.

Test reports and inspection certificates obtained by the contractor must be delivered promptly to the architect. The architect, or the consulting engineer, usually reviews the test or inspection results, and if satisfactory, files the reports with other project documents.

Quality Control Standards

If standards are not otherwise established by law or building codes, the architect usually includes in the specifications the standards by which testing will be conducted and evaluated. Often, industry-wide standards, such as those of the American Society for Testing and Materials (ASTM) are the basis for testing.

Many manufacturers of construction products and materials belong to trade associations that publish standards for the materials and products sold by their members. Such industry-wide standards often serve as the basis for determining whether a particular material or product is satisfactory. However, this does not preclude the architect from requiring higher standards.

Testing

Testing Equipment

The architect does not usually own testing equipment. Testing is usually performed by laboratories or firms that specialize in specific types of tests, such as those for soils, concrete, or welding.

Whether the owner pays directly for tests or requires the contractor to include these costs in the bid price, testing laboratories and companies must be free from any conflict of interest on the project. There should be no business interest between any firm performing tests or inspections and the contractor or any subcontractor, since the owner and the architect rely on test results and reports to determine if the contractor is performing the work in accordance with the contract documents.

On- and Off-Site Testing

Testing is conducted either at the site, in laboratories, or both. Soil testing, for example, requires samples to be taken at the site, and then analyzed in a soils laboratory. The same is often true with concrete; slump tests are conducted at the site, but compression tests require laboratory analysis.

The architect must monitor contract requirements for testing and be aware of construction conditions that might necessitate testing in addition to that called for by the contract documents or by law.

Soil Testing

Soil testing is performed both before design and during construction. The architect and the structural engineer need soil information to locate buildings on the site and to design building foundations. Because of potential liability associated with faulty soil information, it is recommended that the owner, not the architect, hire and pay the geotechnical consultants.

The geotechnical firm obtains core borings at various locations identified by the architect and the consulting engineers. Soil borings are expensive, so the architect and engineer must carefully select the number and location of required borings. If too many borings are specified, the cost to the owner will be excessive. If too few are specified, the architect and structural engineer may not obtain sufficient soils information to proceed properly with the design of the project.

Subparagraph 4.3 of the AIA Owner-Architect Agreement (Document B201) states:

> *The Owner shall furnish services of geotechnical engineers which may include but are not limited to test borings, test pits, determinations of soil bearing values, percolation tests, evaluations of hazardous materials, seismic evaluation, ground corrosion tests and resistivity tests, including necessary operations for anticipating subsoil conditions, with written reports and appropriate recommendations.*

This provision clearly makes it the owner's responsibility to order soil investigations and other tests. These services are not considered design services, but rather the documenting of existing physical site conditions. Therefore, the architect does not obtain these services and is not responsible for their accuracy or completeness. In fact, the AIA Owner-Architect Agreement expressly states that the architect is entitled to rely on the accuracy and completeness of tests and reports furnished by the owner.

AIA form Geotechnical Services Agreement (Document G602) can be used by the owner to request a proposal from a geotechnical engineer for geotechnical services. If the engineer's proposal is acceptable, the document can then be used as the basis for an agreement between the owner and geotechnical engineer for both pre-construction and construction phase services. Pre-construction services include an analysis of the information developed during the investigation and recommendations for foundation design and construction methods. Construction service may include sampling backfill to verify required compaction and inspecting excavations before placing concrete.

The soils report should be made available to the contractor for use and evaluation. This document, however, should not be incorporated into the contract documents by the architect. Because the architect and the consultants do not prepare the soils report, they would be exposing themselves to unnecessary liability by including the report with the contract documents.

The contractor may use the soil information to help determine methods and estimate costs for excavation and underpinning of adjacent structures. Subparagraph 3.2.1 of the AIA General Conditions (Document A201) states:

> *Execution of the Contract by the Contractor is a representation that the Contractor has visited the site, become generally familiar with local conditions under which the Work is to be performed and correlated personal observations with requirements of the Contract Documents.*

However, a bidder is not obligated to have independent soils tests conducted as a part of the bid. If an investigation has been performed, the owner should furnish all bidders with the information contained in the report. Usually, the soils report is made available to bidders with a qualifying statement that the report is provided for information only and that it contains no warranties or other representations.

Subsurface conditions are sometimes different from those described in the soils report. If the contract does not specifically place the responsibility for a change in the work required by these conditions on any particular party and the owner has not withheld information about subsurface conditions, the contractor may be required to pay additional costs for excavations.

In order to provide an equitable solution for both owner and contractor, the AIA General Conditions include a provision to modify the contract sum should concealed conditions be encountered that vary from those indicated in the contract documents. Subparagraph 3.7.4,

Claims for Concealed or Unknown Conditions, allows either the contractor or the owner to request a change in the contract sum if, for example, subsurface conditions encountered during construction result in either more or less work that anticipated by the information shown in the contract documents. Such claims for an increase or decrease in cost must be made by change order within 21 days of first observing the conditions so that the architect can observe these conditions as part of the claim resolution process.

The purpose of this provision is to maximize accuracy in the bid price, as well as fairness to owner and contractor. It prevents the contractor from inflating the bid price as a protection against potential costs for unknown conditions. If these arise during construction, the contractor can request an increase in the contract sum for any additional required work. Similarly, if subsurface conditions are more favorable than anticipated, requiring less work than indicated in the contract documents, the owner can initiate a change order reducing the contract sum. Consequently, whether subsurface conditions result in more or less work, the owner pays only for what is required by the existing conditions.

The bid documents usually contain a request for the contractor to provide *unit costs* for work, such as earth excavation. The cost is quoted in units, such as dollars per cubic yard ($/cy) for both added work and deleted work. Unit costs protect the owner against inflated cost once competitive bidding is over, and allow the owner to compare one contractor's costs to another contractor's and to industry standard unit costs.

Other Testing

Other work that is almost always tested includes concrete and welding, since both are critical to the structural integrity of the project. Concrete is subjected to compression and other tests to determine if it possesses the necessary compressive strength and workability. Welding tests and inspections are conducted to determine if the welding was performed in accordance with industry standards to achieve the structural strength required. Many specifications and laws require that welders be trained and certified.

1 C
2 B
3 C
4 A, B, E
5 D
6 A
7 C
8 D
9 addendum
10 C
11 C
12 B
13 B
14 C
15 D

LESSON 10 QUIZ

1. Shop drawings are prepared by, or on behalf of, the

   A. architect.
   B. owner.
   C. contractor.
   D. surety.

2. A shop drawing, which has been approved by the architect shows that the method for mounting a window frame differs from the method shown in the contract documents. In that case, the contractor must

   A. ignore the contract documents.
   B. inform the architect in writing of the deviation.
   C. inform the owner in writing of the deviation.
   D. test the two different methods to see which works better.

3. A contractor has submitted shop drawings for items for which no shop drawings are called for by the specifications. The architect should

   A. prepare a change order for the work shown in the shop drawings.
   B. send the drawings to the owner for approval.
   C. return them to the contractor without taking any action.
   D. review the shop drawings for conformance with the contract documents.

4. An architect should retain copies of each shop drawing submittal, including multiple submittals, for which of the following reasons? Check all that apply.

   A. To maintain a record of how potential problems were resolved
   B. To protect against possible future claims
   C. To bill the contractor for the time spent reviewing the submittals
   D. To satisfy the owner that every conceivable alternative has been investigated
   E. As a record of the actual field conditions

5. The AIA General Conditions and the Owner-Architect Agreement require both the architect and the contractor to process shop drawings and other submittals within

   A. ten days.
   B. two weeks.
   C. no specific time.
   D. a reasonable time.

6. The architect wishes to make a minor revision of the plans by changing a shop drawing. This is permissible

   A. if the changes do not involve a change in cost or time of performance.
   B. if the changes do not involve a change in time of performance.
   C. at the discretion of the architect.
   D. only if the change is approved by the owner.

7. A specification found in MasterSpec® is used without editing by an architect, eventually resulting in damage to the building. Who is primarily liable in this case?

A. The Construction Specifications Institute (CSI)

B. Professional Systems Division (PSD)

C. The architect

D. The contractor

8. The AIA Owner-Architect Agreement requires the architect to visit the construction site

A. at least once a week.

B. whenever asked by the contractor.

C. as frequently as possible.

D. at appropriate intervals.

9. If changes to the bidding documents are required because of clarifications made during a pre-bid meeting, the architect should issue a(n) ___addendum___.

10. Responsibility for compliance with OSHA standards at the construction site rests with the

A. architect.

B. owner.

C. contractor.

D. insurance company.

11. The architect notices that temporary construction barriers are improperly secured and create a potential danger for workers. What action should the architect take?

I. Report the condition to the local OSHA representative

II. Warn the contractor about the hazard.

III. Stop the work

IV. Inform the owner of the problem

A. I only

B. III only

C. II and IV

D. I, II, and IV

12. An architect has determined that, contrary to the requirements of the contract documents, a cast-in-place concrete slab is out of level. The architect should

A. stop the work.

B. reject the work.

C. ask the owner for instructions.

D. request the concrete subcontractor to remove and replace the defective work.

13. If an architect determines that work requires special testing or inspection, the architect must first

A. secure written permission from the contractor.

B. secure written permission from the owner.

C. ask ASTM for test guidelines.

D. write a change order covering the special testing or inspection.

14. During renovation of an old structure, bracing members not noted in the contract documents are discovered, which will necessitate a major change in the plans. According to the AIA General Conditions, the contractor must request a change in the contract sum to cover the necessary additional work within

A. ten days.

B. two weeks.

C. twenty-one days.

D. a reasonable period.

15. Who should hire and pay the geotechnical engineer?

A. The architect

B. The contractor

C. The structural engineer

D. The owner

LESSON ELEVEN

CHANGE ORDERS AND PAYMENTS

11

CHANGE ORDERS AND CONSTRUCTION CHANGE DIRECTIVES

Contract Document Provisions

Change orders and construction change directives are used to accommodate requirements and conditions not provided for in the contract documents. A change order is defined in the AIA General Conditions (Document A201), Subparagraph 7.2.1 as follows:

> *A Change Order is a written instrument prepared by the Architect and signed by the Owner, Contractor and Architect, stating their agreement upon all of the following:*
>
> *.1 change in the Work;*
>
> *.2 the amount of the adjustment, if any, in the Contract Sum; and*
>
> *.3 the extent of the adjustment, if any, in the Contract Time.*

Changes during construction can result from many unforeseen conditions, such as the discovery of a high water table not indicated in the soils report. In that event, a change order would be issued to add a subsurface drainage system. Another example is the nonavailability of a specified material, such as brick manufactured from a clay deposit that has been exhausted. A change order would be issued to purchase a different brick with a different cost. There could also be changes in an owner's requirements, such as the consolidation of a company's departments. A change order would be issued to revise the office floor plans and related drawings. There might also be changes to correct errors and omissions in the contract drawings. For example, a change order would be issued to add a fire damper in a duct penetration of a fire rated wall, where the drawings did not indicate this requirement.

When a change order is agreed to by both owner and contractor, it becomes a contract modification.

In the absence of total agreement on the terms of a change order, a construction change directive is used. Under the terms of the AIA General Conditions (Document A201), the owner may order changes in the work, within the general scope of the contract documents, by the issuance of a construction change directive. If a construction change directive is signed by the owner and the architect, the contractor is obligated to perform the work described in it. If the contractor is in agreement with the change in contract amount and/or time, the contractor must sign it before performing the work. Once signed, it becomes a contract modification. If the contractor is not in agreement, the construction change directive does not have to be signed. However, the contractor is still obligated under the terms of the contract to perform the work.

Methods to adjust the contract sum found in the AIA General Conditions, Subparagraph 7.3.3 include:

.1 *mutual acceptance of a lump sum properly itemized and supported by sufficient substantiating data to permit evaluation;*

.2 *unit prices stated in the Contract Documents or subsequently agreed upon;*

.3 *cost to be determined in a manner agreed upon by the parties and a mutually acceptable fixed or percentage fee; or*

.4 *as provided in Subparagraph 7.3.7.*

(*Note*: 7.3.7 requires the architect to determine the cost by monitoring the contractor's actual expenditures of time, materials, and overhead.)

Architect's Role

When the owner and contractor are unable to agree on the construction change directive amount by any of the above methods, the cost of the work is determined by the architect. See General Conditions (Document A201), Subparagraph 7.3.7 for a detailed description of this procedure.

The architect's determination of the cost adjustment is based on the estimate of the value of the work, plus a reasonable allowance for the contractor's overhead and profit. The architect may require the contractor to maintain itemized accounting and appropriate supporting data related to the construction change directive work. Usually, the cost of the work is limited to: cost of materials, supplies, and equipment, including sales tax and cost of delivery; cost of labor, including payroll taxes and customary fringe benefits; workers' compensation insurance; bond premiums; rental value of equipment and machinery; and the additional costs of supervision and field office personnel directly attributable to the change.

When both additions and credits to the contract sum are involved in a change, the contractor's overhead and profit, for that change, are calculated on the net increase, if any.

The AIA Owner-Architect Agreement (Document B201) requires the architect to prepare change orders and construction change directives. Subparagraph 2.6.5.1 states:

> *The Architect may authorize minor changes in the Work that are consistent with the intent of the Contract Documents and do not involve an adjustment in the Contract Sum or an extension of the Contract Time. Subject to the provisions of Section 3.3, the Architect shall prepare Change Orders and Construction Change Directives for the Owner's approval and execution in accordance with the Contract Documents.*

Note that change orders and construction change directives are prepared for the owner's approval. The architect does not have the independent authority to make changes that affect either the contract sum or the contract time. On projects involving a construction manager, the responsibility for the preparation of change orders and construction change directives may be assigned to either the architect or the construction manager. However, final approval must still come from the owner. If the architect orders changes without the owner's approval, the architect might be exposed to professional liability claims for changes either not desired by the owner or more expensive than anticipated.

When the contractor discovers a deficiency in the contract documents prepared by the architect, the architect is obligated to remedy the deficiency at her own expense. Such remedies might include revising drawings and processing the change documents. However, the architect is not obligated to pay for the cost of the construction caused by the deficiency unless the owner can prove that the architect was negligent.

If the change is initiated for any other reason, the architect is entitled to additional compensation for revising the drawings and processing the change documents.

For minor changes, the architect may absorb the expense of his services as a courtesy to the owner. For major changes, however, the professional services required to evaluate the change and revise the drawings and specifications may be extensive. Before performing the services, the architect should submit a proposal to the owner for the cost of the additional services anticipated in connection with the change. The owner should provide written approval for these additional services. This procedure accomplishes the following: the fee for the architect's additional services can be evaluated by the owner as a cost associated with the change order, and the architect protects the right to payment for these fees after performing the services. When additional services are performed without proper written authorization, the owner and architect may disagree over the owner's obligation to pay for those services. The owner may claim that the services are minor and should be included in the architect's basic services. Prior written authorization will prevent such misunderstandings.

Subparagraph 3.3.2.3 of AIA Document B201 describe the architect's services in connection with change orders as *Additional Services.* Under Paragraph 3.3 and subparagraph 3.3.2, the architect is obligated to notify the owner that the services will be performed. Upon receipt of the notice, the owner must give prompt written notice to the architect if the owner does *not* want the services to be performed. If the architect does not receive this notice from the owner, the architect can proceed with the additional services and the owner will be obligated to pay for them.

When performing professional services related to changes, the architect must allow sufficient time to properly perform the services. The revised documents must be checked and coordinated with the original contract drawings and specifications.

The architect's consultants have the same rights and responsibilities as the architect with regard to changes and other contractual matters.

Owner's Role

Changes are normal in every construction project. Changes may add to, deduct from, or modify the contract requirements. Owners are permitted to make changes to the work, and are not bound to follow exactly the original contract documents. However, changes must be within the scope of the project as originally contemplated by the owner and contractor. A change cannot be so great that it would impede the contractor's ability to perform the work of the construction contract.

As stated previously, a change order or construction change directive must be signed by, at least, the owner and architect. The architect's signature ensures that he or she is aware of the changes and allows proper administration of the construction contract. Under the AIA General Conditions, the contractor's signature is not required for a construction change directive to be valid. Certain formats for general conditions, often used by governmental agencies, do not recognize the use of a construction change directive. These formats require both owner and contractor's signature before a change is authorized. No distinction is made between authorizing the work and modifying the contract to pay for the work. This approach can restrict the owner's ability to achieve desired changes if the contractor makes unreasonable cost or time demands. Under the approach used in the AIA General Conditions, the owner can have the work accomplished, whether or not there is agreement on price or time. These issues can be resolved separately so that the progress of the work is not delayed.

Whenever changes affect the contract price or the time of performance, construction change directives or changes orders should be issued to document such changes. For example, if the contractor defaults or neglects to properly carry out the work, Paragraph 2.4 of the AIA General Conditions permits the owner to remedy such deficiencies by whatever means are appropriate and reasonable. If the owner hires another contractor to perform this remedial work, the original construction contract can be adjusted by change order to deduct the amount of the payments made to the second contractor. Furthermore, the owner can also deduct by change order the cost of the architect's services required by the original contractor's default.

Usual Procedures

The initial request for a change can originate with the owner, the contractor, or the architect. For example, the owner may request a change to the functional layout of a project. The contractor may claim additional costs for changed conditions or deficiencies discovered in the drawings and/or specifications. The architect may wish to incorporate a new product or system to improve the project. In each case, the owner must issue a construction change directive or a change order if the change will affect either the cost or time of construction.

Usually, the justification, scope, cost, and time associated with proposed changes are fully discussed by all parties before a change order is prepared. Once the architect has sufficient information about the change, including the cost and/or time adjustment requested by the contractor and accepted by the owner, the architect prepares the change order, signs it and forwards it to the owner for signature. After the owner signs the change order, it is sent to the contractor for signature. Once all parties have signed, the change order is distributed to each party for project records.

A Proposal Request (Document G709) is used by the architect to outline the work anticipated for a change order. Once the work is clearly defined, the contractor prepares an itemized quotation for the work. The owner and architect can then evaluate costs and the effect on the contract time in order to decide whether to proceed with the change order. Document G709 states in bold type: *This is not a change order nor a direction to proceed with the work described herein.* The actual authorization to proceed with the work must be made by a construction change directive or a change order.

To enable the architect to order minor modifications to the work when these do *not* involve changes in the contract sum or time, an Architect's Supplemental Instructions, AIA Document G710, can be used.

Examples of minor modifications to the work include:

1. *Issuance of color schedule.*

   Finishes for each room are identified on a room finish schedule in the contract documents. Color selections for the finish materials are usually selected after the material suppliers are awarded a contract.

2. *Revisions to ceiling plan.*

   The layout of lighting fixtures, sprinkler heads, ceiling tile, and diffusers can be modified without change in cost or time provided that the quantity and type are unchanged and the materials have not been fabricated or installed.

3. *Interpretations of the contract documents.*

   A stair is shown in plan with eight treads and in section with nine treads. The architect determines that the plan is correct.

The authority for the architect to order minor changes that do not affect the contract cost or time can be found in Subparagraph 7.4 of the AIA General Conditions, which states:

> *The Architect has authority to order minor changes in the Work not involving adjustment in the Contract Sum or extension of the Contract Time and not inconsistent with the intent of the Contract Documents. Such changes will be effected by written order signed by the Architect and shall be binding on the Owner and Contractor.*

AIA Document G710 allows the architect to order minor changes in written form as required by Subparagraph 7.4. Such minor changes are most often limited to interpretations of the contract documents.

Where time is of the essence, the architect can use a Construction Change Directive, Document G714, to authorize the contractor to proceed immediately even though a change in the contract sum and time may result. As previously mentioned, the owner must sign the construction change directive. Once the contractor's proposal for additional cost or time is satisfactory, the construction change directive is superseded by a change order.

The procedures established by the AIA contract documents assure that all instructions and decisions concerned with modifications to the contract are in writing, in order to allow changes to be made promptly and to avoid disputes. The requirement for written proposal requests, construction change directives, and change orders reduces the possibility of misunderstandings.

Subparagraph 15.1.2 in the AIA General Conditions (Document A201) requires notice of claims to be made within 21 days. This requirement allows the parties to the contract to address the matter while it is fresh in everyone's mind and records are readily available. The architect must initially determine whether claim notices have been made in a timely manner. If an untimely claim is rejected by the architect, he or she should document, in writing, the reasons and state the applicable dates.

Changes

Changes can involve cost, time, or scope, or any combination thereof.

Cost

A change in the cost of construction is usually the most critical element of a change order or construction change directive. Cost changes may either add to, or deduct from, the contract sum. Both cost plus fee and stipulated sum contracts are subject to cost changes. The methods for determining the cost changes are contained in the AIA General Conditions as previously discussed. If the owner and contractor cannot reach agreement on the cost of the work, the architect must make a determination. If the architect's determination is not mutually acceptable, either party may request arbitration.

Time

Changes may affect the time established for construction. If the contractor can perform the work required by a change in sequence with the work of the contract, she may not have to claim any additional time for the change. If the change affects the progress of the work, for example, if delays occur because new materials or equipment must be ordered and delivered, the contractor is entitled to an adjustment in time. The architect may, at times, negotiate with the contractor to determine the length of such time extensions.

It is important that the time aspect of changes be determined when the cost information is developed. If the contract stipulates liquidated damages and the construction is behind schedule, time extensions associated with changes afford the contractor an opportunity to avoid penalties. Liquidated damages require the contractor to pay a significant amount for every day the completion date is delayed, when the delay is caused by the contractor.

The AIA General Conditions require the contractor to claim time extensions within 21 days after the event that caused the claim. Although this provision anticipates claims for time extensions resulting from bad weather, strikes, and other unavoidable delays, the architect should also insist on prompt notice in connection with claims for time extensions associated with changed work. If the claim has to be evaluated long after the fact, there may be no way to determine whether the contractor is entitled to an increase in time, or how many days should be allowed. As with cost, the architect's determination regarding time can be either accepted by both parties or appealed to arbitration by either the owner or contractor.

Scope

There are no restrictions on the type of work covered by a construction change directive or a change order, provided that it is within the scope of the contract documents. Scope can be hard to define. Generally, contractors do not object to changes in the work provided that they are properly compensated, and their ability to meet other commitments is not compromised. If the contractor believes the change in the work is outside the scope of the contract documents, she must promptly advise the architect.

If the contractor refuses to perform a change in the work because she claims it is outside the scope of the contract documents, the architect should review the situation with the owner and the owner's attorney. The owner may either order the contractor to proceed, thereby running the risk of a substantial claim for damages, or have some other contractor perform the work under a separate contract. If the owner chooses the latter course, the original contractor may be liable for breach of contract for refusing to perform work that could have been done by change order or construction change directive. The architect's judgment about the scope of that work will play an important role in determining the outcome if the matter is taken to arbitration.

Proposals

In response to a proposal request, or on her own initiative, the contractor furnishes a proposal indicating revisions to cost and time, if any, for the change in the scope of work. This proposal should provide detailed labor and material costs and the contractor's overhead and profit. Similar proposals from subcontractors are included if the change also affects their work. The general or prime contractor is entitled to mark up the subcontractors' prices to cover administrative costs. Usually, the contract documents fix the percentage for overhead and profit markups for change proposals. The architect and consultants review the contractor's proposal to determine if the cost of labor and materials and time extension are consistent with the intended scope of the change. Usually, the bid documents require the contractor to provide unit costs for labor and materials for certain portions of the work. These unit costs can then be used by the architect to evaluate change proposals.

Documenting Reasons for Changes

For both proper construction contract administration and professional liability loss prevention, it is important that the architect fully document the reasons for changes. Proper documentation requires more than simply completing the construction change directive or change order form to reflect the change in contract amount or time. Whenever a change in the work is proposed, the architect should prepare notes, a memorandum, or a report stating the reasons for the change. This documentation should be retained in the architect's files, and when appropriate, provided to the owner and the contractor.

The contractor or owner may initiate a change without full agreement on the costs and time involved. Sometimes, full agreement is not possible because of unknown quantities of materials, delivery dates, etc. When this occurs, a subsequent claim could be made by the contractor. In an attempt to avoid claims before they occur, the AIA General Conditions call for the issuance of a construction change directive and requires the architect, when there is no agreement between the owner and contractor, to determine the cost of changes on the basis of reasonable expenditures made by the contractor. The architect can require the contractor to keep an itemized accounting in a form prescribed by the architect (A201, Subparagraph 7.3.7). All of these records may be needed to support the reasons for changes in the event of subsequent claims.

Not all claims come from the contractor. An owner may make a claim against an architect for increased construction cost resulting from changes when the changes were caused by the architect's errors and omissions. Accurate documentation will enable the architect to respond to such claims. In some instances, changes allegedly caused by the architect's errors and omissions are the result of the owner's inability to make timely decisions, or the owner's subsequent decisions to modify the project for reasons not previously communicated to the architect. If these reasons are not documented, it may be difficult for the architect to respond to allegations of professional negligence.

Proper documentation is one of the best ways to avoid claims before they arise or to defend against them if they do.

Effect on Bonds

Change orders usually have no effect on the validity of performance and labor and material payment bonds for the project. Bonds assume that construction contracts may involve changes. Contract modifications, such as change orders, do not invalidate bonds, although increases in the contract sum may result in increased bond premium costs to the contractor. Some contractors include these premium increases as part of their charge for the change.

The Performance Bond and Payment Bond (Document A312) does not require that the surety be notified when change orders are issued. If the contractor provides bonds on non-AIA forms, the bonds should be reviewed by the owner's attorney to determine if they contain notice requirements. The owner's attorney should then inform the architect so that the architect can provide the necessary technical support to enable the owner to comply with the surety's requirements. Usually, it is the contractor's responsibility to notify the surety. In fact, most contractors resist having others contact their sureties for reasons other than a default. Contractors consider their relationship with sureties confidential.

AIA Change Order Form

To facilitate the documentation of change orders and to enable proper documentation of amendments to the contract for construction, the AIA has developed a Change Order form, AIA Document G701. This form is usually prepared by the architect, based on information provided by the contractor. For some projects, however, the contractor or the construction manager may prepare the form. Note the space in the upper-right corner of the form for the "Change Order Number." Change orders must be numbered sequentially so there will be complete records of all such changes.

The body of the AIA form is blank to allow the architect to describe the nature of the change. On occasion, it may be necessary to supplement the form with attachments, proposals, and drawings so that the change is fully described. Beneath the description of the change is a space for an accounting of the change, if any, in the contract sum or guaranteed maximum cost. This portion of the form includes the original contract sum, the net change resulting from previously authorized change orders, the contract sum prior to the current change order, the net change resulting from the change order, and the resulting amended contract sum. In addition, there is space to show revisions, if any, to the contract time and the date of substantial completion.

The owner, architect, and contractor must sign the change order for it to be valid. Once signed, the contractor must perform the work called for by the change order, and the change order becomes a contract modification. Copies of the signed change order form should be distributed to the owner, architect, and contractor for their files. If the contractor will not sign the change order because of a disagreement over the change in contract price or time, a construction change directive must be issued if the owner still wants to proceed with the change.

The use of the AIA forms for changes, proposals, and related matters is not mandatory, but highly recommended. If either the owner or architect does not want to use the standard AIA forms, substitute forms should be prepared to standardize record keeping for this important aspect of the project.

PAYMENTS TO THE CONTRACTOR

Contract Provisions

Another important service provided by the architect during the construction phase is processing payments to the contractor for work performed. The architect's construction contract administration responsibilities often involve judgments and determinations about whether the owner is receiving the work for which he is asked to make periodic payments. The standard AIA contract provisions, as well as customary practice, require the architect to evaluate and approve the contractor's applications for payments.

The AIA Owner-Architect Agreement (Document B201), Subparagraph 2.6.3.1, states:

> *The Architect shall review and certify the amounts due the Contractor and shall issue Certificates in such amounts. The Architect's certification for payment shall constitute a representation to the Owner, based on the Architect's evaluation of the Work as provided in Section 2.6.2 and on the data comprising the Contractor's Application for Payment, that, to the best of the Architect's knowledge, information and belief, the Work has progressed to the point indicated and that the quality of the Work is in accordance with the Contract Documents.*

This provision is reinforced by provisions in the AIA General Conditions (Document A201) that bind both the owner and the contractor to accept the architect's determinations. The owner relies on the architect to evaluate whether the contractor's work is in accordance with the contract documents and to certify the amount of money that should be paid for the work with each application for payment. Conversely, the contractor accepts the architect's judgment in determining how much should be paid, because once the architect makes this determination, the owner *must* pay the amount certified by the architect. The mandatory payment provision is found in Subparagraph 9.6.1 of the AIA General Conditions (A201) as follows:

> *After the Architect has issued a Certificate for Payment, the Owner shall make payment in the manner and within the time provided in the Contract Documents, and shall so notify the Architect.*

This provision prevents the owner from arbitrarily withholding payments from the contractor. Since the architect does not have a direct economic interest in the project, he can be objective in deciding 1) how much work has been properly performed, and 2) how much the contractor should be paid for that work. The intent of the AIA General Conditions is that neither the owner not the contractor can take issue with the architect's determinations. Both are bound by the architect's judgments.

The payment provisions in the AIA General Conditions (A201) are described primarily in Article 9. These provisions are correlated with the specific terms, conditions, and amounts of payment for the project set forth in the Owner-Contractor Agreement, Document A101, for a Stipulated Sum Contract, and Document A102, for a Cost Plus Fee Contract. The following describes the interrelation of these contract provisions and the architect's role in certifying payments to the contractor.

Stipulated Sum and Cost Plus Fee Contracts

Of the two basic forms of construction contracts, the *stipulated sum* or *fixed price* contract is more commonly used than the *cost plus fee* contract. Whether the construction contract is competitively bid or negotiated between the owner and contractor, the stipulated sum contract establishes the construction cost of the project before the start of construction. Once the construction cost is agreed to, the owner is obligated to pay that amount. However, as described above, the construction cost may be revised by change order. If the contractor performs the work efficiently, she will probably make a profit. Conversely, if she does not perform efficiently, a financial loss may result. There are many other factors that affect the amount of the contractor's profit or loss on a project, including the bidding climate, the contractor's relationship with subcontractors and suppliers, and the contractor's management and technical experience.

Because the contractor will be paid only the amount stated in the contract, as adjusted by change orders or other contract modifications, the architect must pay careful attention to the payment provisions. If the contractor anticipates a loss on the project, she may submit claims for increases to the contract sum due to alleged negligence by the architect. The architect's best protection against such claims is to have a clear understanding of contract requirements and to enforce those requirements during the construction phase.

The *cost plus fee* contract requires the owner to pay the contractor the actual amount spent by the contractor to perform the work, plus an agreed upon fixed or percentage fee. The contractor is required to furnish the owner with copies of invoices for labor and materials, copies of subcontracts, and copies of any project-related overhead expenses before receiving payment.

Cost plus fee contracts are often used when an owner wants to select a specific contractor for her capabilities, rather than bidding the project competitively. Usually, this type of contract includes a guaranteed maximum price, referred to as a GMP contract. If the contractor spends more than the GMP, the contractor must absorb the loss. If the contractor spends less than the GMP, the owner benefits. Usually, the owner offers the contractor an incentive to perform the work for less than the GMP by sharing the savings with the contractor. For example, if the actual cost is $100,000 less than the GMP, the owner may give the contractor 25 percent of the savings, or $25,000.

In reviewing the contractor's applications for payment, the architect must verify the amounts of actual labor and material invoices, subcontractors' invoices, and allowable overhead expenses. This review is more comprehensive than a review of an application for payment for a stipulated sum contract. Consequently, the architect is usually entitled to increased compensation for this added time and liability.

The AIA Owner-Contractor agreement for stipulated sum contracts (Document A101) incorporates by reference the AIA General Conditions (Document A201) into the contract. The Owner-Contractor Agreement, for Projects of Limited Scope (Document A107) is a stand-alone agreement with its own internal general conditions. It is intended for construction projects of limited scope or medium-to-large sized projects with payments based on a stipulated sum or cost of work plus a fee, either with or without a guaranteed maximum price. Whenever possible, the owner and contractor should use Documents A101 and A201, because they more fully define the legal relationship between the parties.

The payment provisions are found in Articles 4 and 5 of the Owner-Contractor Agreement, Document A101. These relate to Article 9 of the General Conditions, Document A201. A major portion of AIA Document A101 establishes specific payment provisions for the project.

Article 4 of A101 contains the *Contract Sum.* The lump sum bid and accepted alternates are stated here to show how the lump sum or stipulated amount has been determined.

Unit prices, if any, are also listed here. These are stated in dollars per unit for adding or deleting specific portions of the work; for example, x dollars per square foot to add or delete ⅝" drywall.

Article 5 of A101 establishes the timing of payments to the contractor. The AIA contract documents assume that the contractor will apply for and receive payment on a monthly basis as the work progresses. Subparagraphs 5.1.1 through 5.1.5 of A101 state:

> *5.1.1 Based upon Applications for Payment submitted to the Architect by the Contractor and Certificates for Payment issued by the Architect, the Owner shall make progress payments on account of the Contract Sum to the Contractor as provided below and elsewhere in the Contract Documents.*
>
> *5.1.2 The period covered by each Application for Payment shall be one calendar month ending on the last day of the month, or as follows:*
>
> *5.1.3 Provided an Application for Payment is received by the Architect not later than the ______ day of a month, the Owner shall make payment of the certified amount to the Contractor not later than the ______ day of the ______ month. If an*
>
> *Application for Payment is received by the Architect after the application date fixed above, payment shall be made by the Owner not later than ______ days after the Architect receives the Application for Payment.*
>
> *(Federal, state, or local laws may require payment within a certain period of time.)*
>
> *5.1.4 Each Application for Payment shall be based upon the schedule of values submitted by the Contractor in accordance with the Contract Documents. The schedule of values shall allocate the entire Contract Sum among the various portions of the Work. The schedule of values shall be prepared in such form and supported by such data to substantiate its accuracy as the Architect may require. This schedule, unless objected to by the Architect, shall be used as a basis for reviewing the Contractor's Applications for Payment.*
>
> *5.1.5 Applications for Payment shall show the percentage of completion of each portion of the Work as of the end of the period covered by the Application for Payment.*

Subparagraph 5.1.6 describes how progress payments are to be computed. See page 199 for a description of a progress payment.

Subparagraph 5.1.7 of A101 states:

> *5.1.7 The progress payment amount determined in accordance with Section 5.1.6 shall be further modified under the following circumstances:*
>
> *5.1.7.1 Add, upon Substantial Completion of the Work, a sum sufficient to increase the total payments to the full amount of the Contract Sum, less such amounts as the Architect shall determine for incomplete Work, retainage applicable to such work and unsettled claims; and*
>
> *(Section 9.8.5 of AIA Document A201-2007 requires release of applicable retainage upon Substantial Completion of Work with consent of surety, if any.)*
>
> *5.1.7.2 Add, if final completion of the Work is thereafter materially delayed through no fault of the Contractor, any additional amounts payable in accordance with Section 9.10.3 of AIA Document A201-2007.*

Subparagraph 5.1.8 contains space for adjusting the sum withheld for retainage after the work reaches a certain stage, such as 50 percent completion. Retainage is discussed later in this lesson.

The foregoing provisions establish that payment is conditioned upon the submission of an application for payment by the contractor and the issuance of a certificate for payment by the architect. Within an agreed upon number of days thereafter, usually ten days, the owner must make payment for the amount certified, less the agreed upon amount for retainage. At the time of substantial completion, the retainage may be reduced because the risk to the owner for incomplete or incorrect work is also reduced at that point.

Under a cost plus fee contract, the contractor is required to maintain detailed accounting records that are available to the owner upon request. The procedure for payments to the contractor is similar to that used under stipulated sum contracts. The contractor submits a monthly application for payment to the architect. On the basis of these applications for payment, the architect makes determinations about the payments to be made by the owner. Paragraph 12.1.9 of the Owner-Contractor Agreement, Cost Plus Fee (Document A102), states:

> *In taking action on the Contractor's Applications for Payment, the Architect shall be entitled to rely on the accuracy and completeness of the information furnished by the Contractor and shall not be deemed to represent that the Architect has made a detailed examination, audit or arithmetic verification of the documentation submitted in accordance with Paragraph 12.1.4 or other supporting data; that the Architect has made exhaustive or continuous on-site inspections or that the Architect has made examinations to ascertain how or for what purposes the Contractor has used amounts previously paid on account of the Contract. Such examinations, audits and verifications, if required by the Owner, will be performed by the Owner's auditors acting in the sole interest of the Owner.*

Article 7 of Document A102 identifies the items that constitute a contractor's reimbursable costs. The list includes cost of wages, subcontracts, travel, fees, and other customary expenses associated with a construction project. Article 8 of Document A102 identifies those items that are not reimbursable because they are included in the contractor's fee. Among these items are overhead, home office personnel salaries and other costs, costs in excess of the guaranteed maximum sum, if any, and any other item not specifically set forth in Article 7.

* *Substantial Completion*

Substantial completion is defined in the General Conditions (A201), Paragraph 9.8 as:

> *...the stage in the progress of the Work when the Work or designated portion thereof is sufficiently complete in accordance with the Contract Documents so that the Owner can occupy or utilize the Work for its intended use.*

At this point during construction, the contractor is required to prepare a list of items to be completed or corrected in order to achieve final completion. This list, commonly called the *punch list*, must identify all remaining work required by the contract documents, as well as work that is defective or that has been damaged and must be repaired or replaced. Although it may be customary in some instances for the architect to prepare this list, the AIA General Conditions (A201, Subparagraph 9.8.2) requires the contractor to review the status of the project and to prepare the list of items to be completed or corrected. When the architect receives the contractor's list, he or she is required to inspect the work to determine whether it is, in fact, substantially complete, i.e., can the owner use it for the purpose for which it is intended? The failure of the contractor to include an item of remaining work on the punch list does not alter his or her responsibility to perform all the work in accordance with the contract documents. The architect and owner may add items to be corrected or completed to the contractor's list.

Substantial completion is a critical point during the construction process. At substantial completion, the architect is required to inspect the work with a much higher degree of care than during the previous site visits. Prior to substantial completion, the architect is required only to *observe* the progress of the work in order to determine whether it conforms to the requirements of the contract documents. At substantial completion, the architect must *inspect* the work, which means that the work must be looked at very carefully to determine whether the contractor has fully complied with the requirements of the contract.

If the architect agrees with the contractor that the work is substantially complete, the architect prepares the Certificate of Substantial Completion, Document G704, to establish the *date* of substantial completion. The Certificate of Substantial Completion also defines the various responsibilities of both the owner and contractor for the provision of security, maintenance, heat, utilities, and liabilities for damage to the work, as well as the responsibility to provide insurance while the contractor completes the remaining work included in the punch list. The Certificate of Substantial Completion also establishes the time limit for completion of this work. Once the certificate has been prepared, it is submitted to the owner and contractor for signature as evidence of their acceptance of the responsibilities assigned to each of them at that point.

The date of substantial completion is very important. At substantial completion, the owner occupies and uses the project, assumes the responsibilities for insurance, security, utilities, etc., and warranty periods for equipment begin. There may be conflicting interests between the owner and contractor over the date of substantial completion. Unless the owner needs to occupy the project, he or she may prefer to delay the date of substantial completion because less incomplete work will appear on the punch list. He or she will not have to reduce the retainage and thus lose the use of those funds. He or she will not have to assume the responsibility for security, maintenance, and utilities any earlier than absolutely necessary. Conversely, the contractor usually wants to have as early a date of substantial completion as possible. The contractor can then receive retainage funds, and know that there is a finite period of future liability for deficient work. When asked to determine the date of substantial completion, the architect must remain neutral and objective about whether he believes, on the basis of his own professional judgment, the owner can occupy or utilize the work for the purpose for which it is intended and whether the remaining work is of a nature that will not unreasonably impede the owner's occupancy or use of the project.

Final Completion

Both Owner-Contractor Agreements, Documents A101 and A102, contain provisions to define when final payment occurs. Document A101, Subparagraph 5.2.1 states:

> *Final payment, constituting the entire unpaid balance of the Contract Sum, shall be made by the Owner to the Contractor when:*
>
> .1 *the Contractor has fully performed the Contract except for the Contractor's responsibility to correct Work as provided in Subparagraph 12.2.2 of AIA Document A201-2007, and to satisfy other requirements, if any, which extend beyond final payment; and*
>
> .2 *a final Certificate for Payment has been issued by the Architect.*

Subparagraph 5.2.2 states:

> *The Owner's final payment to the Contractor shall be made no later than 30 days after the issuance of the Architect's final Certificate for Payment, or as follows:*

Document A102, Paragraph 12.2 contains similar wording.

Final completion and final payment provisions are also contained in Paragraph 9.10 of the AIA General Conditions (Document A201). Subparagraph 9.10.1 states:

> *Upon receipt of the Contractor's written notice that the Work is ready for final inspection and acceptance and upon receipt of a final Application for Payment, the Architect will promptly make such inspection and, when the Architect finds the Work acceptable under the Contract Documents and the Contract fully performed, the Architect will promptly issue a final Certificate for Payment stating that to the best of the Architect's knowledge, information and belief, and on the basis of the Architect's on-site visits and inspections, the Work has been completed in accordance with terms and conditions of the Contract Documents and that the entire balance found to be due the Contractor and noted in said final Certificate is due and payable. The Architect's final Certificate for Payment will constitute a further representation that conditions listed in Subparagraph 9.10.2 as precedent to the Contractor's being entitled to final payment have been fulfilled.*

When the contractor informs the architect that the work has been completed and submits the final application for payment, the architect must make another inspection with the same degree of care used to determine the date of substantial completion. At the inspection for final completion, the architect determines whether the contractor has completed the project and is entitled to receive final payment.

Subparagraph 9.10.2 of the General Conditions requires the contractor to submit an affidavit that she has paid all payrolls, bills for materials and equipment, and other indebtedness for which the owner and the property might be held liable under lien laws or otherwise. Document G706, Contractor's Affidavit of Payment of Debts and Claims, and Document G706A, Contractor's Affidavit of Release of Liens, may be used for this purpose. The contractor must also submit written consent from her surety company in order to receive final payment. Consent of Surety was discussed in Lesson 9, and should be reviewed to see how performance bonds and labor and material payment bonds concern the architect in determining final payment to the contractor.

The owner may require additional data to verify that all bills have been paid. The owner may require the contractor to submit receipts, as well as releases and waivers of lien, so that the owner will not be liable for the contractor's failure to pay her employees, subcontractors, or suppliers. If any subcontractor or supplier refuses to furnish a release or waiver of lien, the contractor may provide the owner with a bond to indemnify the owner against any liens that might be filed. If a lien is filed for which the owner becomes liable, the contractor remains liable to the owner for all amounts paid by the owner to remove the lien. This latter provision, however, will not help the owner if the contractor is bankrupt or has insufficient assets with which to repay the owner.

Subparagraph 9.10.4 of the General Conditions, Document A201, states that final payment constitutes a waiver of all claims by the owner except those arising from:

> 1. *liens, Claims, security interests or encumbrances arising out of the Contract and unsettled;*
> 2. *failure of the Work to comply with the requirements of the Contract Documents; or*

3. *terms of any special warranties required by the Contract Documents.*

If either the owner or the architect is aware of deficiencies in the work, including those that the contractor may be correcting at the time of final completion, the owner should specifically reserve the right to make a claim for those items not corrected. Alternatively, the owner should not make the final payment.

The same waiver of claims also applies to the contractor. Subparagraph 9.10.5 in the AIA General Conditions states:

> *Acceptance of final payment by the Contractor, a Subcontractor or material supplier shall constitute a waiver of claims by that payee except those previously made in writing and identified by that payee as unsettled at the time of final Application for Payment.*

These provisions require mutual waivers of all claims by the owner and contractor at the time of final payment, except those specifically reserved in writing. They are designed to prevent subsequent claims by either party.

Application for Payment

The application for payment is a document prepared by the contractor for submission to the architect. Usually, the contractor submits an application for payment once a month to request payment, for work performed during the previous month. For small projects, the owner and contractor may agree on a schedule for payments at stipulated stages of the work rather than monthly. If the parties do not adhere to a monthly payment schedule, they must either use non-AIA documents or modify the standard AIA forms.

Subparagraph 9.3.1 of the AIA General Conditions states:

> *At least ten days before the date established for each progress payment, the Contractor shall submit to the Architect an itemized Application for Payment prepared in accordance with the schedule of values, if required under Section 9.2., for completed portions of the Work. Such application shall be notarized, if required, and supported by such data substantiating the Contractor's right to payment as the Owner or Architect may require, such as copies of requisitions from Subcontractors and material suppliers, and shall reflect retainage if provided for in the Contract Documents.*

To facilitate the contractor's preparation of the application for payment, the AIA has prepared Documents G702 and G703.

Document G702, Application and Certificate for Payment, is used to identify the project and parties and to summarize the status of payments under the contract.

In using this document, the original contract sum is adjusted for the net cost of change orders, producing a new contract sum. The value of work completed and stored to date, less retainage, less amounts previously paid, produce the amount due for the current period, which is certified by the contractor and architect.

When the contractor submits an application for payment, the contractor certifies that the work covered by the application has been completed in accordance with the contract documents, all amounts for which previous certificates for payment have been issued have been paid, and that the current payment request shown is now due. If the architect agrees with the amounts shown on the application for payment, he certifies the payment by signing the document. This procedure will be discussed in detail later.

Under the terms of the AIA General Conditions, the contractor warrants that title to all work, materials, and equipment covered by an application for payment passes to the owner either by incorporation into the construction or upon receipt of payment from the owner, whichever occurs first. The title passing to the owner is assumed to be free and clear of all liens, claims, security interests, and encumbrances, whether by the contractor or any other party.

Document G703 is a continuation sheet for Document G702. It is used by the contractor to provide a detailed breakdown of the various components of the work and the scheduled value for each such component for which payment requests are made. Often, this breakdown is based on the divisions in the specifications or the contractor's agreements with subcontractors and material suppliers. Each month, by reviewing the contractor's scheduled value for each item of work, the amounts paid on account of previous applications, the amounts for work in place and stored materials, and the balance to complete each item, less retainage, the architect determines whether the contractor's claims are for the correct percentage of payment for each item of the work. If the architect is in agreement with the amounts claimed by the contractor and the calculations on the application for payment are correct, the architect signs the certificate for payment, Document G702, and forwards it to the owner for payment.

Certificate for Payment

The certificate for payment is issued by the architect to inform the owner of the amount to be paid on account of each of the contractor's applications for payment. Document G702 combines the contractor's application for payment with the architect's certificate for payment. This eliminates potential errors in the amounts certified for payment, which could occur if the various sums were to be transferred to a separate document.

The issuance of the certificate for payment is not a guarantee by the architect that the contractor has properly performed the work. Rather, it is an administrative procedure to permit the orderly processing of payments to the contractor for work performed. The architect is not at the site to observe each worker, nor is the architect a guarantor of the contractor's performance. Therefore, the architect is not expected to know precisely what the contractor has or has not done. The architect is only required to make reasonable professional judgments about the progress of the work and must rely on the documents submitted by the contractor. The owner relies on the architect to determine whether payments requested by the contractor are consistent with the progress of the work. The architect's liability for improper certification would be based on negligent performance of his professional duties and not wrongful acts by the contractor.

Subparagraph 2.6.3 of the Owner-Architect Agreement, Document B201, sets forth the limitations of the architect's duty in regard to the issuance of a certificate for payment.

> *.1 The Architect shall review and certify the amounts due the Contractor and shall issue Certificates in such amounts. The Architect's certification for payment shall constitute a representation to the Owner, based on the Architect's evaluation of the Work as provided in Section 2.6.2 and on the data comprising the Contractor's Application for Payment, that, to the best of the Architect's knowledge, information and belief, the Work has progressed to the point indicated and that the quality of the Work is in accordance with the Contract Documents. The foregoing representations are subject (1) to an evaluation of the Work for conformance with*

the Contract Documents upon Substantial Completion, (2) to results of subsequent tests and inspections, (3) to correction of minor deviations from the Contract Documents prior to completion, and (4) to specific qualifications expressed by the Architect.

.2 *The issuance of a Certificate for Payment shall not be a representation that the Architect has (1) made exhaustive or continuous on-site inspections to check the quality or quantity of the Work, (2) reviewed construction means, methods, techniques, sequences or procedures, (3) reviewed copies of requisitions received from Subcontractors and material suppliers and other data requested by the Owner to substantiate the Contractor's right to payment, or (4) ascertained how or for what purpose the Contractor has used money previously paid on account of the Contract Sum.*

The above contract provisions establish the legal boundaries of the architect's duty and responsibility for the certificate for payment. It clearly states that the architect is making a professional judgment about the progress of the work and that the quality of the work is in accordance with the contract documents.

It does not require the architect to determine whether the contractor has performed the exact amount of work for which he is requesting payment. It simply requires the architect to visit the site, review the data included with the contractor's application for payment, and make a professional judgment to the best of his knowledge, information, and belief in regard to the status of the work. Based on these factors, the architect certifies to the owner how much the contractor should be paid.

Subparagraph 9.5.1 in the General Conditions (A201) gives the architect the right to withhold payments from the contractor. This provision enables the architect to decline to certify payment in whole or in part to reasonably protect the owner. If the architect cannot represent to the owner that the work has progressed as claimed by the contractor, the architect is required to notify the contractor. The architect and contractor must then attempt to agree on a revised amount. If they cannot, the architect must promptly issue a certificate for payment for such amount as the architect can represent to the owner. Usually, the contractor will provide a draft, or *pencil copy*, of the application for payment for the architect's review. Questions and disagreements can be resolved before the typed and signed version is submitted for certification.

The architect can decline to certify payment in whole or in part, or can nullify any prior certificate for payment for the following reasons:

.1 *defective Work not remedied;*

.2 *third party claims filed or reasonable evidence indicating probable filing of such claims unless security acceptable to the Owner is provided by the Contractor;*

.3 *failure of the Contractor to make payments properly to Subcontractors or for labor, materials or equipment;*

.4 *reasonable evidence that the Work cannot be completed for the unpaid balance of the Contract Sum;*

.5 *damage to the Owner or a separate contractor;*

.6 *reasonable evidence that the Work will not be completed within the Contract Time, and that the unpaid balance would not be adequate to cover actual or liquidated damages for the anticipated delay; or*

.7 *repeated failure to carry out the Work in accordance with the Contract Documents.*

If the contractor corrects the work represented by the withheld payments, she is then entitled to receive payment for that work.

The architect is required to issue a certificate for payment within seven days after receipt of the contractor's application for payment, and the owner is required to make payment within seven days after the date established in the contract documents for payment. If the architect fails to issue a certificate for payment through no fault of the contractor, or if the owner fails to make payment of any amount certified by the architect, the contractor is entitled to stop work after giving the owner seven days written notice. After receipt of late payment, the contractor is entitled to compensation for extra costs caused by shutting down, delay, and start-up due to a failure to receive payment on a timely basis.

Stored Materials

Construction contracts may permit the contractor to receive payment for materials stored both on-site and off-site before the materials are incorporated into the work. This provision enables the contractor to order materials as soon as possible without the financial burden of financing the cost of the materials until they are used. Provisions for payment for stored materials should reduce the owner's costs, because the contractor does not have to add the cost of financing materials received before they are needed. The owner may also benefit if the contractor is able to purchase materials before prices increase during periods of inflation. However, the owner assumes risk when making payment for stored materials, particularly when the materials are stored off-site. The architect should alert the owner to these risks so that the owner and the owner's attorney can take legal precautions to protect the owner's interests.

The AIA General Conditions cover payments for stored materials in Subparagraph 9.3.2:

> *Unless otherwise provided in the Contract Documents, payments shall be made on account of materials and equipment delivered and suitably stored at the site for subsequent incorporation in the Work. If approved in advance by the Owner, payments may be similarly made for materials and equipment suitably stored off the site at a location agreed upon in writing. Payments for materials and equipment stored on or off the site shall be conditioned upon compliance by the Contractor with procedures satisfactory to the Owner to establish the Owner's title to such materials and equipment or otherwise protect the Owner's interest, and shall include the costs of applicable insurance, storage and transportation to the site for such materials and equipment stored off the site.*

There are several important concepts in the above paragraph. First, it is mandatory that payment be made for stored materials, unless otherwise provided. If payment is not to be made for stored materials, this must be stated in the Supplementary Conditions. Second, the mandatory payment provisions apply only to materials and equipment stored at the site. The risk associated with payment for on-site stored materials is considerably less than for off-site storage. If payment is to made for materials and equipment stored off-site, additional written approval must be secured from the owner. And, third, for payments for any stored materials and equipment, the owner can require written documentation to protect the owner's title and interest in such property once payment is made.

Owner's Consent

The architect should discuss with the owner and the owner's attorney any changes in the stored materials provisions of the General Conditions. If the owner does not want to pay for materials or equipment until they are incorporated into the work, the owner should be alerted to the probability of higher costs because of the contractor's need to finance the carrying costs for stored materials. Under the provisions of a cost plus fee contract, the contractor may have little incentive to place orders early in order to take advantage of potentially lower prices.

If the owner is willing to pay for materials stored both on-site and off-site, the Supplementary Conditions should include a provision to establish the conditions under which such payments are to be made. The contractor must show how and where the materials are to be stored, how they will be protected, the security precautions taken, how the materials will be transported to the site when needed, and the insurance coverages provided to protect against loss by fire, theft, or otherwise. For items that are not easily identified, such as brick stored in a brickyard, it may be difficult to protect the owner's interest. For example, the brickyard could be subjected to a tax lien, and the premises padlocked by the Internal Revenue Service. Because of such risks, the owner's attorney should be consulted to advise the owner in regard to proper contract protections.

To protect the owner's interests, each application for payment from the contractor containing a request for payment for stored materials must be supported by adequate documentation. If the architect or owner is dissatisfied with the contractor's submittal in regard to stored materials, the contractor may be requested to provide further proof of protection of the owner's interests. If the contractor is unable to provide such proof, the request for this payment may be rejected.

Other Factors

If the architect has concerns about certifying payment for stored materials, he should refer the matter to the owner's attorney. The architect may have to verify whether the materials are actually stored at the site or elsewhere, and the conditions under which they are stored. If the owner intends to pay for materials stored off-site, the architect must be compensated for the time and expense to visit the locations where these items are stored.

If a payment request is submitted for stored materials, the architect must determine whether the owner is required to give written approval for the payment. Documentation, such as bills of sale and copies of insurance policies, must be submitted by the contractor. The evaluation of these documents, however, should be performed by the owner's attorney. In analyzing the contractor's proposals for storage and transportation, the architect must determine if they seem reasonable. He should ask "What if?" about each step of the process. What if the contractor defaults? What if the supplier defaults? What if there is a strike, making delivery impossible when the materials or equipment are needed? How will the contractor deal with that risk? What if there is a fire, theft, or other disaster? Is the warehouse bonded and is proper insurance in force? The architect is not a guarantor of delivery for stored materials, but he or she must exercise a *standard of care* before recommending to the owner that payment be made.

Change Orders and Construction Change Directives

Since change orders and construction change directives may affect the contract sum, they may also affect payments to the contractor.

The architect must maintain a record of all change orders and construction change directives issued by the owner. Claims in process, but not yet agreed to by the owner and contractor, do not affect the contract sum. These claims should be monitored by the architect to assess the owner's total cost exposure. For change orders and claims that are resolved and agreed to by the owner and contractor, the effect on the contract amount can be calculated. The AIA Application and Certificate for Payment, Document G702, provides space for listing approved change orders, additive and deductive, by sequential number, date, and amount. The original contract amount is adjusted by the net sum of these change order amounts. For projects with a large number of change orders, continuation sheets (Documents G703) listing all approved change orders can be attached to each application for payment.

Retainage

Retainage is the process by which the owner withholds money from the contractor as protection against the contractor's potential failure to complete the work according to contract requirements. The construction contract must stipulate the precise terms related to retainage. If no retainage is stipulated, the owner cannot arbitrarily withhold a part of a payment to the contractor, unless the architect certifies that the work is not being properly performed and that the contractor is not entitled to the amount requested in the application for payment.

The provisions related to retainage are normally set forth in the Owner-Contractor Agreement. See Article 5 in Document A101. Although the amount of retainage may vary, the higher the amount of retainage, the more costly the project is likely to be for the owner. Contractors consider retained funds as a "loan" to the owner. They often factor into their bids the interest cost to borrow money equal to that being retained. Thus, if retainages are high, the bid will have to be higher to cover the carrying costs.

In many cases, the owner may withhold 10 percent of each payment as retainage until the work is 50 percent complete, with no additional retainage thereafter. This is equivalent to a 5 percent retainage for the entire contract. Some owners may hold the 10 percent retainage until the work is 90 percent complete. Some owners release retainage on a line-item basis as portions of the work are satisfactorily completed. Retainages are handled in various ways, but whatever approach is used, it must be defined in the contract. The remaining retainages are released to the contractor with final payment.

In the event of a default by the contractor, the retainage can be used by the contractor's surety company to help pay for completing the work. For this reason, as mentioned earlier, retainages never should be reduced during the construction period or released at the end of the work without having the surety's written permission to do so.

LESSON 11 QUIZ

1. The AIA General Conditions requires which of the following parties to sign a construction change directive?

   A. The owner and contractor
   B. The contractor and architect
   C. The architect and owner
   D. The contractor and the architect's field representative

2. Who issues a construction change directive if the work affects either the cost or time of construction?

   A. The owner
   B. The architect
   C. The contractor
   D. The architect's field representative

3. An architect wants to save time by outlining an anticipated change in a construction detail to all the parties involved. What document should the architect use?

   A. Architect's Supplemental Instructions
   B. Construction Change Authorization
   C. Change Order
   D. Proposal Request

4. An architect needs to make an immediate change in construction before the owner and contractor can agree on the cost impact. Which document should the architect use?

   A. Architect's Supplemental Instructions
   B. Construction Change Directive
   C. Change Order
   D. Proposal Request

5. The AIA General Conditions requires that notice of claims regarding change orders must be made within 21 days.

6. When the owner and contractor are unable to agree on the additional cost resulting from a change in the work, the cost of such work is decided by

   A. arbitration.
   B. the architect.
   C. the bonding company.
   D. the mortgagee.

7. Which of the following statements are correct?

   I. The Performance Bond and Payment Bond normally do not require that the surety be notified when change orders are issued.
   II. Contract modifications, such as change orders, do not invalidate bonds.
   III. Increases in the contract sum may result in increased bond premium costs.
   IV. On non-standard bond forms, the owner is normally responsible for notifying the surety when change orders are issued.

   A. I, II, and III
   B. II and III
   C. I and IV
   D. I, II, III, and IV

8. Under a cost plus fee contract, why would the owner prefer the contractor's fee to be fixed?

A. The owner would not, since the cost plus fee contract already guarantees a maximum cost for the work.

B. The owner would not, since a contractor's fee based on a percentage of the cost of work is normally lower than a fixed fee.

C. If the fee is not fixed, reimbursable items, such as overhead, home office personnel salaries, etc., can be charged to the owner.

D. To insure that there is no relationship between the contractor's fee and the cost of the work.

9. Under the AIA General Conditions, what are the architect's responsibilities with regard to the punch list?

A. The architect has no stated responsibilities.

B. The architect must prepare the punch list.

C. If an item is missing from the punch list, the architect is responsible for that omission.

D. The architect must inspect the work from a punch list prepared by the contractor.

10. When does the owner assume the responsibility for insurance, security, utilities, etc. on a construction project?

A. At the date of substantial completion

B. At the date of final completion

C. When the contractor informs the architect that the work has been completed

D. When the owner releases the final retainage

11. When final payment is made pursuant to the AIA General Conditions, it constitutes a waiver of all claims by the owner except those arising from which events? Check all that apply.

A. Unsettled liens

B. Faulty or defective work appearing after substantial completion

C. Undocumented claims made after substantial completion

D. Failure of the work to comply with the requirements of the contract documents

12. The AIA General Conditions requires that the contractor submit to the architect an itemized application for payment how long before each progress payment?

A. Ten days

B. Two weeks

C. At the contractor's discretion

D. When required by the architect

13. The architect's duty and responsibility regarding the certificate for payment include

I. making a professional judgment about the progress of work.

II. verifying that the contractor has properly used the moneys paid to her under the contract.

III. determining that the quality of work is in accordance with the contract documents.

IV. representing to the owner that the work has properly progressed.

A. I, II, and IV

B. I, III, and IV

C. III only

D. II only

14. The owner has failed to make payment in seven days of the amount certified by the architect in a certificate for payment. The contractor has the right to

A. immediately stop work.

B. stop work in seven days.

C. ask for arbitration.

D. request an equal amount from retainage.

15. The architect can decline to certify payment in whole or in part for all of the following reasons, EXCEPT

A. failure of the contractor to properly pay subcontractors, suppliers, or workmen.

B. reasonable evidence that the work cannot be completed for the unpaid balance of the contract sum.

C. reasonable evidence that the work will not be completed on time.

D. part of the payment is for materials stored, but not yet incorporated into the work.

LESSON TWELVE

12

PROJECT FILES, PROJECT COMPLETION, AND PROBLEM AREAS

PROJECT FILES

Contract Documents and Project Records

Neither the AIA Owner-Architect Agreement (Document B201) nor the AIA General Conditions (Document A201) contains specific requirements for the architect to maintain project files. However, it is good business practice to keep records and documents filed in an orderly manner. Lesson 10 covers record-keeping for site visits and for submittals. This lesson will cover some of the other records that an architect normally keeps on file during the project and for an appropriate time thereafter.

There are a number of AIA forms to facilitate gathering and maintaining project data. AIA Document G809, Project Data, is designed to record basic information about the project, such as the project location, the owner's name and address, a basic description of the proposed improvements, the site description, the names of the owner's legal, insurance, and other advisors, the names and addresses of the utility companies that serve the site, relevant financial data, and proposed methods for contracting. By completing this document, the architect is able to keep a permanent record of the owner's team and keep them informed about developments during the course of the project.

AIA Document G807, Project Directory, provides the architect with a format for information about the architect's staff, consultants who are working on the project, and the contractors who are hired by the owner. By updating the document as necessary, the architect can keep the design and construction team informed about developments, and avoid uncertainty about whom to contact regarding problems, meeting schedules, notices, etc.

The AIA also publishes a form for listing subcontractors and others who will be employed on the project by the contractor. AIA Document G705, List of Subcontractors, is filled out by the contractor and sent to the architect. Although the architect usually has no direct contact with subcontractors and other parties under contract to the contractor, the architect needs to maintain a record of the various subcontractors performing work on the project. In the event of claims against the project, this information will save considerable time for the attorneys representing the architect and the owner.

To assist in the administration of the bidding phase, AIA Document G804, Register of Bid Documents, enables the architect to list information including the amount of the bid deposit, if required, the costs for sets of drawings as well as single sheets of documents, and when and where bids are to be submitted. The form also provides for a listing of each recipient of bid sets so that the architect can distribute addenda to each bidder, and monitor bidders' deposits and return of bid sets. The record of the distribution of addenda can be particularly important. Addenda can have a considerable impact on the bids that are received and on the final scope of the contract requirements.

In addition to the basic information about the project, the architect must keep complete copies of all contracts to which she is a party. Files should be established for the owner-architect agreement, amendments, and written authorizations for additional services. Similar files should be established for each contract between the architect and the engineers and other consultants. Records related to the fulfillment of the professional service contract, such as copies of billing statements and owner approvals at milestones such as completion of the schematic design and design development phases and approval of the construction documents for bidding, should also be kept.

Before the start of construction, the architect should obtain a copy of the contract between the owner and the contractor, including the agreement, the general and supplementary conditions, the drawings, specifications, and all addenda. During the construction phase, this file should be updated as change orders are agreed to by the owner and contractor. See Lesson 3 for a detailed review of change orders. If the owner's attorney prepares the owner-contractor agreement, the architect should also obtain a copy for her files.

Construction Administration Procedures

The scope of the architect's services during the construction phase is defined in the AIA Owner-Architect Agreement (Document B201), as well as in the AIA General Conditions (Document A201). Proper performance of construction phase services begins with an understanding of what is contractually required of the architect. The three most important areas requiring proper administrative procedures are: processing of submittals, evaluation and preparation of change orders, and certification of payment applications. Each of these subjects has been covered in detail in earlier lessons.

In addition to recordkeeping related to fulfillment of contractual duties, the architect may be asked to attend meetings and to prepare minutes of those meetings. This duty can include private meetings between the owner and the architect, as well as meetings attended by several parties, such as the owner, the contractor, professional consultants, and others. Often, the architect may be assigned the responsibility of taking notes to document decisions reached at such meetings, and to distribute copies of records to the attendees.

For large or complex projects, the owner may authorize the architect to hire, as a Change in Services, a full-time on-site project representative to provide a higher degree of monitoring of the progress of the work than might otherwise occur. Unlike the basic construction phase services where the architect visits the site only when necessary to observe important aspects of the work or to generally evaluate its progress or quality, the full-time project representative is expected to be at the site whenever work is in progress. The scope of the project representative's duties should be defined in a written attachment to the owner-architect contract.

AIA Document B207 should be given to the contractor to avoid misunderstandings about the limitations on the project representative's duties. For example, under certain circumstances, the representative may not be authorized to make decisions in the field, but must refer matters to the architect.

The architect's full-time project representative should not be confused with the *clerk of the works*, which refers to a person employed by the owner to check on matters at the site and to maintain records of the progress of construction. Because the clerk of the works usually is employed by the owner, his or her presence at the site can create problems. For example, the contractor may be uncertain about whether to consult the architect, the architect's on-site representative, or the owner's clerk of the works. A division of authority can lead to confusion, conflicting instructions, and potential claims against the owner.

Correspondence

Preparing proper correspondence and maintaining correspondence files are extremely important. Copies of correspondence should be filed by topic, as well as chronologically. The chronological file is especially important to show the evolution of the project, in the event of a claim. Construction projects are usually complex and take place over long periods of time, and correspondence that details each step of the project's development may be the only record of what actually happened. Written documentation reduces the likelihood that important items will be overlooked or that misunderstandings will occur. For example, if a letter from the contractor requests information or a decision, the request must be acknowledged and responded to promptly. If the decision or information requested cannot be provided, a letter of explanation should be sent.

Correspondence may be supplemented by photographs, videotape, and/or audio tape to document site conditions, meetings, etc. These should also be filed with the project records.

To expedite the transmittal of other documents and to reduce the need for formal letter writing, the AIA publishes Document G810, Transmittal Letter. The use of this document provides the architect with a written record of what was sent to whom and when it was sent. The form also has space to indicate the purpose of the transmittal. Some architects use transmittal letters, which are printed in multiple copies, to facilitate internal recordkeeping. Others use the AIA form and make file copies after filling in the blanks.

If problems arise during the course of the project, the architect should review correspondence with his or her attorney. When letters are received from the contractor or owner alleging deficiencies in the contract documents or the architect's services, the architect's legal counsel should be asked to review the correspondence and to assist in preparing responses. If actual claims are made against the architect, the architect's professional liability insurance carrier should be notified immediately so that a claims representative or an attorney can be assigned to advise on how to respond. The architect must remember that any written document could be entered as evidence in court during a lawsuit. This does not mean that the architect should avoid written documentation, but only that it should be done carefully and factually. When in doubt, legal or insurance counsel should be consulted.

Architect's Relationship to Owner and Contractor

The AIA General Conditions assigns the architect the role of an intermediary between the owner and contractor. Subparagraph 4.2.4 of AIA Document A201 states, in part, that... *the Owner and Contractor shall endeavor to communicate with each other through the Architect...*

Throughout the General Conditions, the contractor is required to submit documents to the architect, not directly to the owner. For example, this requirement applies to the contractor's application for payment, shop drawings and samples, and various documents during the closeout of the project at the end of the construction phase. All parties to the project expect the architect to maintain copies of these documents and to keep them organized in an adequate filing system.

Shop Drawings and Schedules

The specifications state the requirements for the submission of shop drawings and the preparation of schedules. At the start of the construction phase, the contractor should prepare a schedule for the submission of shop drawings. The purpose of this schedule is to notify the architect when to expect submittals from the contractor, and when approved shop drawings must be returned to the contractor. Other aspects of shop drawings and other submittals are covered in detail in Lesson 11.

The architect's project files should contain a copy of each shop drawing, with the approval stamp and signature of the architect, the architect's consultants, if applicable, and the contractor to show that the shop drawing was processed in accordance with contract requirements. Sophisticated owners may sometimes process shop drawings with their own technical staff. If this is the case, the architect should also maintain a record of the distribution of the shop drawings in accordance with the owner's requirements. The shop drawing schedule must allow time for the shop drawing to be transmitted to, and reviewed by, all appropriate parties.

The architect should retain copies of all shop drawings for a reasonable period of time after the completion of construction. This includes not only the final approved set of shop drawings, but also intermediate submittals that were not approved. The shop drawings help to document the progress of the project. In the event of any claims, a record of what was approved or rejected, and when the action occurred, can play an important role in the resolution of such claims.

When files are cleaned out, old shop drawings are often the first documents to be discarded. As with all project files, nothing should be thrown away until the expiration of the statute of limitations applicable to the architect's services or to the project. The architect should ask if the owner has a need for the project documents before they are destroyed. Clients sometimes rely on the architect to maintain construction records because they do not have the facilities themselves. If the owner plans an expansion or renovation of the project, the architect may supply copies of the drawings. By anticipating the client's needs, the architect can build a strong business relationship with the client.

Record Drawings

Record drawings show field changes that occurred during construction, which vary from the information shown in the working drawings. Record drawings are sometimes referred to as *as-built drawings*. Use of that term should be avoided, since no set of documents can ever show the project exactly as built, and using the term may create a liability exposure.

The owner generally decides whether to require the preparation of record drawings. Since the AIA General Conditions does not cover the preparation of record drawings, if they are required, the procedures are usually indicated in Division One, General Requirements of the Specifications. The recommended procedure is for the contractor to mark up a set of prints in the field showing the changes as they occur during construction. The contractor is responsible for the supervision of the work and is in the best position to know what changes are being made to the work. If changes involve additional or reduced cost or time, they must be documented by the issuance of a change order. Other changes, not involving adjustments in cost or time, can be recorded by supplemental instructions. Some minor changes may simply occur in the field without a change order or supplemental instructions. Field changes should be recorded by the contractor on a set of record prints at the site. This record is invaluable if and when future repairs or modifications to the project are required. The exact location of items, such as underground high voltage lines, must be carefully documented to prevent the possibility of serious injury or death in the event future work is required in the area.

Once the contractor has marked up the changes on a set of prints at the site, the notations should be transferred onto a permanent reproducible medium, such as mylar drafting film or CAD files. The backgrounds used should be the current version of the contract documents, including all revisions issued during construction. If the owner chooses to authorize the architect to transfer the notations, these services are normally considered a Change in Services under the Owner-Architect Agreement. Otherwise, the contract documents should require the contractor to transfer the notations.

Normally, the architect has no obligation to verify if the information provided by the contractor on the marked-up set of record prints accurately represents the installed work because the architect is not continuously at the site during construction and does not control the activities of the construction workers. However, the contractor, whose superintendent is always present at the site, is expected to be aware of all changes made to the construction and is expected to note these changes on a set of record prints.

Warranties

A warranty is a legally enforceable promise made by one party to another about something that will or will not happen or about certain circumstances. There is no legal distinction between the terms *warranty* and *guarantee*. On construction projects, warranties are often required to establish performance standards. If the standards are not met, the breach of the warranty would entitle the owner to recover monetary damages.

The AIA General Conditions requires the contractor to give a broad warranty to the owner about the quality of the work. Paragraph 3.5 of AIA Document A201 states:

> *The Contractor warrants to the Owner and Architect that materials and equipment furnished under the Contract will be of good quality and new unless the Contract Documents require or permit otherwise. The Contractor further warrants that the Work will conform to the requirements of the Contract Documents and will be free from defects, except those inherent in the quality of the Work the Contract Document require or permit. Work, materials, or equipment not conforming to these requirements may be considered defective. The Contractor's warranty excludes remedy for damage or defect caused by abuse, alterations to the Work not executed by the Contractor, improper or insufficient maintenance, improper operation, or normal wear and tear and normal usage. If required by the Architect, the Contractor shall furnish satisfactory evidence as to the kind and quality of materials and equipment.*

If the work is not of good quality, free from faults and defects, or is not in conformance with the contract requirements, the contractor will have breached the warranty and will be liable to the owner for all damages that result.

In addition to the broad warranty contained in Paragraph 3.5 of the AIA General Conditions, the contractor also may be required to provide the owner with specific warranties related to the operation of mechanical systems and other items of equipment. Often, these warranties are given by the equipment manufacturers and are simply passed through by the contractor to the owner. The requirements for warranties are usually specified by the architect in the appropriate sections of the technical specifications. If the architect has any questions about the precise terms of any warranty, he or she must ask the owner's attorney to prepare the appropriate legal language for the specifications.

After the work has been performed, the contractor is required to assemble the manufacturers' warranties and forward them to the architect. Subparagraph 2.6.6.1 of the AIA Owner-Architect Agreement (Document B201) requires the architect to receive and forward these warranties to the owner for review. If there are any discrepancies between the terms of the warranty required by the specifications and the terms of the warranty provided, the owner's attorney should resolve this matter with the contractor. The contractor cannot avoid contractual responsibility by claiming that the manufacturer will not provide the warranty required by the specifications.

PROJECT COMPLETION

Compliance with Contract Documents

When the contractor has fulfilled all the requirements of the contract documents, his obligations are over, subject to remedying defective work that may appear within a year and any liability for breach of contract until the expiration of the statute of limitations. Due to the complexity of most projects, determining project completion can require a great deal of the architect's time and effort. The architect must inspect the work, and she may be required to mediate the competing interests of the owner and contractor. The owner may be anxious to occupy the project, but not if work remains that would interfere with the owner's use of the premises. As the project nears completion, the contractor may minimize the importance of the remaining work because he usually wants to establish the earliest date of substantial completion, get paid, and move on to other projects. At this point, there may seem to be a never-ending series of minor problems that cannot be resolved. The painter may be committed to other work and unable to get back to touch up some walls. A part may be missing for a piece of equipment, and the manufacturer may be unable to promise a delivery date. Whatever the problems, they prevent completion of the project, which can be frustrating for all parties.

The owner relies on the architect to determine whether the contractor's work complies with the requirements of the contract documents. This evaluation is continuous throughout the construction phase. Each of the contractor's applications for payment must be evaluated by the architect to determine compliance with the contract documents for work completed to date. A systematic approach to periodic evaluations of the work in progress facilitates the final evaluation during completion of the project. The remaining work and correction of work should be readily identifiable for inclusion on the contractor's *punch list*, and completion can be verified as the work is performed.

Assembling Documents

The AIA Owner-Architect Agreement (Document B201) requires the architect to forward certain documents to the owner as part of the architect's services related to closing out the project. Subparagraph 2.6.6.1 states:

> *The Architect shall conduct inspections to determine the date or dates of Substantial Completion and the date of final completion; issue Certificates of Substantial Completion; receive from the Contractor and forward to the Owner, for the Owner's review and records, written warranties and related documents required by the Contract Documents and assembled by the Contractor; and issue a final Certificate of Payment based upon a final inspection indicating the Work complies with the requirements of the Contract Documents.*

The contractor's obligation to assemble warranties and documents is defined by the contract documents, both in the general conditions and the specifications. The contractor may prepare a binder to organize operating manuals and other documents that the owner may need to properly maintain equipment and other parts of the project. Warranties required by the contract documents may either be included in the binder or forwarded separately. The architect must know what is required by the contract documents so that documents forwarded by the contractor can be verified as they are received.

If the architect believes that any warranties forwarded by the contractor are not in compliance with the contract requirements, the architect should alert the owner so that the owner can request his or her attorney to review the warranties. The architect should not judge the legal sufficiency of warranties. A warranty may meet the requirements of the contract documents, but the precise language in the warranty may be different from that required by the contract. Conversely, the warranty may not meet contract requirements, and the owner may have to decide whether to accept the warranty as given, or to demand that the contractor furnish the warranty as required. The contractor may claim that the warranty is merely being forwarded from the manufacturer, and the manufacturer will not provide a warranty that meets the contract requirements. In this situation, the owner still has the right to insist that the contractor provide the warranty required by the contract documents. The owner does not have to accept a lesser warranty simply because the contractor is having a problem with one of his suppliers. The architect should limit his involvement in this situation to providing the owner with technical assistance. Leave the legal arguments to the owner's attorney.

Substantial Completion and Final Completion

As part of closing out the construction phase, the architect is required to inspect the project to determine whether the contractor has achieved substantial completion and, subsequently, final completion. Substantial completion means that the work has progressed to the point where the owner can occupy or utilize the work, or designated portions thereof, for the use for which it is intended. The detailed procedures to be followed by the architect in determining substantial completion and final completion are covered in Lesson 10.

Inspections by the Architect

The contractor is required by the AIA General Conditions to prepare a list of items of work to be completed or corrected at the time that he or she claims that the work is substantially complete. In some areas, the local practice is to have the architect prepare this list, called the *punch list*, contrary to the provisions of AIA Document A201. When the contractor prepares the list, the architect has the authority to add items to the list that the contractor may have overlooked. When inspecting the project to determine these two important milestones—substantial completion and final completion—the architect must look at the work with a much higher degree of care than that required during prior site visits, when the architect was only required to determine, in general, whether the work was proceeding in accordance with the contract requirements. Architects should generally avoid the term *inspection* to describe the architect's services. The Owner-Architect Agreement makes an exception in Paragraph 2.6.6 with regard to inspections for substantial and final completion.

Cleaning Up

The AIA General Conditions (Document A201) requires the contractor to keep the premises neat and orderly. Subparagraph 3.15.1 states:

> *The Contractor shall keep the premises and surrounding area free from accumulation of waste materials or rubbish caused by operations under the Contract. At completion of the Work, the Contractor shall remove waste materials, rubbish, the Contractor's tools, construction equipment, machinery and surplus materials.*

The architect may add to these requirements by provisions in the supplementary conditions or in Division One of the specifications. It is common practice to require the contractor to clean and polish all glass, to wax tile floors, to vacuum carpets, and to leave other spaces "broom clean."

Subparagraph 3.15.2 gives the owner the right to clean the premises if the contractor fails to meet his obligations by stating:

> *If the Contractor fails to clean up as provided in the Contract Documents, the Owner may do so and Owner shall be entitled to reimbursement from the Contractor.*

Paragraph 2.4 relates to the owner's right to carry out the work. If the owner chooses to exercise this right, the contractor must be given written notice. If the contractor fails to respond to the owner's notice, the owner can proceed to have the work performed by others. The owner must issue a change order to deduct from the contract amount the cost of such work performed by others. If the owner decides to have the work done by others because the contractor is not performing the work properly, the owner's actions and the amount to be charged to the contractor are both subject to the prior approval of the architect.

The General Conditions give the owner this right because the owner should not be constrained by the contractor's failure to carry out the work properly or to keep the premises reasonably clean and orderly. If the architect believes the contractor is not meeting his responsibilities to keep the premises clean, or has failed to clean up properly at the completion of the work, he or she can recommend to the owner that another party be hired to clean up.

Occupancy by the Owner

The owner is entitled to occupy the project when the work, or designated portions thereof, are substantially complete. Substantial completion, as noted previously, requires that the work be sufficiently complete so that the owner can *occupy or utilize the Work for its intended use.*

On occasion, the work may be sufficiently complete to permit owner occupancy, but the owner does not choose to move in. If this occurs, the architect may be required to make an independent judgment that the work is substantially complete, entitling the contractor to the release or a reduction of retainage, and to shift to the owner responsibilities for insurance, warranties, utilities, maintenance, and so forth. Conversely, the owner may choose to occupy the project *before* the contractor reaches substantial completion. In this case, the contract may claim added costs and/or delays caused by the owner's interference.

The owner's personnel, equipment, furniture, and security requirements may further impede the contractor's operations. The architect may be required to evaluate the validity of the contractor's claim. The AIA General Conditions does not provide for situations where the owner either refuses to take occupancy at substantial completion or desires to move in prematurely. In either case, the architect must try to get the owner and contractor to identify in writing the consequence of their decisions so that each party's legal rights under the contract can be protected.

Release of Liens

Final payment to the contractor is conditioned on the contractor giving the owner a release of liens from the contractor and all subcontractors and material suppliers. Mechanics' liens are covered in detail in Lesson 9. At project completion, the AIA General Conditions requires the contractor to give the owner an affidavit that all payrolls, bills for materials and equipment, and other indebtedness connected with the work have been paid. The contractor also is required to give the owner releases of lien in whatever form the owner may require. If the contractor is unable to get a release of lien from a subcontractor, the contractor can furnish a bond to the owner to indemnify the owner in the event a lien is filed. Because lien-related matters have serious legal consequences and can adversely affect the owner's title to the property, the owner's attorney must become involved in the event that liens are, or may be, asserted against the property.

As part of closing out the project, the architect usually receives from the contractor the Contractor's Affidavit of Payment of Debts and Claims, AIA Document G706 and Contractor's Affidavit of Release of Liens, AIA Document G706A. These forms should be forwarded to the owner for review by the owner's legal counsel. Neither the Owner-Architect Agreement nor the General Conditions requires the architect to make an independent evaluation of these documents. The architect, however, should not issue the final certificate for payment to the owner if he or she has any reason to believe that these documents are not in order, or if the owner, or the owner's attorney, informs the architect that the affidavits or releases of lien are not proper. Final payment to the contractor should be withheld until all such problems are resolved.

If the owner does not require the contractor to furnish a performance bond or a labor and material payment bond, it is even more important to make sure that releases of lien are in order before issuing the final certificate for payment and closing out the project. Without bonds, the assertion of liens can cloud the owner's title to the property, as well as cause the owner to pay twice for the work if the contractor has not disbursed funds received from the owner to the subcontractors and suppliers. In many states, if the contractor goes bankrupt, subcontractors and suppliers to whom the contractor owes money can place a lien on the property and can require the owner to pay them directly, even if the owner has already paid the contractor for the work.

PROBLEM AREAS

Establishing the Contract Requirements

The construction process can be fraught with problems. An important part of the challenge of being an architect is to deal successfully with these problems. In fact, most problems do get resolved to everyone's satisfaction, and relatively few end up in litigation or as formal legal claims. Whether problems are resolved successfully often is directly related to how well the parties to a construction project understand their contractual and other legal obligations.

A construction contract is intended to be a detailed statement of the obligations of the owner and contractor with regard to each other and the project. If the contract, including the owner-contractor agreement, the contract conditions, the drawings, the specifications, and any modifications, is comprehensive and clear, there should be few questions about the rights, duties, and obligations of the contractor to perform the work and the owner to pay for it. Likewise, the role of the architect as the owner's agent during construction should be clearly stated. If problems arise, such as the contractor failing to perform the work in accordance with the contract documents, or the owner not paying for it as required by the contract, the contract should provide guidance for resolving the problems.

From the architect's standpoint, knowledge about the requirements of the contract documents goes beyond the technical content of the drawings and specifications. The architect must also be familiar with the owner-contractor agreement and the general and supplementary conditions of the contract. Problem resolution begins with measuring the facts associated with the problem against contract requirements. Guessing or making uninformed responses magnifies problems rather than solving them.

The AIA General Conditions (Document A201) requires both the owner and the contractor to ask the architect for an initial determination when disputes arise. Subparagraphs 15.1.1 and 15.2.1 state:

> *15.1.1 Definition. A Claim is a demand or assertion by one of the parties seeking, as a matter of right, payment of money, or other relief with respect to the terms of the Contract. The term "Claim" also includes other disputes and matters in question between the Owner and Contractor arising out of or relating to the Contract. The responsibility to substantiate Claims shall rest with the party making the Claim.*
>
> *15.2.1 Claims, excluding those arising under Sections 10.3, 10.4, 11.3.9, and 11.3.10, shall be referred to the Initial Decision Maker for initial decision. The Architect will serve as the Initial*

Decision Maker, unless otherwise indicated in the Agreement. Except for those Claims excluded by this Section 15.2.1, an initial decision shall be required as a condition precedent to mediation of any Claim arising prior to the date final payment is due, unless 30 days have passed after the Claim has been referred to the Initial Decision Maker with no decision having been rendered. Unless the Initial Decision Maker and all affected parties agree, the Initial Decision Maker will not decide disputes between the Contractor and persons or entities other than the Owner.

The legal concepts behind the architect's role as the initial decision maker are covered in Lesson 1. The technical aspects of that role require the architect to take the time to study the contract documents carefully before making a determination. The architect has to step away from his or her role as the preparer of the drawings and specifications and look objectively at those documents and any applicable provisions in the agreement and general and supplementary conditions. After the architect has reviewed the contract documents, he or she is then in a position to analyze the problem presented by the owner or contractor, or both.

When the architect gives a written determination about the problem, it is often helpful to cite the applicable provisions of the contract documents so that the parties can clearly understand the basis for the architect's determination. If the determination is adverse to a party's interests, that party will be less likely to conclude that the architect acted out of bias or self-interest if the architect has reinforced his or her findings with a statement of the requirements of the contract.

Documenting Problems

Factual Information

The most difficult task in analyzing problem areas often is gathering sufficient factual information to enable a proper evaluation to be made in a timely manner. It is imperative that conclusions not be drawn until the architect is satisfied that there is sufficient information available to make valid judgments.

Testing

If the architect suspects that the cause of a problem is noncompliance with the requirements of the contract documents, the architect can order work to be tested. Special testing not called for in the construction contract as part of the work requires written authorization from the owner. If the work subjected to special testing is found to be in compliance with contract requirements, the owner has to pay for the test, as well as for any remedial work to restore the work to its pre-test condition. Conversely, if the test reveals that the work does not comply with contract requirements, the contractor has to pay all related costs, including testing.

If tests are required, the architect establishes the standards applicable to the tests and how the results are to be measured. The architect may rely on industry-wide standards, such as those developed by ASTM, or he or she can establish special standards for unique project requirements. The latter is often the case when the architect determines that the project requires a higher standard than is normally accepted in the industry.

If testing is required, it usually is performed by an independent organization under direct contract to the owner or the contractor. Testing laboratories and companies must be free from any conflict of interest because their work can affect the safety and long-term viability of the project. The architect should not perform tests with her own employees or consultants because of the potential liability. See Lesson 10 for a more detailed discussion of testing.

Causes

The causes of problems can be difficult to determine. Often, when a problem arises, responsibility is quickly assigned before all relevant facts are known or analyzed. The analysis of construction problems should be done carefully, much as the design was developed initially. Problem analysis involves research, investigation, documentation, review, and reporting. The reporting usually takes the form of a written document instead of drawings and specifications. However, since construction moves at a rapid pace, and one decision can affect many aspects of the work, the architect must act expeditiously in making a determination. Paragraph 15.2 of the AIA General Conditions establishes time limits on the architect's response.

Design

When the construction of a building appears to be deficient, the design is often cited as the cause of the problem. The project design is a convenient target for several reasons. First, it provides the only record of what was intended to go into the project. The drawings and specifications are readily available to be analyzed. What actually went into the construction may not be easy to determine. For example, it may be impossible, after a structural collapse, to determine whether the contractor properly placed the reinforcing steel before the concrete was poured. Secondly, after problems arise, it may be convenient to say that a different or better design could have avoided the problem. When analyzing a problem, the real issue is not whether the design could have been done differently, but whether the design was appropriate for the project requirements and in accordance with the architect's duties to the owner.

When problems arise, care must be taken not to conclude prematurely that a faulty design caused the problem. There may have been intervening causes. For example, an underdesigned structural system might have collapsed, not because of the underdesign, but because of severe overloading by the contractor that would have caused a collapse regardless of the design. A curtain wall may appear to have leaked, not because of faulty detailing or quality of construction, but because the owner's employees did not close the windows during a heavy rainstorm. In every case, careful investigation is required before conclusions can be drawn.

For the architect to be liable for faulty design, it must be shown that the architect had a *duty* in regard to that aspect of the design in the first place. The architect would not be liable for a failure to properly design the structural system if, for example, the owner retained the architect merely to sketch a floor plan. Likewise, the architect would not be liable for the malfunctioning of an elevator system if the owner contracted directly with an elevator manufacturer for the design of the system. If the architect had a duty in regard to the design of the part of the project under investigation, the next step would be to determine if the architect had exercised reasonable care. Was there adequate investigation? Did the architect have the background and experience to proceed? Should she have utilized consultants? What were the constraints imposed by the owner and the budget? What were the trade-offs and options considered? All of these factors bear on whether the architect used *reasonable care* in reaching the design decisions that led to the actual design.

The law does not hold an architect liable simply because a design does not work or because it could have been executed differently and avoided the problem. After a problem arises, most architects would probably admit that they might have done things differently had they known of the potential problems beforehand. The test of liability is whether the architect was *negligent* in proceeding with the design. Did the architect fail to meet the ordinary *standard of care* when she performed the services? The fact that the design does not work is not conclusive in establishing the architect's liability. Therefore, any analysis of a construction problem that focuses on the design must be weighed against the legal standards applicable to design liability.

Construction

If the design and the contract documents have been prepared properly, the contractor is expected to construct the project in accordance with the contract documents. A failure by the contractor to do so is a breach of the contract with the owner.

Any analysis of construction-related problems must begin by establishing what the contract documents require. The contractor's performance must be measured against contract requirements. Once they have been determined, the investigation can focus on the work performed by the contractor and how it deviated, if at all, from the contract requirements. Unlike an analysis of the design that can involve many subjective factors, the evaluation of the contractor's work usually can be measured objectively against the standards established by the contract documents.

Field investigations can involve photographing the failed portions of the work, installing gauges to measure deflections, taking concrete core samples to test for compressive strength, cutting samples from the roofing material to analyze chemical composition, and various other technical evaluations. In each instance, the investigation must be carefully documented, including who performed the test, the date of the test, weather conditions, and other relevant factors. Photographs must be dated and the photographer identified. These precautions are necessary so that the investigation results can be used as evidence in litigation.

Maintenance

Problems can arise with a completed project if the owner or occupant fails to properly maintain it. Maintenance can involve such things as keeping metal roofs painted and changing filters on HVAC equipment. All buildings have ongoing maintenance requirements, and it is the owner's responsibility to see that maintenance is carried out on a proper schedule. The architect should alert the owner at the end of the construction phase about those aspects of the project for which there are special maintenance requirements. Paragraph 9.8 of the AIA General Conditions shifts the responsibility for maintenance from the contractor to the owner on the date of substantial completion.

If the architect is called back to the project because of a problem caused by a lack of proper maintenance, the architect should be prepared to recommend to the owner both corrective work to remedy the immediate problem as well as what can be done to prevent recurrences in the future. Services provided after the completion of construction are considered additional services according to Article 3 of the Owner-Architect Agreement.

If the problem is design-related, an architect might not seek additional compensation for the services required to deal with the problem because of the potential for alienating the client. An architect has to use good judgment in responding to a client's notification about problems with the project so that the owner and architect can work as a team to solve the problems.

Other Problem Areas

In addition to design deficiencies, poor construction quality, failures to adhere to contract requirements, and improper or inadequate maintenance, other problems may occur after the project is complete. Theses include normal wear and tear, extraordinary weather conditions, and unanticipated changes in usage. Whatever the apparent cause of problems, the architect must approach the investigation of the problem with a high degree of professionalism and objectivity. This is particularly true when the architect is requested to investigate problems on a project designed by another architect. In this situation, the architect is performing the role of *expert witness*. The architect must gather all available facts, retain objectivity when analyzing the facts, and know the proper standard against which to measure the circumstances under investigation. Having done so, the architect will have fulfilled her obligations to the client, as well as to the profession, regardless of the outcome.

X 1 B
2 A, B, D
3 A
4 B
5 C
6 Contractor
7 A
8 B
9 D
10 B
11 C
12 A
13 D
14 C
15 A

LESSON 12 QUIZ

1. Which AIA document requires that the architect maintain project files?

   A. Owner-Architect Agreement
   B. General Conditions
   C. Project Data
   D. None of the above

2. Which three of the following are important administrative tasks of the architect? Check all that apply.

   A. Processing of shop drawings and samples
   B. Evaluation and preparation of change orders
   C. Coordinating the work of contractors and subcontractors
   D. Certifying payment applications

3. A Project Representative is selected, employed, and directed by the

   A. architect.
   B. owner.
   C. contractor.
   D. clerk of the works.

4. A clerk of the works is generally employed by the

   A. architect.
   B. owner.
   C. contractor.
   D. project representative.

5. Nothing from the project file should be thrown away until

   A. a reasonable period of time after the completion of construction.
   B. two years have passed.
   C. the expiration of the statute of limitations.
   D. the bonding company gives its approval.

6. Record drawings noting the exact location of construction elements in a set of drawings when a change is made are the responsibility of the as built record dwg.

7. The architect's duties and responsibilities in connection with manufacturers' warranties are limited to which of the following?

   A. Forwarding the documents from the contractor to the owner
   B. Verifying that the warranties are legally sufficient
   C. Furnishing evidence to the owner that the kind and quality of materials and equipment are satisfactory
   D. Negotiating with the manufacturer in the event that the work is not of good quality or is defective

8. A contractor is obligated to leave a project "broom clean," but fails to do so. What should the architect do?

   A. Hire a professional cleaning firm and charge its fee to the contractor
   B. Have the owner issue a change order to have the work performed by others
   C. Stop work, thus stopping final payment
   D. Nothing

9. If a contractor is unable to get a release of lien from a subcontractor, the contractor can

A. deduct the amount from the owner's final payment.

B. have the architect negotiate a settlement with the subcontractor.

C. draw upon the performance bond.

D. furnish a new bond to the owner to indemnify the owner.

10. If an owner feels that a project is not progressing fast enough, he or she should

A. ask the contractor for an explanation.

B. ask the architect to request an explanation from the contractor.

C. bring in an independent management consultant.

D. fire the contractor.

11. Work that was subjected to special testing at the request of the architect is found to be in compliance with the contract requirements. Who must pay for the test?

A. The architect

B. The contractor

C. The owner

D. The testing company

12. If special testing is required, which is not called for in the construction documents, who establishes the standards applicable to the tests?

A. The architect

B. The owner

C. An independent testing agency

D. The engineer responsible for the work in question

13. An affirmative answer to which of the following questions could render an architect liable?

I. Did the architect fail to meet the ordinary standard of care when he or she performed services?

II. Does the design work poorly?

III. Could the design have been executed differently and the problem avoided?

IV. Did the architect lack care in designing a defective component or system, which was constructed or manufactured in accordance with that design?

A. I only

B. III only

C. II, III, and IV

D. I and IV

14. Documentation for a construction field investigation should include all of the following, EXCEPT

A. who performed the test.

B. the date of the test.

C. the architect's role in supervising the work in question.

D. weather conditions.

15. Whose responsibility is it to see that maintenance is carried out on a proper schedule after substantial completion?

A. The owner

B. The architect

C. The subcontractor

D. The manufacturer

Part II

The Graphic Vignette

LESSON THIRTEEN

THE NCARB SOFTWARE

INTRODUCTION

There is a wide variety of programs used by candidates at the firms in which they work. Therefore, an essential part of every candidate's preparation is to practice using the examination's computer tools. Candidates can download this software from the NCARB Web site (*www.ncarb.org*). This program contains tutorials and sample vignettes for all the graphic portions. Spend all the time necessary to become familiar with this material in order to develop the necessary technique and confidence. You must become thoroughly familiar with the software.

The drafting program for the graphic portions is by no means a sophisticated program. While this may frustrate candidates accustomed to advanced CAD software, it is important to recall that NCARB aimed to create an adequate drafting program that virtually anyone can use, even those with no CAD background at all.

VIGNETTE SCREEN LAYOUT

Each vignette has a number of sections and screens with which the candidate must become familiar. The first screen that appears when the vignette is opened is called the Vignette Index and starts with the Task Information Screen. Listed on this screen are all the components particular to this vignette. Each component opens a new screen when the candidate clicks on it with the mouse. A menu button appears in the upper left corner of any of these screens that returns you to the Index Screen. Also available from the Index Screen is a screen that opens the General Test Directions Screen, which gives the candidate an overview of the procedures for doing the vignettes. Here are the various screens found on the Index Screen:

- **Vignette Directions** (found on all vignettes)—describes the procedure for solving the problem
- **Program** (found on all vignettes)—describes the problem to be solved
- **Tips** (found on all vignettes)—gives advice for approaching the problem and hints about the most useful drafting tools
- **Code**—gives applicable code information if required by the vignette
- **Sections**—typically found on the Stair Design Vignette and shows a section through the space in which the stair will be located

- **Lighting Diagrams**—found on the Mechanical and Electrical Plan vignette to show light fixture distribution patterns

The beginning of each vignette lesson in this study guide provides a more detailed description of each vignette screen.

To access the actual vignette problem, press the space bar. This screen displays the problem and all the computer tools required to solve it. Toggle back and forth between the Vignette Screen and one of the screens from the Index Screen at any time by simply pressing the space bar. This is not as convenient as viewing both the drawing and, say, the printed program adjacent to each other at the same time. Thus, it is a procedure that the candidate must become familiar with through practice. Also, some vignettes are too large to be displayed all at once on the screen. In this case, use the scroll bars to move the screen up and down or left and right as needed. The Zoom Tool is also helpful.

THE COMPUTER TOOLS

There are two categories of computer tools found in the ARE graphic portions:

- Common Tools
- Tools specific to each vignette

The Common Tools, as the name implies, are generally present in all the tests and allow a candidate to draw lines, circles, and rectangles, adjust or move shapes, undo or erase a previously drawn object, and zoom to enlarge objects on the screen. There is also an on-screen calculator and a tool that lets you erase an entire solution and begin again.

Vignette-specific tools include additional tools that enable the candidate to turn on and off layers, rotate objects, and set elevations or roof slopes. In addition to these extra tools, each vignette also includes specific items under the draw tool required for the vignette, such as joists or skylights. Become an expert in the use of each tool.

Each tool is dependent on the mouse; there are no "shortcut" keys on the keyboard. Press the computer tool first to activate it, select the item or items on the drawing to be affected by the tool, and then re-click the computer tool to finish the operation. Spend as much time as required to become completely familiar with this drafting program. The Common Tools section of the practice vignettes, available from NCARB, is particularly useful for helping you become familiar with the computer tools. Three things to note: the left mouse button activates all tools; there is no zoom wheel on the mouse, nor an associated tool on the program; and the shift key activates the Ortho Tool.

The standard computer tools and their functions are shown in Figure 13.1.

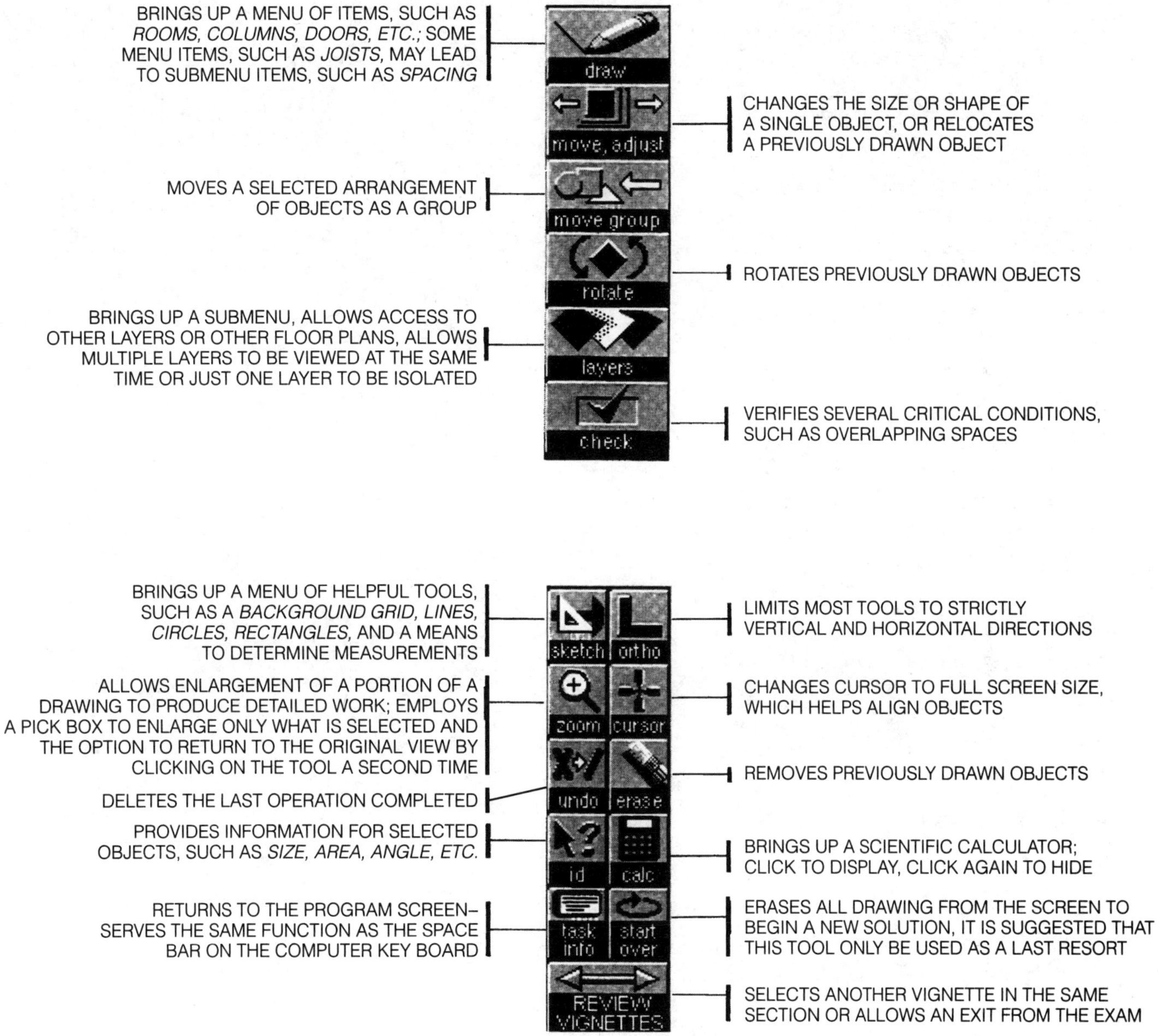

Figure 13.1 Standard Computer Tools

TAKING THE EXAM

INTRODUCTION

Preparation for the ARE usually begins several months before taking the actual exam. The first step is to submit an application for registration with your state board or Canadian provincial association. Most, but not all, registration boards require a professional degree in architecture and completion of the Intern-Development Program (IDP) before a candidate is allowed to begin the exam process. Since the processing of educational transcripts and employment verifications may take several weeks, begin this process early. The registration board will review a candidate's application to determine whether he or she meets the eligibility requirements.

SCHEDULING THE EXAM

The exams are available to eligible candidates at virtually any time, since test centers are open nearly every day throughout the year. However, it is the responsibility of the candidate to contact a test center to schedule an appointment. This must be done at least three days prior to the desired appointment time, but it is probably more sensible to make an appointment a month or more in advance. It is not necessary to take the test in the same jurisdiction in which you intend to be registered. Someone in San Francisco, for example, could conceivably combine his or her test-taking with a family visit in Philadelphia.

FINAL PREPARATION

Candidates are advised to complete all preparations the day before their appointment, in order to be as relaxed as possible before the upcoming test. Avoid last minute cramming, which in most cases does more harm than good. The exams not only test design competence, but also physical and emotional endurance. You must be totally prepared for the strenuous day ahead, and that requires plenty of rest and composure.

One of the principal ingredients for success on this exam is confidence. If you have prepared in a reasonable and realistic way, and if you have devoted the necessary time to practice, you should approach this exam with confidence.

EXAM DAY

Woody Allen once said that a large part of being successful was just showing up. That is certainly true of the licensing examination, where you must not only show up, but also be on time. Get an early start on exam day and arrive at the test center at least 30 minutes before the scheduled test time. Getting an early start enables you to remain in control and maintain a sense of confidence, while arriving late creates unnecessary anxiety. If you arrive 30 minutes late, you may lose your appointment and forfeit the testing fee. Most candidates will begin their test session within one-half hour of the appointment time. You will be asked to provide a picture identification with signature and a second form of identification. For security reasons, you may also have your picture taken.

THE EXAM ROOM

Candidates are not permitted to bring anything with them into the exam room: no reference materials, no scratch paper, no drawing equipment, no food or drink, no extra sweater, no cell phones, no digital watches. You are permitted to use the restroom or retrieve a sweater from a small locker provided outside the exam room. Each testing center will have its own procedure to follow for such needs. A calculator is provided as part of the drafting program.

Scratch paper will be provided by the testing center. The candidate might wish to request graph paper, if available.

Once the candidate is seated at an assigned workstation and the test begins, he or she must remain seated, except when authorized to leave by test center staff.

Exam room conditions vary considerably. Some rooms have comfortable seats, adequate lighting and ventilation, error-free computers, and a minimum of distractions. The conditions of other rooms, however, leave much to be desired. Unfortunately, there is little a candidate can do about this, unless, of course, his or her computer malfunctions. Staff members will try to rectify any problem within their control.

EXAM ROOM CONDUCT

NCARB has provided a lengthy list of offensive activities that may result in dismissal from a test center. Most candidates need not be concerned about these, but for those who may have entertained any of these fantasies, such conduct includes:

- Giving or receiving help on the test
- Using prohibited aids, such as reference material
- Failing to follow instructions of the test administrator
- Creating a disturbance
- Removing notes or scratch paper from the exam room
- Tampering with a computer
- Taking the test for someone else

BEGIN WITH THE PROGRAM

You can either solve the vignettes in the order they are presented or build confidence by starting with one that looks easier to you. Only you know what works best for you; the practice software should give you a sense of your preferred approach.

The vignette solution begins with the program. Read the entire program carefully and completely, and consider every requirement. During this review, identify the requirements, restrictions, limitations, code demands, and other critical clues that will influence the solution. Feel free to use scratch paper to jot down key points, data, and requirements. This will help ensure that you understand and meet all the requirements as you develop your solution.

The vignette problem has two components: the written program and a graphic base plan. Both components are complementary and equally important; together they completely define the problem. Candidates should not rush through a review of the program and base drawing in an attempt to begin the design sooner. It is more important to understand every constraint and to be certain that you have not overlooked any significant detail. Until you completely understand the vignette, it is pointless to continue.

GENERAL STRATEGIES

The approach to all vignette solutions is similar: Work quickly and efficiently to produce a solution that satisfies every programmatic requirement. The most important requirements are those that involve compliance with the code, such as life safety, egress, and barrier-free access.

Another important matter is design quality. Strive for an adequate solution that merely solves the problem. Exceptional solutions are not expected, nor are they necessary. You can only pass or fail this test, not win a gold medal. Produce a workable, error-free solution that is good enough to pass.

During the test session, candidates will frequently return to the program to verify element sizes, relationships, and specific restrictions. Always confirm program requirements before completing the vignette so that oversights or omissions may be corrected while there is still time to do so. Candidates must always keep in mind the immutability of the program. That is, never—under any circumstances—modify, deviate from, or add anything to the program. Never try to *improve* the program. Lastly, taking time at the end of each section to review all the vignettes can help to eliminate small errors or omissions that could tip the balance between a passing and failing grade.

Candidates should have little trouble understanding a vignette's intent; however, the true meaning of certain details may be ambiguous and open to interpretation. Simply make a reasonable assumption and proceed with the solution.

THE TIME SCHEDULE

The most critical problem on the exam is time, and you must use that fact as the organizing element around which any strategy is based. The use of a schedule is essential. During the preparation period, and especially after taking a mock exam, note the approximate amount of time that should be spent on each vignette solution. This information must then become a candidate's performance guide, and by following it faithfully, the candidate will automatically establish priorities regarding how time will be spent.

It is important to complete the vignette in approximately the time allotted. Forget the details, do not strive for perfection, and be absolutely certain to finish the test. Even the smallest attempt at solving a vignette will add points to your total score.

Vignettes have been designed so that a reasonable solution for each of the problems can be achieved in approximately the amount of time shown in the *ARE Guidelines.* These time limits are estimates made by those who created this test. In any event, a 45-minute-long vignette may not necessarily take 45 minutes to complete. Some can be completed in 30 minutes, while others may take an hour or longer. The time required depends on the complexity of the problem and your familiarity with the subject matter. Some candidates are more familiar with certain problem types than others, and since candidates' training, experience, and ability vary considerably, adjustments may have to be made to suit individual needs.

Candidates who are aware of the time limit are more able to concentrate on the tasks to be performed and the sequence in which they take place. You may submit an imperfect solution, but you *will* complete the test. Only solve what the program asks you to solve, and don't use real-world knowledge, such as specific building code requirements.

TIME SCHEDULE PROBLEMS

It is always possible that a candidate will be unable to complete a vignette in the time allotted. What to do in that event? First, avoid this kind of trouble by adhering to a rigid time schedule, regardless of problems that may arise. Submit a solution even if it still has problems or is incomplete.

Candidates are generally able to develop some kind of workable solution in a relatively short time. If each decision is based on a valid assumption and relies on common sense, the major elements should be readily organized into an acceptable functional arrangement. It may not be perfect, and it will certainly not be refined, but it should be good enough to proceed to the next step.

MANAGING PROBLEMS

There are other serious problems that may arise, and while each is potentially fatal, they must be managed and resolved. Consider the following:

- The candidate has inadvertently omitted a major programmed element.
- The candidate has drawn a major element too large or too small.
- The candidate has ignored a critical adjacency or other relationship.

The corrective action for each of these issues will depend on the seriousness of the error and when the mistake is discovered. If there is time, one should rectify the design by returning to the point at which the error occurred and begin again from there. If it is late in the exam and time is running out, there may simply be insufficient time to correct the problem. In that case, continue on with the remainder of the exam and attempt to provide the most accurate solutions. The best strategy, of course, is to avoid critical mistakes in the first place, and those who concentrate and work carefully will do so.

WORKING UNDER PRESSURE

The time limit of the graphic portions creates subjective as well as real problems. This exam generates a unique psychological pressure that can be harmful to performance. While some designers thrive and do their best work under pressure, others become fearful or agitated under the same conditions. It is perfectly normal to be uneasy about this important event; and although anxiety may be a common reaction, it is still uncomfortable.

Candidates should be aware that pressure is not altogether a negative influence. It may actually heighten awareness and sharpen abilities. In addition, realize that, as important as this test may be, failure is not a career-ending event. Furthermore, failure is rarely an accurate measure of design ability; it simply means that you have not yet learned how to pass this difficult exam.

EXAMINATION ADVICE

Following is a short list of suggestions intended to help candidates develop their own strategies and priorities. We believe each item is important in achieving a passing score.

The *ARE Guidelines,* available from the NCARB Web site, also lists suggestions for examination preparedness.

- **Get an early start.** Begin preparation early enough to develop confidence by the time you are scheduled to take the exam. Arrive at the exam site early and be ready to go when the test begins.
- **Complete all vignettes.** Incomplete solutions risk failure. Complete every problem, even if every detail is not complete or perfect.
- **Don't modify the program.** Never add, change, improve, or omit anything from a program statement. Never assume that there is an error in the program. Verify all requirements to ensure complete compliance with every element of the program. If ambiguities exist in the program, make a reasonable assumption and complete your solution.
- **Develop a reasonable solution.** Since most vignettes generally have one preferred solution, solve the problem in the most direct and reasonable way. Never search for a unique or unconventional solution, because on this exam, creativity is not rewarded.
- **Be aware of time.** The strict time constraint compels you to be a clock-watcher. Never lose sight of how much time you are spending on each vignette. When it is time to proceed to the next problem, quit and move on to the next vignette.
- **Remain calm.** This may be easier said than done, because this type of experience often creates stress in even the most self-assured candidate. Anxiety is generally related to fear of failure. However, if a candidate is well prepared, this fear may be unrealistic. Furthermore, even if the worst comes to pass and you must repeat a division, all it means is that your architectural license will be delayed for a short period of time.

BUILDING SECTION VIGNETTE

INTRODUCTION

The Building Section vignette is a one-hour-long problem that tests your ability to produce a complete building section. Candidates are provided with floor plans of a two-story structure together with architectural, structural, and mechanical requirements. The program also includes a description of materials and construction methods, as well as vertical dimensions, all of which must be coordinated with the floor plans to produce the required section.

VIGNETTE INFORMATION

The Building Section vignette is presented on a series of computer screens that include the following:

- **Vignette Directions**—This screen directs candidates to establish a grade line and draw a complete schematic section that corresponds to the section cut line shown on the plans. One is advised to produce a section that accurately reflects the dimensions, structural relations, and spatial relations shown in the plan.
- **Program**—This screen lists the information one must have in order to produce the building section, such as the basic materials and construction methods, separation wall requirements, ceiling heights, parapet heights, and the amount of space required between a ceiling and the structure above to accommodate the mechanical and lighting components. For example, the program may state that nine inches of clear space is required between the bottom of ducts and the finished ceiling to permit the installation of recessed light fixtures.
- **Tips**—The suggestions on this screen are especially helpful to those unfamiliar with the exam's computer program. For example, only one floor level at a time may be viewed, and it is selected by clicking on the *layers* tool. One may view either level of the architectural plan only, the plan with *either* the structural framing or mechanical ductwork overlaid, or the plan with *both* the structural and mechanical elements included. One is also advised that both the joist depth and spacing must be

selected before the joists can be added to the section; and to relocate the joists, one must use the *move group* tool, not the *move, adjust* tool that is used to relocate a single joist. Finally, candidates are warned that unless they draw a grade line, their solution will not be graded.

Much of this information may have little meaning to a first-time test taker, but a few moments after starting this vignette, it all makes sense and is quite helpful. This screen also discusses the benefits of the *zoom* tool, which makes possible more accurate drawings; the *full-screen* cursor, used to align elements; and the *sketch* tool, which is used to establish heights and clearances.

- **Work Screen**—This screen, on which you are required to draw your solution, is accessed by pressing the space bar on the computer keyboard. This screen contains plans of the building's two levels (viewed one at a time, never together) with the section cut line indicated on both plan levels so that one always knows the location and direction of the required section. Candidates are required to construct the building section directly over the floor plan, and the complete array of computer tools enabling one to do this is typically located on the left side of this screen.

DESIGN PROCEDURE

The Building Section vignette tests a candidate's ability to create a schematic building section based on the given floor plans and construction information. When one clicks on the *draw* tool, which is the top tool icon at the left side of the work screen, an on-screen list appears that includes each required element of the section. Clicking on *slab on grade*, for example, enables one to draw a slab in section, which is automatically the proper thickness. When other elements, such as *joists* and *ducts*, are brought up on the screen, they are accompanied by sublists from which one must also choose the depth, width, or spacing. Incidentally, each element is color coded, so that it is relatively easy to distinguish among elements, as well as differentiate the section drawing you create from the existing floor plan over which it is superimposed. The left side of the section should be placed along the line that is designated "exterior face of section."

The development of your solution should proceed logically from the foundation level to the roof. For example, one's first step must be to establish the grade line, which will typically be horizontal. If a grade line is not horizontal, clear contour lines will be provided. Note: it will be important for the candidate to be comfortable with how contour lines affect the grade line in section view. This should be located directly over the floor plans, just above or just below the section cut line, so elements intersected by the cut line can be more easily identified. Beginning at the grade line, you would then establish the floor slab and footings at their appropriate elevations. Next, you would establish exterior walls, the first story ceiling height, the necessary interstitial space between the ceiling and the floor above, and the second floor level. Similarly, you would establish the ceiling height of the second story, create the necessary roof construction, and add any required parapets above the roof level. When the exterior building envelope is completed, you would add the minor construction details, such as interior partitions and mechanical elements.

Two further details of this vignette bear mention. First, the sites on which structures sit are generally level, as are ceilings and roof surfaces. Therefore, roof slope and drainage do not present problems. Secondly, bearing walls and other fire-rated walls must extend from a floor slab to the slab above. Other non-bearing partitions extend from a floor slab to approximately the ceiling line. This allows the ceiling space to remain unobstructed and suitable for use as a return-air plenum.

DESIGN APPROACH

The Building Section vignette is principally a straightforward exercise in following directions. One does not even have to know or remember every required construction element, since each appears on screen with a simple click of the *draw* tool. If you incorporate the given dimensions, materials, and treatments, you will produce a passing solution. On the other hand, any deviation from specific directions or any display of originality can only lower your final grade. This is not a vignette that requires creativity; rather, one must display a rigid adherence to the explicit instructions provided. Your approach, therefore, must be to carefully construct a building section, one detail at a time, and to use the programmed information as a check list of components that must be incorporated. In all this, one must always be aware of the section cut line location. It is equally important to establish all vertical dimensions and clearances at the minimum values that satisfy the requirements of the program.

PROBLEM AREAS

Even the most elementary problems, however, can cause trouble to those who are inexperienced. For example, the program requires a specific frost depth, and the candidate must determine the depth of footings relative to this frost depth. However, there is not complete agreement as to what that footing depth should be. While it is common practice to place the bottom of the footing at the frost depth line, it is also permissible for the *top* of the footing to coincide with the frost line. Both solutions are generally acceptable, but if footings are shown excessively deep, a candidate's score will be lowered.

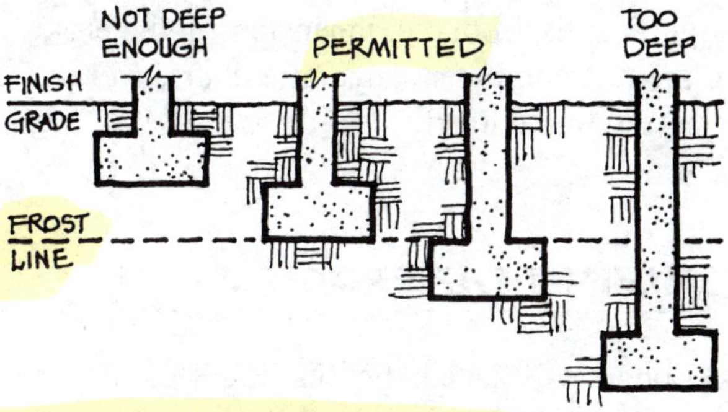

FOOTING/FROST LINE RELATION

Figure 15.1

Interior footings, which do not support exterior walls, are not affected by frost, since they are protected because of their location. Normally, the tops of interior footings align with the bottom of the ground floor slab. Interior footings are not required for non-load bearing walls. Interior load-bearing walls should have interior footings. The load-bearing wall will typically extend to the footing with the slabs abutting it.

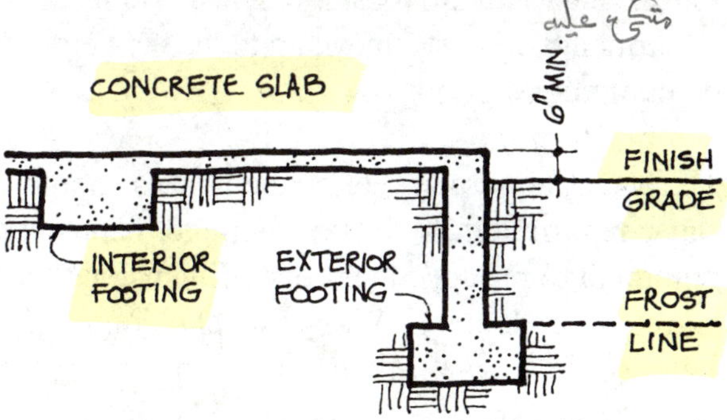

TYPICAL FOOTING DEPTHS

Figure 15.2

Another problem concerns the depth of the space between a ceiling and the floor or roof line above, often referred to as *interstitial space*. Within this total depth, one must allow sufficient space for mechanical ducts, light fixtures, structural members, and any other elements of the construction assembly. However, one does not want to provide more space than is required, which would reduce the structural efficiency of the building. The amount of interstitial space is important because it affects the overall story heights. If your solution deviates significantly from the dimensions in the program, the computer program will discover the mistake immediately.

SOLVING THE PROBLEM

As previously stated, the Building Section vignette must be solved logically, one element at a time. Exterior walls are aligned directly over the given plan, which establishes the width of your section. Heights are developed from the lowest floor level upward, with each incremental height based on a specific program requirement. At each step along the way, candidates must verify every detail given in the program. For example, if the section cut line passes through a 12-inch-deep HVAC duct, but at another location in the same room a duct is 18 inches deep, you must be certain that your interstitial space allows for the duct with the greater depth. This must be confirmed before you move up to the next element of your section.

The accurate portrayal of element sizes should cause no problem, since the computer is programmed to produce the proper dimensions for all required elements. When you click on floor slab or exterior wall, for example, the computer program automatically draws the slab or wall at the suitable thickness. On the other hand, when you choose the size and spacing of joists, they must match exactly the joist sizes shown on the plans.

FINAL DRAWING

Because most of your building section is projected directly from the given floor plans, you must be certain to follow the path of the section cut line very carefully. With the profusion of wall lines, framing lines, and duct lines, it is easy to lose one's way and make mistakes.

Candidates may be surprised by the simplicity of the final section. Although the vignette will generally take the full one hour allowed, the resulting drawing is relatively simple. It is also devoid of textures, room names, notes, dimensions, and elevations, although one may verify most vertical dimensions by using the *sketch* tool. The completed section, however, represents a multitude of important details and individual decisions. When your building section is completed, you should return to the program to verify every piece of information listed to be certain no detail has been omitted.

KEY COMPUTER TOOLS

- **Layer Tool**—This is a helpful tool in many vignettes, but the candidate will find it especially helpful in the Building Section Vignette.
- **Draw Tool**—Note that in this vignette, a 4-inch deck is automatically drawn on top of joists one selects.
- **Ortho Tool**—This tool helps the candidate keep elements in line with each other—particularly columns.
- **Zoom Tool**—Again, one will find this an essential tool in almost any vignette. In the Building Section Vignette it is particularly useful in assuring accuracy.
- **Sketch Tool**—Use this tool to help set floor elevations and clearance space requirements.
- **Full Screen Cursor**—Along with the Ortho Tool, the Full Screen Cursor helps the candidate keep elements lined up accurately.

VIGNETTE BUILDING SECTION

Introduction

Our example for this problem is based on a Building Section vignette created for a paper-and-pencil mock exam. Following are a detailed solution and an explanation, which illustrate the procedure one must follow to achieve a passing solution. Although the discussion appears lengthy, the actual process takes little time. Candidates are urged to follow the sequence of activities so that they may apply these principles to any building section vignette they may encounter on the exam.

The Exam Sheet

Shown are the Building Section program and plans.

The program describes a structure with masonry-bearing walls, steel joists, and a flat roof. The design and construction information are presented in a logical sequence, beginning with foundation information and proceeding with floor and roof sections, ceiling heights, and parapets. Also included in the program requirements is information about mechanical and lighting elements to be considered in preparing the section drawing.

Below the program, we see a horizontal line labeled *LOWER FLOOR - EL: 48'-0"*. This is the established level at which we must begin our solution, and in fact, one of our first tasks is to plot the existing grade relative to this lower floor line.

We may also review the partial upper and lower floor plans of the given office building. Candidates should be immediately aware of several important details. First, there is a finished grade differential of about two feet across the width of the structure along the section cut line. Next, there is a two-story Conference Room with a roof framed perpendicular to other roof framing in the structure. Finally, a portion of the Director's Office on the upper floor overhangs the Secretary's Office below. None of these conditions is particularly unusual, nor do any of them present a serious problem. However, with just one hour to complete the section, one must be aware of any individual element that could slow you down.

Developing the Framing

The program states that the minimum ceiling height is 9'-0", except at the Conference Room, where the ceiling height is 16 feet, and that the structural system includes steel joists. On the lower floor plan, we note that the floor framing consists of 24-inch-deep joists spaced 3'-0" on center. Before drawing the joists in the section, however, we must determine at what elevation to place them. This level must consider the joist depth, the space below the joists to accommodate mechanical ducts, and the space allowed for lighting fixtures between the ducts and the ceiling. In this case, the duct cut by the section line is 12 inches deep, and the required allowance for lighting fixtures is 9 inches. Thus, the bottom of the upper level floor slab must be 24" (joist) + 12" (duct) + 9" (lighting) = 45 inches above the ceiling line.

Developing the Foundation

We begin our solution by projecting downward and drawing the right and left exterior walls, which should be the same thickness and distance apart as they appear in the lower floor plan. We have now established the right and left limit lines of the structure at the lower floor level. At the left side, we see that the existing

VIGNETTE BUILDING SECTION

Using the lower floor line shown, draw a schematic building section corresponding to the section cut lines indicated on the floor plans. Draw only the elements intersected by the section cut line, as well as joists in elevation immediately adjacent to the cut line. Your section should accurately reflect the dimensions, structural relationships, and all other components indicated in the program and plans.

Program Requirements

- The structural system consists of steel joists on masonry-bearing walls with continuous 12" × 24" concrete spread footings and concrete slab on grade.
- All floor and roof slab sections should be shown 6 inches thick. Ceilings may be shown with a single line.
- All ceilings and roofs are flat.
- The Conference Room ceiling height is 16 feet. The ceiling height of all other spaces is 9 feet.
- Ceiling spaces will be used as return-air plenums, and all such spaces must be the minimum dimension required to accommodate structural, mechanical, and lighting components.
- All ducts must be placed below joists.
- Provide 9 inches of clear space minimum between the bottom of ducts and finish ceilings to accommodate light fixtures.
- The frost depth is 3 feet below natural grade.
- Parapets must extend 3 feet above roof decks at roof edges and at any change in roof elevation.
- All windows are 4-1/2 feet high with heads 7 feet from the floor.

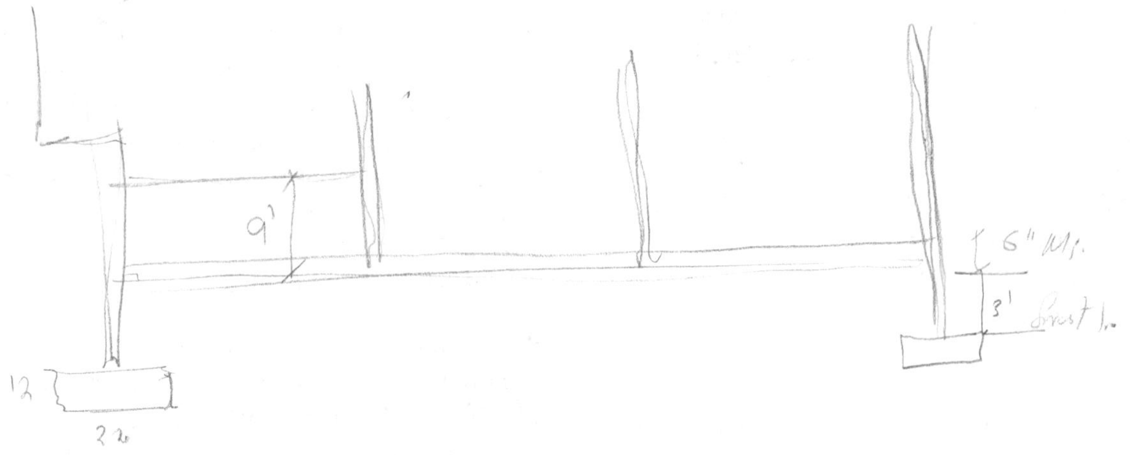

9 + 15

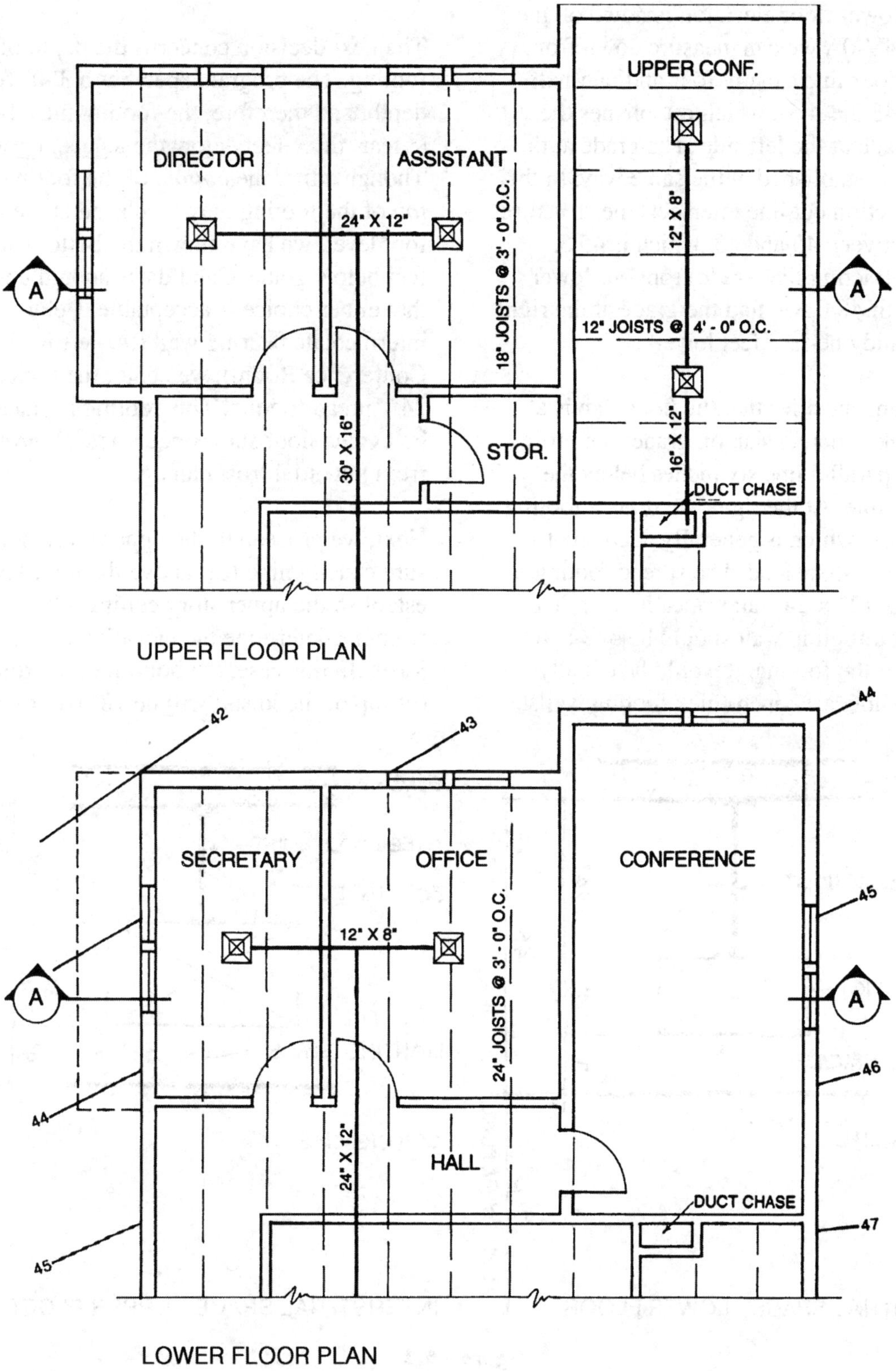
UPPER CONF.
DIRECTOR
ASSISTANT
24" X 12"
12" X 8"
18" JOISTS @ 3' - 0" O.C.
12" JOISTS @ 4' - 0" O.C.
A
A
30" X 16"
16" X 12"
STOR.
DUCT CHASE
UPPER FLOOR PLAN
42
43
44
SECRETARY
OFFICE
CONFERENCE
12" X 8"
24" JOISTS @ 3' - 0" O.C.
45
A
A
46
44
24" X 12"
HALL
DUCT CHASE
47
45
LOWER FLOOR PLAN

exterior grade at the section cut line is midway between contours 43 and 44, or elevation 43.5. Since our lower floor elevation is noted on the section as 48'-0", we can measure down from the finish floor line exactly four and one-half feet (48 – 43.5 = 4.5), which establishes the existing grade at the left side. The grade at the right side is established in the same way. In this case, the section cut line intersects the structure midway between 45 and 46, which is 45.5. Again, subtracting that grade from our lower floor level of 48.0, we find the grade at the right to be two and one-half feet lower.

The program specifies that the floor slab is a 6-inch-thick concrete slab on grade; therefore, we draw a parallel line six inches below the given floor line. At the right, we draw a footing wall 9" thick, which is generally adequate to support a two-story load. The spread footing is specified as 12" × 24", and since it is a rule of thumb that a footing wall should be one-half the width of the footing, it would be equally correct to show a 12-inch-thick footing wall. Either way, the 24-inch-wide continuous spread footing is centered below the footing wall.

The next decision concerns the depth of this footing. The program specifies a 3'-0" frost depth, and therefore, the footing must be at least three feet below the adjacent grade. Though either the *bottom* of the footing or the *top* of the footing may be placed at the three-foot level, we have shown the bottom at three feet below grade. Candidates should understand that either choice is acceptable. Below the intermediate bearing wall (the left wall of the Conference Room), we center the typical 12" × 24" spread footing. This footing is placed just below the floor slab, since it is well protected from potential frost damage.

Next, we proceed to the upper level and measure exactly nine feet above the floor level to establish the upper story ceiling. Once again, we must determine the elevation of the roof joists. In this case, the bottom of the roof slab (or top of the joists) will be 18" (roof joists) +

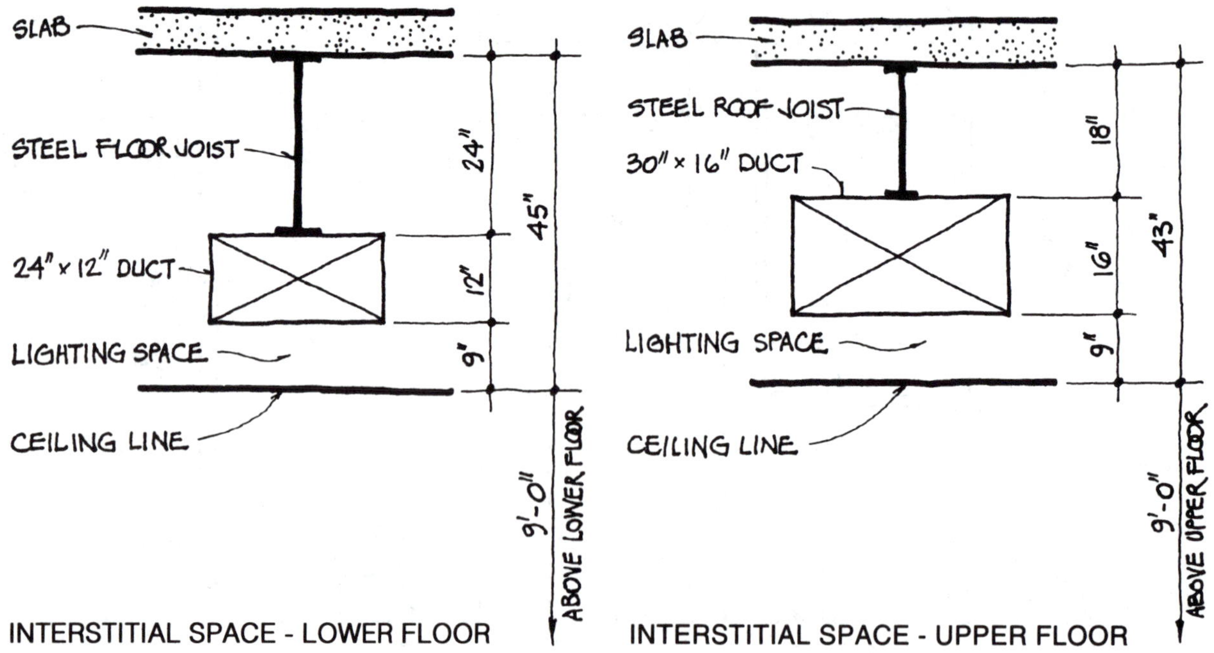

Figure 15.3

16" (duct) + 9" (lighting) = 43 inches from the ceiling line. We have assumed that the lower and upper floor ceilings have no dimension, since the program requires ceilings to be shown with a single line.

We next add the six-inch-thick roof slab and project the exterior walls on both sides of the offices to form parapet walls extending three feet above the roof deck. The program states that all roofs are flat, and since there are no dotted lines on the upper floor plan, we have no roof overhangs to deal with.

Conference Room

The section through the Conference Room is drawn similarly. However, the space is higher, and the framing members run perpendicular to those in the other areas. We establish the ceiling line 16 feet above the lower floor slab on grade. From this point, we must allow 9" for lighting, 12" for ducts, and 12" for the joists, which totals 33 inches of space between the ceiling line and the top of the joists (bottom of the roof slab).

Candidates may wonder why we allow 12 inches for ducts when the section cut line actually passes through an 8-inch duct. If you refer to the upper floor plan, you will see that the duct running from the duct chase into the Conference Room is 12 inches deep up to the first register. This is the depth that one must allow for ducts. If the depth were governed by the 8-inch-deep duct, a break in the room's ceiling would be necessary where the 8-inch duct becomes a 12-inch duct. Therefore, the ceiling depth must always allow for the deepest duct in a space, which is not necessarily the duct that is cut by the section line. Just as before, we add the roof slab and extend the exterior wall three feet above the roof slab to form the final parapet.

The Overhanging Floor

A planning element is rarely included in a vignette for absolutely no reason. That certainly applies to the portion of the upper floor that overhangs the lower floor, as indicated in both floor plans. The overhang is actually not much of a problem, but it was included to test a candidate's understanding that the exterior walls for the two building stories do not align.

Furthermore, the upper wall may stop at the upper floor and expose the bottom of the floor slab or, as shown in our suggested solution, the space may be enclosed down to the ceiling line of the lower level. These alternatives are illustrated in Figure 15.4.

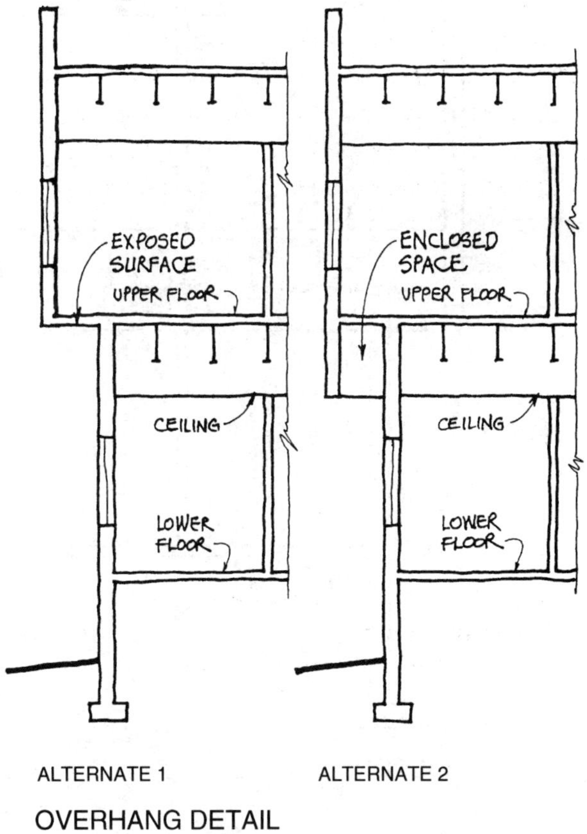

Figure 15.4

Other Elements

We have nearly completed the building section, but a few details remain. First, we must draw the non-bearing interior partitions through which the section line passes. On the lower floor, such a partition separates the Secretary and Office spaces, and aligned directly above, a partition separates the Director and Assistant spaces. These two partitions run from the floor level to the ceiling line. They do not extend to the slabs above, because the program states that ceiling spaces are used as return-air plenums, and they must, therefore, remain unobstructed. On the other hand, if a partition were fire-rated, such as an exit corridor wall, the partition would have to extend above the ceiling to the underside of the floor above.

Next, we must indicate the HVAC ducts cut by the section line. This is a relatively mechanical process, but one must be absolutely certain to identify the correct duct. For example, one can see on the lower floor plan the horizontal duct labeled 12" × 8"; however, it is the vertical duct

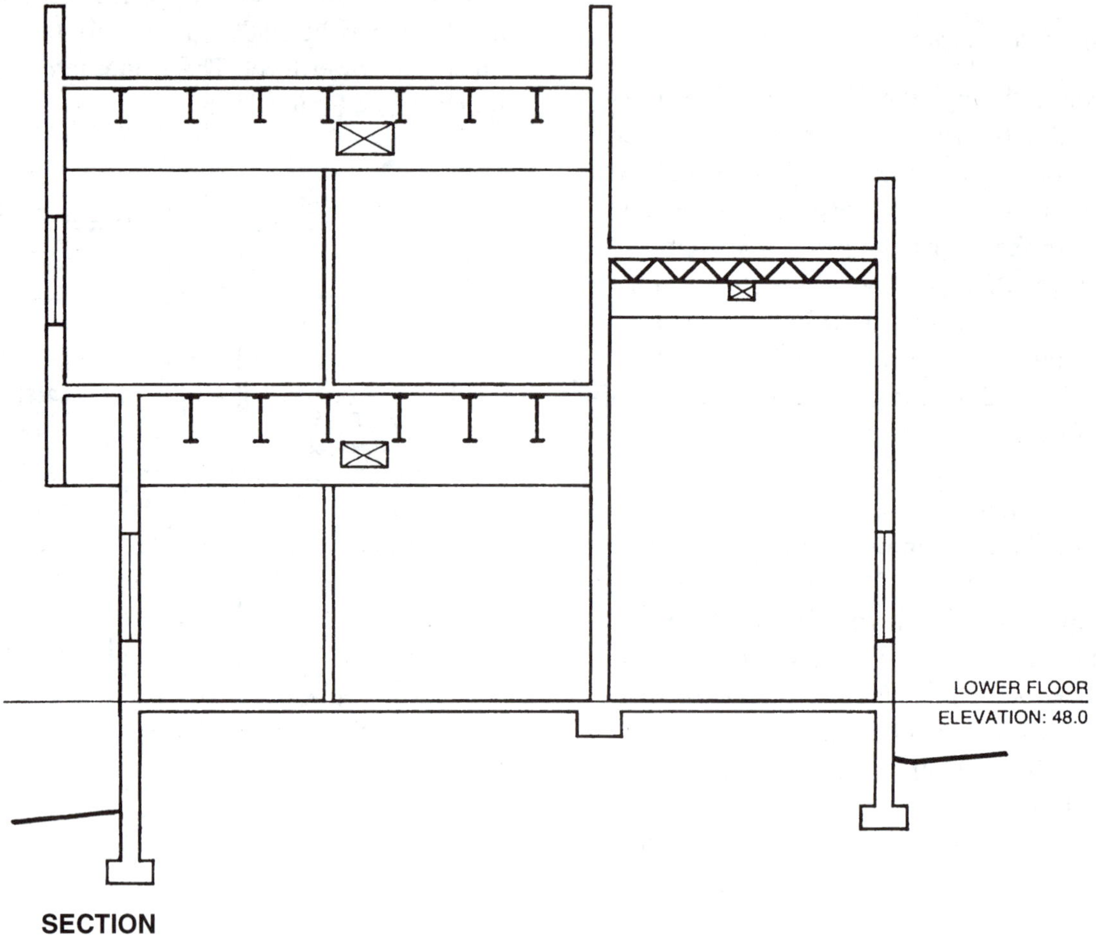

BUILDING SECTION VIGNETTE SUGGESTED SOLUTION

Figure 15.5

labeled 24" × 12" through which the section line actually passes. On the actual exam, with the profusion of lines that make up the floor plans, one must remain focused on the problem to avoid such errors.

The program mentions that windows are 4-1/2 feet high with heads 7 feet from the floor. The section line cuts through three windows on the two levels, and each of these is drawn identically. Again, there is nothing complicated about this requirement, but it must be acknowledged and drawn correctly.

The program mentions the clear space required for lighting components, but no details about the actual fixtures are furnished. Thus, we must assume that lighting fixtures are not included in the building section.

Final Presentation

At this point we have satisfied every major requirement, and our building section is complete. There are no further details that remain, since it is unnecessary to identify spaces or provide notes for clarification. Additionally, no dimensions are required to be shown. In the actual exam, the computer program automatically computes the critical elevations and grades one's solution accordingly.

Before leaving this problem, we should verify that all elements intersected by the section cut line are represented on the drawing. These would include all the foundation members, walls, slabs, framing members, and ducts. Any element that occurs beyond the cut line does not have to be shown, except for joists in elevation immediately adjacent to the cut line.

Conclusion

The Building Section vignette is relatively uncomplicated. One might discover minor problems involving the footing depth, partition height, mechanical ducts, and to some extent, the framing; but, this is essentially a fairly simple problem. On the actual computer exam, more serious problems might involve insufficient familiarity with the computer tools and not having enough time to complete the vignette within the strict time limit. That is why we stress the importance of practicing with the exam software until one develops the necessary technique and speedy performance needed to succeed.

The successful strategy for problems such as this is to keep one's eye on the complete solution and not become distracted by the multitude of details of which it is composed. Since time is of the essence, one must solve the larger aspects of the problem and complete the less important details as time allows. Those who are familiar with conventional construction methods will, of course, work faster than those who are not. And for those who have less construction knowledge, we advise further study on the subject.

(MasterFormat 95 Numbering)

SECTION 07160—BITUMINOUS DAMPPROOFING

Part 1—General

1.01 RELATED DOCUMENTS

A. Drawings and general provisions of Contract, including General and Supplementary Conditions and Division 1 Specification Sections, apply to this Section.

1.02 SUMMARY

A. Section Includes:

1. Substrate preparation.
2. Bituminous dampproofing.

B. Related Sections:

1. Unit masonry: Section 04200.
2. Building insulation: Section 07210.
3. Flashing: Section 07600.

1.03 SUBMITTALS

A. Product Data: Submit technical product information and installation instructions which demonstrate that products comply with project requirements.

1.04 DELIVERY, STORAGE, AND HANDLING

A. Deliver dampproofing materials to project site in factory-sealed containers.

B. Store materials in dry, well-ventilated space.

1.05 SITE CONDITIONS

A. Install dampproofing only when site weather conditions are acceptable per manufacturer's recommendations.

B. Ventilation: Provide sufficient ventilation during application and curing of dampproofing to prevent buildup of toxic or flammable fumes.

Part 2—Products

2.01 DAMPPROOFING MATERIALS

A. Fibrated Dampproofing: Cold-applied, asphalt emulsion semi-mastic of spraying or brushing (medium) consistency, meeting the requirements of ASTM D 1227, Type IV; asbestos free.

1. Products: Provide one of the following:

a. "A-H Semi-Mastic Emulsion, Asbestos Free"; Anti Hydro Company, Inc.

b. "No. 220 AF Fibrated Emulsion Dampproofing"; Karnak Corporation.

c. "Hydrocide 700B Semi-Mastic"; Sonneborn Building Products Division/ChemRex, Inc.

d. "Sealmastic, Type 2"; W.R. Meadows, Inc.

2.02 INSTALLATION ACCESSORIES

A. Detailing Mastic: Asphalt-based plastic roof cement, trowel consistency, meeting the requirements of ASTM D 4586; asbestos free.

Bituminous Dampproofing

07160-1

Part 3—Execution

3.01 EXAMINATION

A. Verify that surfaces are smooth, sound, clean, and dry, and that elements which will penetrate dampproofing have been completed and are rigidly installed.

3.02 PREPARATION

A. Remove ridges and projecting rough areas.

B. Fill cracks, holes, and irregularities with detailing mastic as recommended by membrane manufacturer.

3.03 INSTALLATION

A. General: Comply with dampproofing manufacturer's instructions for handling, preparation, application, and protection of dampproofing materials.

B. Dampproofing: Apply dampproofing to entire exterior face of concrete masonry back-up wythe (inner wythe) for all cavity wall construction and other locations indicated.

1. Apply 2 coats of bituminous dampproofing at manufacturer's recommended coverage rate.
2. Apply coatings using application method best suited for obtaining full, uniform coverage of surfaces to be coated.

C. Coordinate dampproofing work with installation of the following materials:

1. Brick masonry.
2. Masonry flashings: Ensure watertight flashing installation.
3. Cavity wall insulation.

3.04 PROTECTION AND CLEANING

A. Take measures required to protect completed dampproofing after installation.

B. Clean spillage and soiling from adjacent surfaces using cleaning agents and procedures recommended by the manufacturer of the surface.

END OF SECTION 07160

EXAMPLE OF SPECIAL CONDITIONS

SPECIAL CONDITIONS

1. Soil Borings

A. Subsurface soil investigations have been made at the site, and logs of the test holes are available in Architect's office to assist in ascertaining character of material to be encountered. Contractor shall make his own interpretation of the data, since Owner or Architect or Soils Engineer in no way guarantees adequacy or accuracy of the data or that data are representative of all conditions to be encountered.

2. Maintenance of Traffic

A. The Contractor shall maintain traffic on all streets adjacent to or leading to the site. Where construction operations interfere with the free movement of traffic, controls, flagmen, or similar devices to efficiently control traffic movement shall be provided.

3. Protections of Finishes

A. The various materials, work, equipment, and finishes provided by the several trades are all to be protected from other operations or work so that all items are in perfect condition at the time project is turned over to the Owner. The final responsibility for this protection rests with the General Contractor even though various sections of the specifications may contain specific comments or precautions about protection.

4. Reference Documents

A. Building Code: Reference to *Code,* or to *Building Code,* or to specific code sections, not otherwise identified, means Uniform Building Code—1997 Edition together with additions, changes, amendments, and interpretations in force on the date of the contract. Nothing in the drawings or in these specifications is to be construed as requiring or permitting work that is contrary to code requirements.

B. Standard Specifications: Standard specifications, codes, rules, and regulations referred to in these specifications by basic name of designation only, shall be considered to be of the latest issue with all amendments, as of the date of these specifications. Whenever a date of issue is shown, that particular issue shall govern. Whenever the initials only of a Society or Association are used, the following organizations are referred to:

NBFU	National Bureau of Fire Underwriters
NFPA	National Fire Protection Association
ASTM	American Society for Testing and Materials
AASHTO	American Association of State Highway and Transportation Officials
AWWA	American Water Works Association
AWS	American Welding Society
ACI	American Concrete Institute
AISC	American Institute of Steel Construction

C. State Highway Specifications: Wherever in these specifications reference is made to *State Highway Specifications,* it shall be understood to refer to the Standard Specifications of the State of California, Department of Transportation, with all additions and revisions thereto.

5. Examination of Site

A. Data in these specifications and on the drawings are as accurate as possible, but are not guaranteed. The Contractor shall verify locations, levels, distances, and features of the site and related improvements that may affect the work. No allowance will be made in his behalf for any extra expense resulting from failure or neglect in determining the conditions under which work is to be performed.

6. Telephones

A. The Contractor shall provide and pay for all necessary temporary telephones. Telephones shall be in continuous service available at all times, free and unrestricted, to Architect's and Owner's representatives for calls in direct connection with the work. All such temporary telephones shall be removed upon completion of the work.

7. Tests and Inspections

A. All tests and inspections required by these Specifications will be performed by a person or testing laboratory employed by the Contractor with the prior approval of the Architect.

8. Layout

A. Principal lines, levels, and control stakes shall be established by a Registered Civil Engineer or Licensed Surveyor employed by the Contractor, except that upon submission of adequate proof by the Contractor that the Contractor is capable of performing this work and his assumption of full responsibility for its accuracy, the requirements may be waived by the Architect. Such lines and points shall be marked and maintained as required by the Architect for construction and inspection purposes.

9. Shop Drawings

A. Five copies of material lists, schedules, and brochures as required under the various sections shall be submitted to the Architect for approval. One reproducible copy of all drawings shall be submitted sufficiently in advance of the work to allow for selection of colors and patterns and coordination of the work shown with related work. Drawings shall be clearly marked with the name of the project and name of the Contractor. These requirements apply to original submittals and to any resubmittals that may be necessary, and shall be as further specified in General Conditions Article 4. The reproducibles will then be forwarded to the Contractor for the Contractor's use.

10. Dust Control

A. The work includes dust control as required to abate any dust nuisance on and about the site which is the result of construction activities. Dust control shall consist of the application of water by means of approved sprinkling equipment to the extent and in the amounts required at any time, including weekends and holidays, that dust control is necessary. The use of chemicals, oil, or other palliatives will not be permitted.

11. Cleaning

A. The Contractor shall at all times during the course of this contract keep the buildings, the Owner's premises, and the adjoining premises, including streets and driveways, free from accumulations of waste materials and rubbish caused by his employees or work or by the employees or work of his subcontractors. Rubbish shall not be buried on the Owner's premises. At completion of the work, or prior thereto if so directed, the Contractor shall remove from the buildings and the premises all tools, appliances, surplus materials, debris, temporary structures, temporary construction, and rubbish, and shall be responsible for clean-up of the work as well as work under other contracts affected by his work. Cleaning, polishing, sealing, waxing, and all other such finish operations noted on the drawings or required in the specifications shall be taken to indicate the required condition at the time of acceptance of work under the contract. At completion of work, the Contractor shall sweep and clean thoroughly, shall clean all glass, removing all paint, stains, etc. therefrom without scratching or injuring the glass, and shall leave the work bright, clean, and polished. All marks, stains, fingerprints, dust, dirt, paint, drippings, and the like shall be removed throughout the building; waxed work shall be polished, plumbing fixtures shall be washed clean, all hardware and other unpainted metals shall be cleaned and polished, all equipment and paint work shall be cleaned and touched up if necessary, and all temporary labels, tags, and paper coverings shall be removed, all to the approval of the Architect and the Owner. Finally, the exterior of the building, the grounds, approaches, railings, fences, equipment, planting, etc. shall be similarly clean and in good order at time of final acceptance of the building, with paint fresh, coatings unbroken, hardware clean and polished, and planting well established, neatly trimmed, and in good condition. If the Contractor, upon request by the Architect, does not attend to such cleaning with responsible promptness, the Owner may cause such cleaning to be done by others and charge the cost of the same to the Contractor or deduct the said cost from payments still due the contractor under the contract.

12. Safety

A. Precaution shall be exercised at all times for the protection of persons (including employees) and property. The safety provisions of applicable laws, building codes, and construction codes shall be observed. Machinery, equipment, and all hazards shall be guarded or eliminated in accordance with the safety provisions of the latest edition of the manual of Accident Prevention in Construction, published by the Associated General Contractors of America, to the extent that such provisions are not in contravention of applicable law. The Contractor shall protect all hazards with adequately constructed guard rails or barricades and shall provide lanterns, warning lights, and the like, as necessary. The Contractor shall eliminate all attractive nuisances from the work and from the site. To this end, the Contractor shall so dispose, store, guard, and protect the premises and all work, materials, equipment, and both permanent and temporary construction as to preclude the unauthorized use thereof by children or others and particularly to eliminate possible consequent injury to all unauthorized persons. In no case shall the Owner or Architect be responsible for construction means, methods, techniques, sequences, or procedures or for safety precautions and programs in connection with the work, nor shall the Owner or Architect be responsible for Contractor's failure to employ proper safety procedures.

13. Scaffolding and Hoists

A. The Contractor shall furnish and maintain hoists, staging, rigging, scaffolding, and runways required in the prosecution of the work under this contract. Such temporary work shall be erected, equipped, and maintained in accordance with statutes, laws, ordinances, rules, or regulations of the State of California or other authorities and insurance companies having jurisdiction and shall be approved by the State of California.

14. Dewatering

A. The Contractor shall furnish and maintain all pumps or other dewatering devices which may be required by the work under this contract.

15. Construction Fence

A. If no suitable fence exists, Contractor shall erect an 8-foot-high chain link fence around the site, with necessary gates. Remove at project completion. If suitable fencing exists, the Contractor shall be responsible for any additional fencing, gates, or relocation for construction purposes, and for any fees or permit costs during the construction period.

16. Job Sign

A. Furnish and erect two 96″ × 96″ job signs painted and lettered to identify the Project, the Owner, the Architect, the Contractor, and four Consulting Engineers. Mount on posts and brace as directed by the Architect before commencing.

17. Insurance

A. As per Article 11 of the General Conditions, the Contractor's insurance limits of liability for bodily and personal injury, occupational sickness or disease, or death shall not be less than $500,000.00 for each person, and subject to that limit for each person, not less than $1,000,000.00 for each occurrence. The limit for property damage liability shall not be less than $500,000.00.

18. Index of Drawings

A. The list of drawings is shown on Sheet 1 - Sheet Index.

19. Project Office

A. Provide and maintain, for the duration of the Contract, a project office, complete with heat, light, ventilation, and convenience outlets. Office shall be of sufficient size for Contractor's personnel and operations and shall provide desk space for use of the Architect and inspection personnel.

20. Temporary Toilets

A. Provide temporary toilet facilities for all personnel employed on the project. Toilets shall be maintained in a clean and sanitary condition at all times. Remove at project completion.

21. Construction Water and Power

A. Contractor shall make arrangements for all water and power required for the project. Provide all temporary lines and pay all bills. Remove temporary facilities at project completion.

22. Premises and Fees

A. All permits, assessments, and fees shall be paid for by the General Contractor including but not limited to:

1. Electric permit and service
2. Plumbing permit and service
3. Building permit
4. Street cut fees
5. Sewer connection charges
6. Water connection fees

23. Utilities

A. Contractor to notify all applicable utility companies, including those listed on the accompanying Drawings, and to verify all utility grades, locations, and crossings before beginning construction. Start of construction without a written report to the Architect of any discrepancies assumes acceptance of the site conditions by the Contractor and claims for extra work involving these utilities will be the responsibility of the Contractor.

B. The Contractor is expected to use reasonable caution when trenching in the vicinity of lines shown, and in the event of damage to them, will be responsible for their prompt repair or other acceptable temporary means of maintaining service until permanent repairs can be made.

C. All work on this project shall be so conducted as to permit utility companies to maintain their services or install additional facilities without interruption.

GLOSSARY

The following glossary defines a number of terms, many of which have appeared on past exams. While this list is by no means complete, it comprises much of the terminology with which candidates should be familiar. You are therefore encouraged to review these definitions as part of your preparation for the exam.

A

Addenda Statements or drawings that modify the basic contract documents after the latter have been issued to the bidders, but prior to taking of bids.

Addition (to the contract sum) An amount added to the contract sum either by an accepted additive alternate or by change order.

Advertisement for Bids Published public notice soliciting proposals for a construction project.

Agent One who acts on behalf of another.

Alternate Bid Proposal by bidder for amount to be subtracted from, or added to, the base bid if the corresponding change in the work is accepted.

Americans with Disabilities Act (ADA) Federal civil rights legislation that establishes the right of persons with disabilities to equal access to sites and buildings, and sets design guidelines for its implementation.

Application for Payment Contractor's written request for payment of amount due on account of work completed and/or materials suitably stored on the site. Also called Payment Request.

Approved Equal Material or method that is approved by the architect as being equivalent to what was originally specified.

Arbitration An alternative to litigating a dispute, in which one or more arbitrators hears the evidence and renders a decision.

As-Built Drawings See Record Drawings.

B

Barrier-free Provisions Regulations that provide for accessibility to buildings and sites for persons with disabilities.

Base Bid The sum of money stated in the bid for which the bidder offers to do the work, not including any alternate bids.

Basic Services The architectural services normally required for a building project, usually consisting of schematic design, design development, construction documents, bidding or negotiation, and construction contract administration.

Bid A proposal by a contractor to do the work required by the contract documents for a stipulated sum of money.

Bid Bond A surety bond guaranteeing that the bidder will sign a contract, if offered, in accordance with his or her proposal.

Bidding Documents The invitation to bid, instructions to bidders, bid form, and the contract documents.

Bond See Surety Bond.

Bonus and Penalty Clause A provision in the construction contract for payment of a bonus to the contractor for completing the project prior to a specific date, and for a charge (penalty) against the contractor for failing to complete the project by that date.

C

Cash Allowance An amount included in the contract sum to cover the cost of certain items not specified in detail. Hardware and other finish items are often handled in this manner.

Certificate for Payment A statement by the architect informing the owner of the amount due the contractor on account of work completed and/or materials suitably stored.

Certificate of Occupancy A document issued by the governing authority stating that a building complies with applicable laws and permitting occupancy for its designated use.

Certificate of Substantial Completion A document prepared by the architect stating that the work is substantially complete, thereby establishing the date of substantial completion. It generally fixes the time within which the contractor must complete the unfinished work listed.

Change in Services (of the architect) Professional services rendered by the architect, upon the owner's request, in addition to the basic services identified in the owner-architect agreement. Also called Extra Services.

Change Order A written order to the contractor, prepared by the architect and signed by the owner, contractor, and architect, which authorizes a change in the work, the contract sum, or the contract time.

Clerk of the Works The owner's job site representative.

Completion Bond A bond obtained by the contractor or owner, which guarantees that the project will be completed free of liens.

Comprehensive Services The architect's basic services expanded to include certain additional services, such as programming, land use studies, etc.

Construction Budget The sum established by the owner as available for construction of the project. See Project Budget.

Construction Change Directive A document, signed by the owner and the architect, which authorizes a change in the work, the contract sum, or the contract time. Used when there is not complete agreement on the terms of a change order. See Change Order.

Construction Documents Working drawings and specifications.

Construction Management (CM) Management services performed by the architect or others, over and above normal architectural services, that contribute to the control of time and cost in the construction of a project.

Consultant An engineer or other specialist retained by the architect to provide specified professional services to the architect that the architect is required to provide to the owner under the owner-architect agreement.

Contingency An amount of money set aside in a budget to cover unanticipated expenses.

Contract Documents Working drawings, specifications, addenda, general conditions of the contract, supplementary conditions, and the owner-contractor agreement.

Contract Sum The total amount payable by the owner to the contractor, as stated in the owner-contractor agreement, for performing the work under the contract documents.

Contract Time The period of time within which the work must be completed, as established in the contract documents.

Cost Plus Fee Contract An agreement under which the contractor, or the architect, is reimbursed for his or her costs and, in addition, is paid a fee for his or her services.

Critical Path Method (CPM) A project management procedure in which all events and operations are charted in a way that establishes the optimum sequence and duration of operations.

CSI MasterFormat A systematic listing of construction trades, materials, systems, and administrative requirements published by the Construction Specifications Institute (CSI) for the purpose of organizing construction specifications.

D

Date of Substantial Completion The date, certified by the architect, when the construction is sufficiently completed, in accordance with the contract documents, so that the owner can occupy the project or specified area of the project for the intended use.

Davis-Bacon Act A federal law that requires the Department of Labor to set prevailing wages for government-financed construction projects.

Deposit for Bidding Documents A deposit of money required of each bidder as security for the bidding documents, to ensure the return of the documents by unsuccessful bidders.

Direct Expense Expense items directly incurred by, or attributable to, a specific project.

Direct Personnel Expense (DPE) Salaries and wages attributable to a specific project, plus benefits, such as employment taxes, insurance, sick leave, holidays, vacations, pensions, and similar contributions and benefits.

Division One The *General Requirements* Division of the specifications that establishes the administrative and procedural duties of the contractor, architect, and owner during construction.

Due Care The requirement that a professional exercise reasonable ability and judgment in a specific circumstance, the absence of which constitutes negligence. Also called Standard of Care.

E

Errors and Omissions Insurance See Professional Liability Insurance.

Estimate A forecast of probable costs, as opposed to a firm bid.

Express Warranty A legally enforceable promise made by the warrantor.

Extra An item of work involving additional cost. See Addition (to the contract sum).

Extra Services See Change in Services (of the architect).

F

Fast-Track Construction A construction technique by which construction on each element of a project is begun as soon as the design for that element is completed, without waiting for completion of the design for the entire project. Its principal objective is to shorten the overall construction time.

Final Completion The completion of all work in accordance with the terms and conditions of the contract documents.

Float The extra time available for a construction activity above its estimated time duration, without causing any delay of project completion.

G

General Conditions The part of the contract documents that states the rights, responsibilities, and relationships of the parties involved, usually by means of a standard document published by the American Institute of Architects.

General Contract The agreement between the owner and the contractor for the construction of a project.

Guarantee A legally enforceable assurance of the quality of materials and labor furnished for a project, or of the length of time that a project or a part thereof will perform satisfactorily. Also called Warranty.

Guaranteed Maximum Cost The amount established by agreement between owner and contractor as the maximum cost of performing specified work. Also called Guaranteed Maximum Price (GMP). See Upset Price.

H

Hold Harmless Clause See Indemnification.

I

Incentive Clause Clause in a cost plus fee contract between the owner and contractor in which the savings between the guaranteed maximum cost and the actual project cost are proportionally shared.

Indemnification A contractual obligation whereby one party agrees to guarantee another party against loss or damage from specified liabilities. Also called Hold Harmless Clause.

Indirect Expense Overhead expense, that is, expenses indirectly incurred and not chargeable to a specific project.

Instructions to Bidders Instructions in the bidding documents for preparing and submitting bids for a project. Also called Notice to Bidders.

Insurance Coverage by contract (insurance policy), whereby one party (the insurance company) agrees to indemnify or reimburse another (the insured) against loss from a specified hazard. See Liability Insurance and Professional Liability Insurance.

Invitation to Bid An invitation to a selected list of contractors soliciting bids for a project.

L

Labor and Material Payment Bond A bond guaranteeing to the owner that the contractor will pay for all labor and materials used for the project.

Letter of Intent A letter signifying intention to enter into a formal agreement and setting forth the general terms thereof.

Liability Insurance Insurance that financially protects the insured against liability on account of bodily injury or property damage sustained by another.

Liquidated Damages A sum chargeable against the contractor as reimbursement for damages suffered by the owner because of the contractor's failure to complete the work within a specified time.

Lowest Acceptable Bona Fide Bid The lowest bid that complies with all the stipulated requirements.

Lump Sum Contract See Stipulated Sum Contract.

M

Means of Egress The elements that comprise an exit to the outside, including aisles, corridors, doors, and stairways.

Mechanic's Lien A claim on property by those who furnish material or labor for the construction of a building. Clear title to the property cannot be obtained until the claim is settled.

Multiple of Direct Personnel Expense Agreement An agreement providing for payment for professional services based upon the direct personnel expense multiplied by an agreed factor.

N

Notice to Bidders See Instructions to Bidders.

O

Observation A term sometimes used for on-site examination of the contractor's work by the architect to determine in general if it is proceeding in accordance with the contract documents.

Option A choice given to the contractor to provide specified alternates without prior approval of the architect.

Overhead Expense See Indirect Expense.

P

Payment Request See Application for Payment.

Penalty See Bonus and Penalty Clause.

Percentage Agreement An owner-architect agreement under which the professional fee is based upon a percentage of the construction cost of the project.

Performance Bond A bond that guarantees to the owner that the contractor will perform the work in accordance with the contract documents.

Post-Completion Services Additional services rendered to the owner by the architect following actual completion of the project.

Prime Contractor Any contractor on a project who has a contract directly with the owner.

Professional Liability Insurance Insurance that financially protects an architect against claims for damages resulting from professional negligence. Also called Errors and Omissions Insurance.

Program A written statement of the owner's conditions and requirements for the project.

Progress Payments Payments made to the contractor during progress of the work on account of work completed and/or materials suitably stored.

Project Budget The sum established by the owner as available for the entire project, including the construction budget, land cost, equipment cost, financing cost, cost of professional services, and contingency allowances. See Construction Budget.

Project Manual The manual prepared by the architect for a project, including the technical specifications, bidding instructions and forms, general conditions, supplementary conditions, special conditions, and other legal and administrative documents.

Project Representative The architect's representative at the site, who assists in the general administration of the construction contract for a project.

Punch List A list of items to be corrected or completed, which is provided by the contractor and usually expanded by the architect based on a detailed inspection of the work prior to substantial completion.

R

Record Drawings Drawings revised to show changes made during construction. Sometimes erroneously called As-Built Drawings.

Release of Lien A legal document signed by a supplier of material or labor for a project, which releases his or her mechanic's lien against the property. See Mechanic's Lien.

Retainage An amount withheld from each payment to the contractor in accordance with the terms of the owner-contractor agreement.

S

Sample Material or assembly submitted for the architect's approval prior to manufacture or delivery to the project.

Schedule of Values A statement furnished to the architect by the contractor reflecting the amounts to be allotted for the principal divisions of the work. It serves as a guide for reviewing the contractor's periodic applications for payment.

Separate Prime Contract One of several owner-contractor agreements for a project, each of which provides for constructing a major portion of the work (general construction, electrical, mechanical, etc.) by a different contractor.

Shop Drawings Drawings prepared by contractor, subcontractor, manufacturer, or supplier, showing how specific portions of the work shall be fabricated and/or installed.

Sick Building Syndrome A term used to describe poor indoor air quality, which can lead to discomfort and possibly disease.

Single Prime Contract A contract for building construction under which one prime contractor is responsible for the entire project, in contrast to having separate contracts. See Separate Prime Contract.

Special Conditions Part of the contract documents, other than general and supplementary conditions, describing unique conditions of a project.

Specifications Part of the contract documents, comprising written descriptions of materials, construction systems, and workmanship.

Standard of Care See Due Care.

Statute of Limitations An ordinance that specifies the period of time within which legal action must be brought in order to obtain legal relief for damage or injury.

Stipulated Sum Contract An agreement under which the architect or contractor is paid a specific amount as the total fee for services performed. Also known as Lump Sum Contract.

Subcontractor One who has a contract with a prime contractor to perform a portion of the work.

Submittal A shop drawing, project data, or sample submitted by the contractor to the architect for review prior to incorporation in the work.

Subrogation The substitution of one entity for another with regard to legal rights.

Substantial Completion As defined in the AIA General Conditions, completion of a project to the point where the owner can occupy all or designated portions of the work for the purpose for which it is intended.

Superintendent The contractor's representative at the site.

Supervision Direction of the work by the contractor's personnel. Supervision is not the responsibility of the architect.

Supplementary Conditions Part of the contract documents, prepared by the architect, which may modify provisions of the general conditions of the contract.

Surety An individual or company that guarantees to make good to another party the debt, default, or failure to perform of a third party.

Surety Bond An agreement under which one party (the surety or bonding company) guarantees to make good to another party (the obligee or owner) the debt, default, or failure to perform of a third party (the principal or contractor). See Bid Bond, Labor and Material Payment Bond, and Performance Bond.

T

Total Quality Management (TQM) An approach to the delivery of goods or services in which quality is determined by customer satisfaction and conformance to requirements.

Trade Discount The difference between the list price and the actual price paid.

U

Unit Price An amount, stated in dollars per unit, provided by the contractor with his or her bid for adding or deleting specific portions of the work.

Upset Price Same as Guaranteed Maximum Cost.

V

Vapor Barrier A membrane that prevents the passage of water vapor through a wall or roof.

W

Waiver of Lien A document by which one relinquishes the right of mechanic's lien against the property of another. See Mechanic's Lien and Release of Lien.

Warranty See Guarantee.

Work All materials and/or labor required for a project.

Working Drawings The part of the contract documents, prepared by the architect, that graphically illustrates the construction required for the project.

X

XCU The exclusion from insurance coverage for liability arising out of (X) explosion or blasting, (C) collapse of or structural damage to a building, and (U) underground damage caused by mechanical equipment.

Part III

ARE Strategic Planning

LESSON SIXTEEN

16

ARE STRATEGIC PLANNING

GOAL SETTING

Setting specific goals in order to reach an accomplishment is like identifying a destination prior to leaving. If you know exactly where you are going, you are able to focus on how you will reach your destination, stay on course, and get back on course if any obstacles get in the way.

Goals also give you something against which to measure your progress, to see if what you are doing is moving you closer or farther from your ultimate accomplishment. If your actions are moving you closer to your goal, then keep doing what you're doing. If your actions are moving you away from you goal, then you will need to adjust what you are doing to get back on track.

Your Specific Licensure Goals

In what state, or states, do you want to practice architecture?

When are you going to be eligible to start taking ARE exams?

By what date do you want to be licensed?

GETTING STARTED

Select the order in which you will take the exams.

There are numerous opinions on what is the optimum order for taking the exams. Recently, one Kaplan candidate passionately argued for ordering exams based on pass rates with a bell curve model, where the easiest exams were taken first and last and the hardest exams were taken in the middle. Our research suggests many candidates choose exam order based primarily on personal convenience and availability of study materials, seminars, and study partners. Occasionally, a candidate will insist on a systematic but whimsical approach, such as taking the exams in alphabetical order.

Kaplan recommends carefully considering several factors to determine the order that is best for you, including the following:

- *Personal areas of strength*—Many candidates start with subjects they are familiar with, or strongest in, in order to become familiar with the exam format first.
- *Timing concerns*—Leaving the hardest exams until the end can be problematic if the candidate does not pass and has to wait six months before retaking the exam.
- *Grouping of exams with closely related content*—While Kaplan strongly recommends delineating which exam a candidate is studying for at any given time, there is a benefit in studying related subject areas sequentially. Studying should be easier as the candidate progresses and builds on previous knowledge.

Sample exam order strategy.
Following is a sample order of common exam divisions, as well as an explanation of why this order might be considered:

1. Schematic Design
2. Structural Systems
3. Building Design & Construction Systems
4. Building Systems
5. Construction Documents & Services
6. Site Planning & Design
7. Programming, Planning & Practice

This order starts with a division this candidate thinks will be "easier" in order for him to become familiar with taking the exams. After that, the candidate has front-loaded the divisions he feels will be more "challenging" in order to allow for the possibility of retaking one of these exams and still meeting the overall licensure target date. Divisions with related content, such as Site Planning & Design and Programming, Planning & Practice, are taken sequentially. Programming, Planning & Practice was intentionally left for the end as the content touches so many disparate general topics the candidate found it disorienting, particularly as he had been out of school for a long time. Building Design & Construction Systems also includes information from other exams; however, this candidate thought the content was easy enough to take sooner.

The downside of this order is that Schematic Design, the first division, does not allow a student to become familiar with the multiple-choice exam format prior to the harder divisions. This candidate decided to compensate by spending additional time reviewing for multiple-choice section exam format prior to the second division.

Alternative sample exam order strategy.

Following is an alternate exam order strategy and rationale based around a different candidate's strengths and weaknesses:

1. Site Planning & Design
2. Programming, Planning & Practice
3. Construction Documents & Services
4. Structural Systems
5. Building Systems
6. Building Design & Construction Systems
7. Schematic Design

This order also starts with an exam the candidate believes will be "easier," so she has time to become more familiar with the exam format and the general process of taking these exams. This order also groups related subject areas sequentially. However, this candidate, as a recent graduate, decided to take Programming, Planning & Practice and Construction Documents & Services earlier, as much of this information was still fresh in her mind. Memorization and quickly navigating the multiple-choice format were her strengths, so she also elected to put the vignette-heavy exams later to allow more time for her to become familiar with the NCARB software.

The downside of this order is putting Building Design & Construction Systems, a division with historically low pass rates, so close to the end and increasing the chances the candidate will be delayed in reaching her licensure date.

YOUR DIVISION ORDER

Division	Order
Building Design & Construction Systems	
Building Systems	
Construction Documents & Services	
Programming, Planning & Practice	
Schematic Design	
Site Planning & Design	
Structural Systems	

This order makes sense for you because...

Set aside scheduled study time.

This is one of the biggest challenges for busy professionals, but one that is often necessary for success.

Establish a routine.

Unfortunately, ad hoc studying is often incomplete and uneven studying. A little organization and dedication can go a long way.

Identify areas for improvement.

You should not just review areas you know you are weak in. You need to at least test yourself on your perceived areas of strength. There are things you think you know, but do not. You need to find those areas for improvement, too.

Increase your speed.

This exam, like most exams, is timed. Speed counts. It is not just a matter of knowing the material, but of recalling it and applying it in the exam format quickly. Speed is something that should be planned for and practiced.

Don't forget to learn something.

Too many people try to cram or rely on educational tools with poor instructional design. Most people forget exam preparation should involve actual learning about the exam format, the best way to approach problems, personal areas of strength and weakness and of course the content being tested.

Sample Daily Breakdown

Daily study routine: 1 hour morning + 1 hour lunch + 2 hours evening = 4 hours/day

Sample Study Schedule

Week	Activity	Hours
1	Diagnostic quizzes and test banks	20
2	Multiple-choice areas of weakness	20
3	Multiple-choice areas of weakness and NCARB sample problems	20
4	Seminar review	16
5	Flashcards and mock exams	20
6	NCARB software and vignette sample problems	20
7	NCARB software and vignette sample problems	20
8	Multiple-choice and vignette areas of weakness	15
9	Exam	

It is important to remember that the study tools used often have different strengths and weaknesses from an educational standpoint. These strengths and weaknesses should be considered when developing your study plan.

Seminars and study groups.

The time constraints of most seminars and study groups do not allow for an in-depth review of all the content. Instead, they are great opportunities to ask others about the exam format, the best way to approach problems, and how to address personal areas of strength and weakness.

Reference books.

The value of reading or rereading about *all* of the exam content topics in depth varies significantly by student. However, reference books are almost always extremely helpful once an area of weakness has been identified.

Quizzes, test banks, and problems and solutions books.

These allow you the opportunity to pinpoint areas where you need improvement.

Mock final exams and NCARB sample problems.

These are designed to simulate the ARE in a way that can teach you about the exam format and allow you to practice for speed as well as accuracy.

Flashcards and terminology and definitions mobile applications.

These are great for ad hoc studying to ensure the main concepts are understood and can be recalled quickly. They should always be supplemented with study tools that review the exam format and the exam required concept application.

NCARB vignette software.

You need to learn how to use this software prior to taking the exam. You do not have time to learn to use it during the exam. Many people do not find the software very intuitive. Practice, practice, practice!

Sample Study Breakdown

To ensure that your studying is appropriate for the exam, consider which portion of your preparation will be "active studying," such as information gathering, reading, and memorization, and which will be "active practice," such as diagnostic testing, mock examinations, flashcards, and so on.

- 60 percent active study—information gathering, reading, and memorization
- 40 percent active practice—flashcards, mock exams, and answering approaches

MULTIPLE-CHOICE QUESTION CONSIDERATIONS

It helps to recognize that there are distinct question types within the multiple-choice part of the exams. These include:

- Checking all that apply
- Comparative analysis
- Terminology and definitions
- Diagrams, principles, materials, definitions, and details
- Scenarios
- Questions with partially correct answers that appear to be correct but have minor errors
- Selecting the most accurate statement: code, public health, safety, welfare, and so on

Historically, the most common question type has been comparative analysis. "Check all that apply" questions are newer.

Many candidates fall victim to questions seeking the "best" answer. In these cases, it may appear at first glance as though several choices are correct. Remember the importance of reviewing the question carefully and interpreting the language literally. Consider the following example:

1. Which of these cities is located on the east coast of the United States?

   A. Boston

   B. Philadelphia

   C. Washington, DC

   D. Atlanta

At first glance, it may appear that all of the cities could be correct answers. However, if you interpret the question literally, you'll identify the critical phrase as "on the east coast." Although each of the cities listed is arguably an "eastern" city, only Boston sits on the Atlantic coast. All the other choices are located in the eastern part of the country but are not coastal cities.

GENERAL TEST-TAKING TIPS

- Arrive in plenty of time—you do not need the additional stress.
- Bring layered clothing—the room temperature at testing facilities can vary, and you should not let that distract you.
- Breathe slowly and deeply—this will help you remain calm, focused, and confident.

MULTIPLE-CHOICE TIPS

Pace yourself.

Each division allows candidates at least one minute per question. You should be able to comfortably read and reread each question and fully understand what is being asked before answering.

Read carefully.

Begin each question by reading it carefully and fully reviewing the choices, eliminating those that are obviously incorrect. Interpret language literally, and keep an eye out for negatively worded questions.

Decide on a strategy.

Have a strategy in mind for how you will ensure that you answer all the questions in the allotted time. Some people answer every question before moving on, while others use the three-pass method.

Review difficult questions.

The exam allows candidates to review and change answers within the time limit. Utilize this feature to mark troubling questions for review after completing the rest of the exam.

Use your instincts.

If you continue to be unsure of the answer after returning to the question a second time, it is usually best to stick with your first guess. Unless you have misread the question, statistics show that the first answer you choose is usually correct.

Answer every question.

All unanswered questions are considered incorrect, so answer every question. If you are unsure of the correct answer, select your best guess and/or mark the question for later review. If unfamiliar with terminology, look for something similar.

Be cautious of wording.

Words such as *never* or *always* suggest a situation where the conditions "always" or "never" exist, which rarely happens in the practice of architecture. Look for strong words such as *guarantee* or *eliminate.*

Use scratch paper liberally.

Sketching helps many candidates think more clearly. Do not hesitate to sketch out a vignette strategy if it will improve your solution and still meet the time restrictions.

Double-check your math calculations.

This is a good idea for the complex formulas and the simple problems as well. Small miscalculations can be easily avoided with a little extra attention.

Utilize reference materials available to you.

Some divisions include reference materials accessible through an on-screen icon. These materials include formulas and other reference content that may prove helpful when answering questions in these divisions. Note, however, that candidates may *not* bring reference material with them to the testing center.

The Three-Pass Method

Pass #1. Read all questions, answering only those you can answer quickly. This gives you an opportunity to see all questions and gauge time-management requirements.

Pass #2. Answer all remaining questions, except for those where you clearly do not have the knowledge base.

Pass #3. Guess at remaining questions and try to eliminate any nonviable answers to increase your odds. Remember that all unanswered questions are marked incorrect and count against you.

VIGNETTE CONSIDERATIONS

Style doesn't count.

Vignettes are graded on their conformance with program requirements and instructions. Don't waste time creating aesthetically pleasing solutions and adding unnecessary design elements.

Pace yourself, and don't panic.

Speed is often important, but each vignette allows candidates ample time to complete a solution within the time allotted. It is typically very helpful to already have a strategy for how you will approach each vignette type.

Read carefully.

Review instructions and requirements. Quickly make a list of program and code requirements to check your work against as you proceed through the vignette.

CONSTRUCTION DOCUMENTS & SERVICES

EXAM OVERVIEW

According to NCARB, the Construction Documents & Services division covers the "application of project management and professional practice knowledge and skills, including the preparation of contract documents and contract administration."

CONSTRUCTION DOCUMENTS & SERVICES		
Intro Time	:15	
MC Testing Time	**2:00**	**100 items**
Scheduled Break	:15	
Intro Time	:15	
Graphic Testing Time	**1:00**	**Building Section**
Exit Questionnaire	:15	
Scheduled Appointment Time	**4:00**	

Multiple-Choice Topic Areas

Multiple-choice topic areas will likely include the following:

1. Codes and regulations.

You will need to be able to incorporate building codes, specialty codes, zoning, and other regulatory requirements in construction documents and services.

Building and zoning codes should be incorporated into the construction documents and services. For example, The International Building Code (IBC) requires that mixed-occupancy buildings either be separated or nonseparated with fire barriers.

- International Building Code
  - Occupancy Use Group (chap. 3)
  - Construction Type (chap. 6)
    - Type I through V
    - "A" = One hour
    - "B" = Not rated
  - Egress and Exiting (chap.10)
- Fire and Life Safety Codes
  - Fire Resistance (chap. 8)
  - Hazardous materials
- Energy Code
  - R values of roof and wall assemblies
  - Total amount of allowable glazing
- Mechanical and Electrical Codes

2. Environmental issues.

You will need to be able to incorporate sustainable design principles, adaptive reuse concepts, alternative energy systems, new material technologies, and hazardous material mitigation in construction documents.

Remember when considering the adaptive reuse of an existing building, issues of designing the building envelope to be energy efficient should be incorporated into the construction documents.

All decisions of hazardous materials removal must also be addressed.

3. Construction drawings and project manual.

You will need to be able to prepare and coordinate construction drawings, including building systems, product selection, and constructability, as well as prepare, coordinate, and review general and supplementary conditions and technical specifications.

You will need to consider compatibility with design intent. Program requirements must be balanced with practical considerations such as code requirements, cost, and material limitations. Consider the following subjects:

- Structural integrity
- Safety
- Durability and maintainability
  - Exterior and interior
- Costs of materials
- Tolerances of materials
- Conformance to industry standards
- Resistance to moisture and weathering
- Thermal resistance
- Coordination with project manual and specifications

4. Project and practice management.

You will need to be able to prepare estimates of probable construction cost, and to consider cost implications of design decisions.

Notes

You will need to be able to prepare and manage the project schedule and coordinate all contract documents, including those of consultants.

You will need to be able to establish project delivery method and provide contract administration documentation and services.

You will need to be able to review and administer professional services and construction contracts, and to consider issues pertaining to practice, including risk management.

Key content areas include cost, scheduling and coordination, project delivery, and contracts and legal issues.

One of the best places to start studying for this portion of the exam would be to review the *AIA Architect's Handbook of Professional Practice.*

Building Section Vignette

This problem tests the understanding of basic building form and the impact of structural, mechanical, and lighting components and other constraints on the building section.

The information typically provided is *a program; floor plans indicating architectural, structural, and mechanical components; and a section cut line.*

The goal is to produce a building section from an indicated datum with the supplied building components (i.e., footings, foundations, floors, walls, ceilings, joists, and decks).

The solution is analyzed for accuracy and completeness based on the program requirements and plans.

1. *Strategy.*

- Determine final floor and roof elevations based on ceiling heights provided, and floor/ceiling and roof/ceiling assembly depths.
- Determine MINIMUM assembly depths required to accommodate structural, mechanical, and lighting requirements.

2. *Analysis.*

- Determine ceiling height (D1) for each floor level in the program.
- Determine clearance for lighting (D2) for each level in the program.
- Determine the greatest combination of joists and ducts (D3) per floor that defines the MAXIMUM assembly depth. Add *deck depth* to D3.
- Study entire floor plan for this combination, starting at vertical chases. Do not assume that it is at the cut line.

3. *Calculation.*

- Add D1 + D2 + D3 to determine the MINIMUM floor-to-floor height for both floors indicated.

4. *Draw section.*

- Locate GRADE LINE on section in Work screen.
- Zoom in to set a view where you can see only the entire section.
- Use rectangles on sketch layer as guides for heights of all elements.
- Draw exterior footings to below frost depth, interior footings under slab.
- Draw finish grade 6" below slab.
- Add exterior walls with Draw tool, *aligning with correct floor plan.*
- Add required parapets.

Notes

- Add ground floor slab, joist/deck combinations for second floor and roof.
- The floor deck is added to selected joist.
- Select joist depth, spacing, and direction that are cut with the cut line.
- Draw interior bearing walls and rated walls to underside of deck.
- Draw non-load-bearing walls up to established ceiling line.
- Draw correct ducts size and location that are cut by cut line in each mechanical floor plan.
- Draw any other required elements.

Notes

YOUR CONSTRUCTION DOCUMENTS & SERVICES SELF-STUDY PLAN

Your target test date is:____________

Your study breakdown.

____ percent active study—information gathering, reading, memorization

____ percent active practice—flashcards, mock exams, answering approaches

Your daily breakdown.

Daily study routine ___ hour(s) morning +

___ hour(s) lunch + ___ hour(s) evening =

____ hours/day

Your study tools.

Your study schedule.

Week	Activity	Hours

Preparation plan.

My areas to improve.

My areas of strength.

My exam strategy.

Multiple-choice strategy.

Vignette strategy.

BIBLIOGRAPHY

The following list of books is provided for candidates who may wish to do further research or study in Construction Documents and Services. Most of the books listed below are available in college or technical bookstores, and all would make welcome additions to any architectural bookshelf. In addition to the course material and the volumes listed below, we advise candidates to review regularly the many professional journals, which are available at most architectural offices.

Architects and Engineers—Their Professional Responsibilities
Acret
McGraw-Hill

Architect's Handbook of Professional Practice (three volumes)
The American Institute of Architects

Architectural and Engineering Law
Tomson and Coplan
Reinhold

Architectural Graphics Standards
Ramsey, Sleeper, and Hoke
Wiley

Architectural Working Drawings
Liebing and Paul
John Wiley & Sons

Avoiding Liability in Architecture, Design and Construction
Cushman, R.
John Wiley & Sons

Building Contracts for Design and Construction
Hauf
John Wiley & Sons

Building Construction Cost Data
Means, R.S.

Case Histories in Construction Law—A Guide for Architects, Engineers, Contractors, Builders
Jabine
Cahners Publishing Company

Construction Bonds and Insurance Guide
Rothschild
American Institute of Architects

Construction Law in Contractors' Language
Stokes
McGraw-Hill

Design Cost Analysis for Architects and Engineers
Swinburne
McGraw-Hill Book Company

Dictionary of Architecture and Construction
Harris, Cyril
McGraw-Hill Book Company

Legal Aspects of Architecture, Engineering and the Construction Process
Sweet
West Publishing Company

Legal Pitfalls in Architecture, Engineering and Building Construction
Walker and Rohdenburg
McGraw-Hill

Life Cycle Cost Analysis: A Guide for Architects
The American Institute of Architects

Life Cycle Cost Analysis 2: Using It in Practice
Haviland
The American Institute of Architects

National CAD Standard
National Institute of Building Sciences
American Institute of Architects
Construction Specifications Institute

The Professional Practice of Architectural Detailing
Wakita and Linde
John Wiley & Sons

The Project Resource Manual
CSI Manual of Practice, 5th Edition
Construction Specifications Institute

Working Drawing Handbook, a Guide for Architects and Engineers
McHugh
Van Nostrand Reinhold Company

Lesson One

1. **D** When preparing specifications, the architect must review the manufacturer's literature to determine if the product specified is appropriate for the application (A). Any specification must at least meet the minimum standards established by the local building code (B). As with all professional services, the architect must rely on his or her own professional judgment and experience when preparing specifications (C). Since A, B, and C are all correct, D is the answer.

2. **C** Although the drawings and specifications organize information, the general contractor divides the construction trades in accordance with trade union rules (III) and business judgment (IV). See page 11.

3. **B** Dimensions (A), quantities (C), and configurations (D) can be determined from the drawings, but the level of quality (B) is indicated in the specifications.

4. **A** Shop drawings show installation details for the actual product to be used in the project (A). Equipment operating data (B) are indicated in product data submittals. Color and texture (C) are shown on schedules and specifications prepared by the architect and may be repeated on shop drawings. Standard of workmanship (D) is stated in the specifications.

5. **C** A Project Manual contains the technical specifications and other contractual and administrative documents, such as those in choices II, III, and IV, but it does not include the drawings (I), although it may contain a list of the drawings.

6. **A** Specifications that describe the desired end result (A) are *performance specifications*. Statements B, C, and D are all true regarding *descriptive specifications.*

7. **A** See page 15 for a discussion of cash allowance specifications.

8. **D** All the factors shown influence the selection of a mechanical system.

9. **B** *Reference specifications* refer to standard specifications that are incorporated into the project specifications by reference, not by actual text. See page 15.

10. **C** *Descriptive specifications* explain all components of the specified items in detail (C). Statements A, B, and D correctly describe *performance specifications.*

11. **B** Level of quality, desired performance, and installation methods (A, C, and D) are contained in the specifications, but quantities (B) are determined from the drawings.

12. **D** The warranty (A), code compliance (B), and maintenance costs (C) must be comparable for a product to be approved as a substitute for a specified product. Installation equipment is not considered by the architect, although if the in-place cost of the product is reduced, the savings may accrue to the owner if the substitution is approved.

13. **B** *Master specifications* are generally edited by *eliminating* information that does not apply to the project, not by *adding* information.

14. **B** *Proprietary specifications* list one or more acceptable products by name. An *open* proprietary specification allows equivalent products not listed to be provided, while a *closed* proprietary specification allows only those products listed to be provided.

15. **C** Plumbing drawings indicate the size of pipes by notation, not graphically. (C is the incorrect statement we are looking for.) Statements A, B, and D correctly describe plumbing drawings.

Lesson Two

1. **D** The architect has the authority to reject work under subparagraph 2.6.2.2 of the AIA Owner-Architect Agreement (Document B201). Answer A is partially true, but caulking is only one part of a waterproofing system. In answer B, durability is one of several important considerations in selecting finish materials. Equally important are appearance and cost. Samples of manufactured products are required when the product is part of a critical assembly (C is incorrect).

2. **A** In correct answer A, concrete may be shop fabricated (precast), but more commonly it is cast-in-place (job fabricated). In incorrect answer B, cabinets are usually shop fabricated. Stairways are usually shop fabricated (C is incorrect). In incorrect choice D, air conditioning equipment is usually manufactured.

3. **B** Wood has a low coefficient of thermal expansion. In addition, the connections in wood framing are generally flexible enough to accommodate any thermal movement that may occur. Concrete is very susceptible to cracking, especially during curing (A is incorrect). Metal has a relatively high coefficient of thermal expansion. Also, paving, gutters, and curtain walls are exposed to the exterior, where, unlike wood framing, they are subject to extreme temperature variations (C and D are incorrect).

4. **C** See pages 41 and 42 for a discussion of *acoustical control.*

5. **C** Answer A is incorrect because shop drawings are usually prepared by fabricators. The AIA General Conditions state that shop drawings do not supersede the Contract Documents (B is incorrect). D is also incorrect because the AIA General Conditions specifically place the responsibility for checking dimensions and quantities on the contractor.

6. **B** All five methods tend to reduce the transmission of vibration and/or sound.

7. **A** Locating property lines (I) requires the greatest degree of accuracy because of legal concerns. These and column center lines (II) are always established on site before any construction occurs, and are used as references for layout of all subsequent work. Dimensions of cabinet work designed to fit into an existing opening (V) are also critical. Exposed concrete (IV) requires greater accuracy than concealed concrete (III), which is normally hidden by finish materials.

8. **A, B, C, and D** A moment (A) is created when wind acting on a building tends to overturn the building. Wind typically acts horizontally or laterally (B). Wind can also create uplift, especially under overhangs and above roofs (C). The massing of a building may cause uneven distribution of lateral forces, thereby introducing torsion (D).

9. **B** Impact noise of a material (A) is rated by an Impact Isolation Class (IIC). Amplification and reverberation (C and D) are functions of the absorptive characteristics of a material rated by a Noise Reduction Coefficient (NRC). The correct answer, therefore, is B.

10. **C** See discussion of *electrolysis* on page 38.

11. C Answer A is incorrect, since building and fire codes are intended to protect buildings from fire damage. In incorrect answer B, life safety codes are used in conjunction with, not instead of, building codes. Life safety codes primarily address egress (D is incorrect).

12. C Mock-ups generally do not produce time or cost savings (III and V are incorrect). For new or innovative systems (I and II), mock-ups allow testing under actual job site conditions. For repetitive systems (IV), mock-ups allow design refinements.

13. D Answer A is incorrect, since building codes are minimum, not ideal, standards. In incorrect answer B, the fire resistive requirements of building codes are intended to allow the evacuation of occupants within a specified period of time. Zoning ordinances regulate land use, while the type of construction required by building codes is a function of use as well as area, height, and fire resistive properties of materials and assemblies (C is incorrect).

Lesson Three

1. A, B, C, and D Life-cycle costing considers cost of the building throughout its life. Some aspects of a sustainable design may cost more initially but may result in lower life-cycle costs due to lower operating and manufacturing costs.

2. B Sustainable designs may or may not have lower initial costs.

3. D Refer to page 59.

4. C

5. B U.S. Green Building Council, The American Society of Heating, Refrigeration and Air Conditioning Engineers (ASHRAE) provides established standards by which building performance is measured, but the USGBC sponsors the program.

6. C Refer to the complete list of all six categories on page 56. Recycling is a sustainable practice included within the rating system, but it is not one of the six categories.

7. A Volatile organic compounds.

8. D

Lesson Four

1. D Subparagraph 3.12.4 of the AIA General Conditions specifically states that shop drawings (III) are not contract documents. The architect's rights and responsibilities during the construction phase are referenced in the AIA General Conditions of the Contract for Construction, but the contractor is not a party to the Owner-Architect Agreement (IV). Subparagraph 1.1.1 of the AIA General Conditions specifically includes the specifications, addenda, and supplementary conditions as part of the contract documents (I, II, and V).

2. Performance bond A performance bond guarantees proper execution of the work by the contractor. Conversely, a payment bond guarantees payment to the subcontractors by the prime contractor.

3. C The various types of construction insurance are discussed on pages 65 to 67. Both the owner and contractor (IV and V) are required to purchase liability insurance by Article 11 of the AIA General Conditions.

4. A See page 69 for information about liquidated damages. The AIA Owner-Contractor Agreement instruction sheet advises the parties to include *the amount of damages due for each day lost* in

Paragraph 3.3 of the Agreement or in the Supplementary Conditions.

5. A Subparagraph 3.11 of the General Conditions requires the contractor to maintain a set of record drawings (A) at the site *marked currently to record field changes and selections made during construction*. A field sketch (C) is used to revise or clarify information shown in the contract documents based on field conditions.

6. A See pages 6-8 for information regarding the Project Manual. The General Conditions (II) are either a standard form, such as AIA Document A201, or written by the owner's attorney. An architect should never prepare a contract to which the architect is not a party, such as the Owner-Contractor Agreement (IV). The Specifications (I) are prepared by the architect, but not necessarily the Supplementary Conditions (III).

7. D Article 2 of the AIA General Conditions requires the owner to obtain easements and provide copies of construction documents (IV and V). Article 9 requires the owner to pay the contractor (II). The contractor, not the owner, is required to pay the subcontractors (III), and access to the site as an owner's responsibility is implied but not explicitly stated (I).

8. B Under the definitions of time in the AIA General Conditions (8.1.1), time is the period allotted for substantial completion of the work (B is correct, C is incorrect). Subparagraph 8.1.4 defines *day* as calendar day (A is incorrect). The date of the contract is often the start of construction time, but may be otherwise if noted in Paragraph 3.1 of the Owner-Contractor Agreement, or if a *notice to proceed* has a different starting date (D is also incorrect).

9. A Subparagraph 9.8.4 of the AIA General Conditions states that warranties commence on the date of substantial completion, unless a different commencement date is stated in the certificate of substantial completion (A is correct). See page 73 for a discussion of warranties.

10. C See page 70, and subparagraph 9.5.1 of the AIA General Conditions for payment nullification reasons. All of the reasons listed, except for the lender's refusal to release funds (II), are legitimate.

11. C See pages 66 to 67. The owner must pay the contractor as stipulated in the contract documents, not necessarily within 21 days after the certificate for payment.

12. A See page 174 and Article 10 of the AIA General Conditions for the general contractor's responsibility for site safety.

13. D Since substantial and final completion are significant events for the contractor and owner, the architect is required to make detailed inspections prior to certifying completion. Other site visits are not considered *inspections*, since they are less detailed.

14. B See page 74 and Paragraph 9.10 of the AIA General Conditions for prerequisites for final payment. Answer B is the incorrect answer we are looking for, because there is no certificate of final *completion*, only a final certificate for *payment*.

15. A and B Subparagraph 3.12.5 of the AIA General Conditions requires the contractor to approve shop drawings and 4.2.7 requires the architect to do likewise (A and B). The owner (C) is not involved in the shop drawing process. Subcontractors may prepare the shop drawings (E is incorrect). The engineers may review and stamp shop drawings, but the AIA General Conditions do not recognize the engineers as entities independent of the architect (D is incorrect).

Lesson Five

1. **D** See pages 84 and 85, and Paragraph 5.1 of AIA Document B201 for definition of *construction cost*. Land acquisition costs and the architect's fee (III and IV), while part of the owner's *project budget*, are specifically excluded from *construction cost* in paragraph 5.1.

2. **B** See pages 87 to 89 for a description of the subsystems method of estimating. Answer B is correct because subsystems estimates include all components of a particular subsystem (e.g., for a brick cavity wall: brick, concrete masonry units, insulation, flashing, mortar, joint reinforcement, lintels, brick ties, etc.). Thus, the total for a brick cavity wall can easily be compared to the total for a metal and glass curtain wall, on a cost per square foot basis.

3. **B** See pages 88 through 92 for a description of detailed estimates of construction cost.

4. **A, B, C, and D** See subparagraphs 5.6 and 5.7 of AIA Document B201 for the four options available to the owner if the budget is exceeded by the lowest bid. These options are those in answers A through D. The owner has no claim against the architect if the budget is exceeded.

5. **D** See pages 86 and 87 for adjustments that may be necessary to the cost per square foot estimate.

6. **B** See subparagraph 5.5 of AIA Document B201 for the architect's authority to modify scope and quality to meet the owner's budget. Answers C and D are incorrect because the architect is not able to control price directly.

7. **D** The architect is not able to control the cost of labor, materials, or the contractor's bidding methods. Subparagraph 5.2 of AIA Document B201 relieves the architect from responsibility for these factors.

8. **C** II through V all may affect the contractor's bid prices. See page 93 for a discussion of these factors. Funding is the owner's responsibility (I is incorrect). Errors and omissions may result in change orders after the contract is awarded (VI is incorrect).

9. **B** Subparagraph 5.2 of AIA Document B201 states that the architect has no control over bidding, market, or negotiating conditions (B is correct). There is no National Construction Cost Index (A is incorrect). Estimates of construction cost are basic services in AIA Document B201 (C is incorrect). The architect never guarantees the ultimate construction cost (D is incorrect).

10. **D** See pages 86 and 87 for a discussion of construction costs based on area/volume estimates.

11. **A** See page 88 for items included in *indirect costs*. Excavation is a direct cost (B is incorrect). Architect's fees and FF&E (furniture, furnishings, and equipment) are owner's costs (C and D are incorrect).

Lesson Six

1. **A** Activities B and E both terminate at event 3, after which activity C may begin. See page 105.

2. **D** The critical path is the path with the longest total required time. See page 107.

3. **C** Critical path 1 – 2 – 3 – 4 – 5 has a total time of 2 + 2 + 3 + 1 = 8 days.

4. **C** Total construction cost has little to do with the construction schedule. See pages 103-105.

5. **Fast track** Fast track scheduling is appropriate in situations that require minimum construction time. Time savings are achieved by overlapping the production of construction documents with construction activities, with certain portions of the project documented and built in phases.

6. **B** Certain phases of design and production, such as client approval and project bidding, have fixed times. Therefore, the 25 percent reduction in time would probably come from the construction drawing phase. This would likely lower the overall quality of the construction documents. The construction budget and time would be unaffected.

7. **A** A shortened time schedule may reduce some fixed overhead expenses, such as rent, but it would undoubtedly lead to higher costs because of overtime work, additional hired help, and/or work that is subcontracted to others. Because of higher labor costs, profit would decrease, and the documents would be adversely affected. See page 125.

8. **D** See pages 99–102.

9. **C** Complexity is generally more critical than size, while cost and quality rarely affect scheduling.

10. **A** Project cost would be increased because of inefficiencies resulting from additional labor and overtime work. See page 125.

Lesson Seven

1. **Construction management** The construction management delivery method involves an owner who hires a construction manager to work with the architect to resolve constructability and cost issues during the design phase.

2. **C** The design/build entity is typically responsible for the design and construction of a project, based upon requirements established by the owner and issued in the request for proposal.

3. **C** An architect acts as an agent for the owner in the design/award/build and the construction management delivery methods (II and III are the correct answers). An architect acts as a vendor responsible for the cost and construction of a project in a joint venture with a contractor and in the design/build delivery method (I and IV are incorrect).

Lesson Eight

1. **C** A pre-bid conference is held to inform the bidders about unique circumstances of the project not readily apparent in the bid documents (answer C is correct). Subcontractors submit bids just prior to the general contractors' bids (answer A is incorrect). An addendum may be issued at the conference, but subsequent addenda may be issued within several days of the bid date (answer D is incorrect).

2. **D** A private owner may choose any of the listed methods to request proposals. In general, competitive bidding among pre-qualified bidders results in the best combination of price and quality.

3. **A, B, and D** Shop drawings (C) are submitted by the successful contractor *after* the construction contract has been awarded.

Drawings and instructions to bidders (A and B) are issued with the initial bid documents. Addenda are issued during the bid period (D) and are considered part of the bid documents.

4. A Late bids and non-standard bid forms generally do not materially affect the outcome of competitive bidding and may be overlooked by a private owner (I and II are correct). Incomplete bids make it impossible to determine the low bidder (answer III is incorrect). Omission of a performance bond is an indication that the contractor is not capable of performing the work (answer IV is incorrect).

5. C The number of bidders generally has no effect on the length of the bid period, except when distribution of many sets of bid documents to many bidders consumes excessive time (answer C is correct). See pages 124 and 125 for a discussion of time allowed for bidding.

6. C See pages 137 and 138 for a discussion of bid security. A certified check is an acceptable form of bid security (answer C is correct). A performance bond is an indication of the bidder's ability to perform the work (answer A is incorrect). Covenants and indemnities afford some protection, but are not as liquid as a cashier's check or a bid bond (answers B and D are incorrect).

7. A See page 137 for a discussion of the list of subcontractors. B, C, and D are all correct. A is the incorrect statement and therefore the answer to this question: the purpose of the list of subcontractors is to provide the owner and architect the opportunity to review the list and make reasonable objections.

8. A Bidding documents are issued to general contractors only (answer A is correct). Sub-bidders and material suppliers obtain the bid documents from, and submit their bids to, the general contractors (answers B, C, and D are incorrect).

9. B The lowest bid is the lowest total of the base bid (C) and accepted additive and deductive alternates (A and D). Unit prices (B) cannot be combined with the base bid and alternates to determine the low bidder.

10. C See the glossary for definitions of *change in services*, *change orders,* and *addenda* (answers A, B, and D are incorrect). This question correctly defines *alternates* (answer C is correct).

Lesson Nine

1. B Subparagraph 2.6.2.2 of the AIA Owner-Architect Agreement specifically gives the architect the authority to reject work that does not conform to the contract documents. The other three actions would expose the architect to professional liability. Choices A and D are specifically excluded from the architect's services in Subparagraphs 2.6.1.2 and 2.6.2.1. Only the owner can stop the work (see Paragraph 2.3 of the AIA General Conditions).

2. D Article 2.6 of the AIA Owner-Architect Agreement describes the architect's services during the construction phase. The AIA General Conditions is part of the contract documents between owner and contractor, but also establishes the architect's rights and responsibilities for the contractor's information. See Paragraph 4.2 of the AIA General Conditions. The Owner-Contractor Agreement does not relate to the architect's status.

3. C See pages 143 and 144 for a discussion of professional liability coverage.

4. D See pages 144 to 146 for a discussion of liens. Lien laws vary considerably from state to state, but general contractors, subcontractors, and material suppliers always have lien rights.

5. **A and B** Retainage on progress payments to the contractor is intended for the owner's protection against the contractor's failure to pay subcontractors or to correct deficient work. See Lesson 11 for a discussion of retainage.

6. **D** See Subparagraph 9.10.2 of the AIA General Conditions for submittals required from the contractor prior to final payment.

7. **Arbitration** See Paragraph 15.4 of the AIA General Conditions for a discussion of arbitration.

8. **C** See pages 148-149 for a general discussion of bonds. The architect is not a party to construction bonds. The owner and contractor both need protection from risk, and the surety provides that protection for a fee.

9. **A** See page 149 for a discussion of bid bonds. Bid bonds are intended to protect the owner from a bidder who fails to sign a contract for the bid price. The owner can collect damages from the surety with greater certainty and expediency than by the other options listed.

10. **D** A performance bond amount of less than 100 percent indicates that the surety does not believe that the contractor can perform the work according to the surety's standards. The surety may believe that the project is too big for the contractor, or that the contractor has too many commitments to other projects.

11. **B** See pages 151 to 155 and the AIA General Conditions paragraphs 11.1 and 11.2 for a discussion of the owner's and contractor's liability insurance requirements during construction. The Owner-Architect Agreement may require the architect to carry professional liability insurance, but this is not a requirement of the construction contract.

12. **B** Supervision of construction is excluded from an architect's professional liability insurance. Subparagraph 2.6.1.2 of the AIA Owner-Architect Agreement and Subparagraph 4.2.2 of the AIA General Conditions both state that the architect does not supervise construction, since this is the contractor's responsibility.

13. **B** See Paragraph 9.9 of the AIA General Conditions for the requirements for partial occupancy by the owner prior to substantial completion.

14. **D** See Paragraph 11.3 of the AIA General Conditions and page 154 for the requirements for property insurance, which is commonly referred to as builder's risk insurance.

15. **A** Subparagraphs 2.6.2.4 of the AIA Owner-Architect Agreement and 4.2.13 of the AIA General Conditions both state that the architect's decisions regarding aesthetic effect are final if consistent with the intent expressed in the contract documents.

Lesson Ten

1. **C** See AIA General Conditions (Document A201) Subparagraph 3.12.1 for the contractor's responsibility to prepare shop drawings (C). A subcontractor, manufacturer, supplier, or distributor under contract to the prime contractor may also prepare shop drawings. The architect reviews shop drawings (A is incorrect). The owner and surety have no direct involvement in the shop drawing process (B and D are incorrect).

2. B See AIA General Conditions Subparagraph 3.12.8 for the contractor's responsibility to inform the architect when shop drawings deviate from the contract documents.

3. C An architect should not review shop drawings that are not required by the contract drawings because such reviews increase the architect's administrative expenses and create opportunities for error. Architects should only require submittals which are necessary for quality control.

4. A and B Architects should retain copies of shop drawings, transmittals, and a log in the event a contractor claims that he or she was unable to complete the work on schedule due to the architect's delay in processing the shop drawings (B). Multiple submittals may suggest a potential problem, and retaining copies of each shop drawing submittal conveys the nature of the problem and how it was resolved (A). Actual field conditions are documented in as-built drawings and are not related to shop drawings (E is incorrect).

5. D The AIA General Conditions and Owner-Architect Agreement require processing of shop drawings within a reasonable time (correct answer D). This flexibility allows specific job conditions, such as sequencing and work load, to determine the processing time. At the pre-construction meeting, the contractor may provide a submittal schedule coordinated with the construction schedule to indicate the lead times necessary for delivery of materials that must be approved before they can be ordered.

6. A Minor revisions can be made by the architect on the shop drawings. However, the contractor may submit a claim for additional time and/or cost if he or she believes the revisions are beyond the scope of work indicated on the contract documents. The contractor would then request a change order for the revisions.

7. C The architect is responsible for providing the services required by the owner-architect agreement. Retaining a consultant or incorporating information prepared by others does not relieve the architect of his or her professional liability.

8. D Subparagraph 2.6.2.1 of the Owner-Architect Agreement states that the architect must visit the site at *intervals appropriate to the stage of construction.* In a similar manner to the flexibility allowed for shop drawing processing time, this provision recognizes that each project has unique conditions that can best be addressed by the project team.

9. Addendum The AIA Instructions to Bidders (Document A701) states that changes to the bidding documents can only be made by a written addendum. Addenda issued prior to the signing of the contract for construction become part of the contract. Conversely, change orders modify the contract *after* the signing of the contract.

10. C Under OSHA, employers are responsible for the workplace safety of their employees. Therefore, on a construction site, the contractor (C) is responsible for OSHA compliance. Subparagraph 4.2.2 of the AIA General Conditions states that the architect will *not have control over...safety precautions and programs in connection with the Work....*

11. C Although architects are not responsible for job site safety or OSHA compliance (I is incorrect), common sense requires that an obviously unsafe condition should be reported to the party who has direct responsibility for correcting the condition, the contractor (II). The owner, not the architect, has the right to stop the work (III is incorrect). However, the architect should inform the owner of the problem (IV).

12. **B** Subparagraph 4.2.6 of the AIA General Conditions gives the architect authority to reject work (B), but only the owner can stop the work (A is incorrect). An architect should never communicate directly with a subcontractor (D is incorrect).

13. **B** The architect must request the owner's permission to order special testing (B). If the test results indicate that the work tested complies with the contract documents, the owner must pay for the test. If the work does not comply, then the contractor pays the testing expense.

14. **C** Subparagraph 15.1.2 of the AIA General Conditions states that *Claims by either party must be initiated within 21 days after occurrence of the event giving rise to such Claim....*

15. **D** Subparagraph 4.3 of the Owner-Architect Agreement requires the owner to furnish the services of geotechnical engineers. The architect and structural engineer are entitled to rely on the accuracy and completeness of the information provided by the geotechnical engineers in their design.

Lesson Eleven

1. **C** According to AIA Document G714, the architect and owner are required to sign a construction change directive. The contractor can choose not to sign it if he or she disagrees with the terms of the construction change directive. The duties of an architect's field representative do not include signing construction change directives.

2. **A** The owner issues a construction change directive in accordance with the owner's right under the AIA General Conditions to order changes in the work. The architect may prepare the actual document and also must sign the document, but the architect does not have the right to order changes in the work that affect cost or time.

3. **D** A Proposal Request notifies the contractor of an anticipated change and requests a proposal from the contractor stating the cost and/or time impact of the change (correct answer D). The Architect's Supplemental Instructions would be the appropriate document only if there is no change in the cost or time (A is incorrect). A Change Order is the last document in the process, not the first (C is incorrect). A Construction Change Authorization does not exist as an AIA document (B is incorrect).

4. **B** A Construction Change Directive (B) is used when an immediate change is necessary but the owner and contractor cannot agree on the cost. The Architect's Supplemental Instructions does not involve cost (A is incorrect). A Proposal Request does not direct the contractor to proceed with the change (D is incorrect). A Change Order is valid only when signed by the owner, architect, and contractor (C is incorrect).

5. **21 days** Subparagraph 15.1.2 of the AIA General Conditions states that claims by either party must be initiated within 21 days from which the claim stems.

6. **B** See pages 186 to 188 and the AIA General Conditions Subparagraph 7.3.7 for a detailed description of the architect's role in determining appropriate cost adjustments.

7. **A** The Performance Bond and Payment Bond (Document A312) does not require that the surety be notified when change orders are issued (I). Non-AIA bond forms, however, *may* require such notification. If so, this is usually the contractor's responsibility (IV is incorrect). Change orders usually do not invalidate bonds (II), although increases in the contract sum may result in increased bond premium costs (III).

8. **D** Under a cost plus fee contract, the contractor is paid for the actual amount spent to perform the work plus a fee. If the contractor's fee is based on a percentage of the cost of construction, there is no incentive for the contractor to control expenditures (A and B are incorrect, D is correct). Overhead and home office personnel salaries are not reimbursable (C is incorrect).

9. **D** Subparagraph 9.8.2 of the AIA General Conditions states that *the Contractor shall prepare and submit to the Architect a comprehensive list of items to be completed or corrected*, which is called a *punch list*. The architect's responsibility is to make an inspection, not prepare the list.

10. **A** The AIA Certificate of Substantial Completion (Document G704) states: *The responsibilities of the Owner and Contractor for security, maintenance, heat, utilities, damage to the Work, and insurance shall be as follows....* This is followed by a space to be filled in, and usually the owner accepts these responsibilities at this time.

11. **A, B, and C** See AIA General Conditions Subparagraph 9.10.4.

12. **A** The AIA General Conditions Subparagraph 9.3.1 stipulates the ten-day period. During this period, the architect may visit the site to verify that the progress of the work is consistent with the contractor's application, and negotiate any adjustments.

13. **B** See AIA General Conditions Subparagraph 9.4.2 for the representation made by the architect when certifying payment.

14. **B** See AIA General Conditions Subparagraph 9.7 for the contractor's rights when the owner fails to make a payment within the required time.

15. **D** See AIA General Conditions Subparagraph 9.5.1 for the reasons an architect may decline to certify payment. Subparagraph 9.3.2 requires payment to be made for materials suitably stored (D is the incorrect statement we are looking for).

Lesson Twelve

1. **D** None of the AIA documents listed requires the architect to maintain project files. However, it is good business practice to do so.

2. **A, B, and D** Tasks A, B, and D are all included in the architect's services during the construction phase under the AIA Owner-Architect Agreement (Document B201). Coordinating the work of contractors and subcontractors (C) is specifically excluded and could expose the architect to professional liability claims.

3. **A** A Project Representative is employed by the architect.

4. **B** A clerk of the works is employed by the owner. See page 211 for an explanation of the difference between a project representative and a clerk of the works. Neither of these should be confused with the contractor's superintendent.

5. **C** Nothing should be thrown away until the expiration of the statute of limitations applicable to the architect's services or to the project.

6. **Contractor** Record drawings, which show field changes during construction that vary from the information in the working drawings, are generally done by having the contractor mark up the changes on a set of prints at the site.

7. **A** See page 214 and Subparagraph 2.6.6.1 of the AIA Owner-Architect Agreement for the architect's responsibilities regarding warranties, which are limited to forwarding the documents from the contractor to the owner.

8. **B** Subparagraph 3.15.2 of the AIA General Conditions states that the owner may clean up and be entitled to reimbursement from the contractor who has failed to clean up as required.

9. **D** If a subcontractor refuses to furnish a lien release to the contractor, there is a strong possibility that the subcontractor intends to file a lien. Therefore, to protect the owner, the general contractor can furnish a new bond.

10. **10. B** If an owner believes that a project is not progressing fast enough, he or she should ask the contractor for an explanation through the architect, as provided by Subparagraph 4.2.4 of the AIA General Conditions. If the contractor is not responsive, the owner may terminate the contract in accordance with Article 14.

11. **C** See Paragraph 13.5 of the AIA General Conditions for the owner's, architect's, and contractor's responsibilities for special testing. In this case, the owner must pay for the test.

12. **A** See Paragraph 13.5 of the AIA General Conditions for the owner's, architect's, and contractor's responsibilities for special testing. If standards are not otherwise established by the building code, the architect usually includes in the specifications the standards by which testing will be conducted and evaluated.

13. **D** Professional liability, or negligence, is determined on the basis of the ordinary standard of care (I). If the design works poorly (II), or if the problem could have been avoided by a different design (III), the architect would not be liable, as long as he or she used the ordinary standard of care. If the architect was not careful (IV), that could render him or her liable, since he or she failed to meet the ordinary standard of care.

14. **C** Field reports should include all necessary facts, including who performed the test, the date of the test, and the weather conditions. An architect should never *supervise* the work. C is therefore incorrect and the answer to this question.

15. **A** Once the owner accepts the work as substantially complete, he or she becomes responsible for maintenance. If the work is properly maintained, and a defect becomes apparent within the warranty period, then the contractor is responsible for correcting the defect. In some cases, the owner may obtain a maintenance contract from the subcontractor who installed the work, which is separate from the construction contract.

EXAMINATION DIRECTIONS

The examination on the following pages should be taken when you have completed your study of all the lessons in this course. It is designed to simulate the Documentation portion of the Construction Documents & Services division of the Architect Registration Examination. Many questions are intentionally difficult in order to reflect the pattern of questions you may expect to encounter on the actual examination.

You will also notice that the subject matter for several questions has not been covered in the course material. This situation is inevitable and, thus, should provide you with practice in making an educated guess. Other questions may appear ambiguous, trivial, or simply unfair. This too, unfortunately, reflects the actual experience of the exam and should prepare you for the worst you may encounter.

Answers and complete explanations will be found on the pages following the examination, to permit self-grading. **Do not look at these answers until you have completed the entire exam**. Once the examination is completed and graded, your weaknesses will be revealed, and you are urged to do further study in those areas.

Please observe the following directions:

1. The examination is closed book; please do not use any reference material.
2. Allow about two hours to answer all questions. Time is definitely a factor to be seriously considered.
3. Read all questions *carefully* and mark the appropriate answer on the answer sheet provided.
4. Answer all questions, even if you must guess. Do not leave any questions unanswered.
5. If time allows, review your answers, but do not arbitrarily change any answer.
6. Turn to the answers only after you have completed the entire examination.

GOOD LUCK!

EXAMINATION ANSWER SHEET

Directions: Read each question and its lettered answers. When you have decided which answer is correct, blacken the corresponding space on this sheet. After completing the exam, you may grade yourself; complete answers and explanations will be found on the pages following the examination.

1. Ⓐ Ⓑ Ⓒ Ⓓ
2. Ⓐ Ⓑ Ⓒ Ⓓ
3. Ⓐ Ⓑ Ⓒ Ⓓ
4. Ⓐ Ⓑ Ⓒ Ⓓ
5. Ⓐ Ⓑ Ⓒ Ⓓ
6. Ⓐ Ⓑ Ⓒ Ⓓ
7. Ⓐ Ⓑ Ⓒ Ⓓ
8. Ⓐ Ⓑ Ⓒ Ⓓ
9. Ⓐ Ⓑ Ⓒ Ⓓ
10. Ⓐ Ⓑ Ⓒ Ⓓ
11. Ⓐ Ⓑ Ⓒ Ⓓ
12. Ⓐ Ⓑ Ⓒ Ⓓ
13. Ⓐ Ⓑ Ⓒ Ⓓ
14. Ⓐ Ⓑ Ⓒ Ⓓ
15. Ⓐ Ⓑ Ⓒ Ⓓ
16. Ⓐ Ⓑ Ⓒ Ⓓ
17. Ⓐ Ⓑ Ⓒ Ⓓ
18. Ⓐ Ⓑ Ⓒ Ⓓ
19. Ⓐ Ⓑ Ⓒ Ⓓ
20. Ⓐ Ⓑ Ⓒ Ⓓ
21. Ⓐ Ⓑ Ⓒ Ⓓ
22. Ⓐ Ⓑ Ⓒ Ⓓ
23. Ⓐ Ⓑ Ⓒ Ⓓ
24. Ⓐ Ⓑ Ⓒ Ⓓ
25. Ⓐ Ⓑ Ⓒ Ⓓ
26. Ⓐ Ⓑ Ⓒ Ⓓ
27. Ⓐ Ⓑ Ⓒ Ⓓ
28. Ⓐ Ⓑ Ⓒ Ⓓ
29. Ⓐ Ⓑ Ⓒ Ⓓ
30. Ⓐ Ⓑ Ⓒ Ⓓ
31. Ⓐ Ⓑ Ⓒ Ⓓ
32. Ⓐ Ⓑ Ⓒ Ⓓ
33. Ⓐ Ⓑ Ⓒ Ⓓ
34. Ⓐ Ⓑ Ⓒ Ⓓ
35. Ⓐ Ⓑ Ⓒ Ⓓ
36. Ⓐ Ⓑ Ⓒ Ⓓ
37. Ⓐ Ⓑ Ⓒ Ⓓ
38. Ⓐ Ⓑ Ⓒ Ⓓ
39. Ⓐ Ⓑ Ⓒ Ⓓ
40. Ⓐ Ⓑ Ⓒ Ⓓ
41. Ⓐ Ⓑ Ⓒ Ⓓ
42. Ⓐ Ⓑ Ⓒ Ⓓ
43. Ⓐ Ⓑ Ⓒ Ⓓ
44. Ⓐ Ⓑ Ⓒ Ⓓ
45. Ⓐ Ⓑ Ⓒ Ⓓ
46. Ⓐ Ⓑ Ⓒ Ⓓ
47. Ⓐ Ⓑ Ⓒ Ⓓ
48. Ⓐ Ⓑ Ⓒ Ⓓ
49. Ⓐ Ⓑ Ⓒ Ⓓ
50. Ⓐ Ⓑ Ⓒ Ⓓ
51. Ⓐ Ⓑ Ⓒ Ⓓ
52. Ⓐ Ⓑ Ⓒ Ⓓ
53. Ⓐ Ⓑ Ⓒ Ⓓ
54. Ⓐ Ⓑ Ⓒ Ⓓ
55. Ⓐ Ⓑ Ⓒ Ⓓ
56. Ⓐ Ⓑ Ⓒ Ⓓ
57. Ⓐ Ⓑ Ⓒ Ⓓ
58. Ⓐ Ⓑ Ⓒ Ⓓ
59. Ⓐ Ⓑ Ⓒ Ⓓ
60. Ⓐ Ⓑ Ⓒ Ⓓ
61. Ⓐ Ⓑ Ⓒ Ⓓ
62. Ⓐ Ⓑ Ⓒ Ⓓ
63. Ⓐ Ⓑ Ⓒ Ⓓ
64. Ⓐ Ⓑ Ⓒ Ⓓ
65. Ⓐ Ⓑ Ⓒ Ⓓ
66. Ⓐ Ⓑ Ⓒ Ⓓ
67. Ⓐ Ⓑ Ⓒ Ⓓ
68. Ⓐ Ⓑ Ⓒ Ⓓ
69. Ⓐ Ⓑ Ⓒ Ⓓ
70. Ⓐ Ⓑ Ⓒ Ⓓ
71. Ⓐ Ⓑ Ⓒ Ⓓ
72. Ⓐ Ⓑ Ⓒ Ⓓ
73. Ⓐ Ⓑ Ⓒ Ⓓ
74. Ⓐ Ⓑ Ⓒ Ⓓ
75. Ⓐ Ⓑ Ⓒ Ⓓ
76. Ⓐ Ⓑ Ⓒ Ⓓ
77. Ⓐ Ⓑ Ⓒ Ⓓ
78. Ⓐ Ⓑ Ⓒ Ⓓ
79. Ⓐ Ⓑ Ⓒ Ⓓ
80. Ⓐ Ⓑ Ⓒ Ⓓ

FINAL EXAMINATION

1. A good approach to preparing construction drawings is to

   A. draw each item fully each time it appears in the drawings so that there will be no misunderstandings.

   B. draw no details, but rather rely on contractors' shop drawings to provide all specifics.

   C. draw items fully in only one location and provide references to that location throughout the drawings.

   D. show complete details in all sections, although plans may be diagrammatic.

2. The AIA documents state that an architect is to provide a preliminary estimate of construction cost during the schematic design phase and to revise it, as necessary, during subsequent phases. Which of the following statements are true? Check all that apply.

   A. Initial cost estimates are generally based on building area or volume multiplied by historic cost factors.

   B. Once an initial cost estimate is submitted, architects should not subsequently revise it, to avoid giving the impression of having erred initially and thereby becoming liable to the owner.

   C. If area or volume calculations are used initially, they must be used in all subsequent cost estimates.

   D. The cost estimate submitted during the construction documents phase is as detailed as a contractor's quantity take-off.

3. In the specifications for a project, two different qualities of carpeting are inadvertently indicated to be installed in the same space. A flooring subcontractor notices this inconsistency during the bidding period. The proper procedure for the sub-bidder to follow, under the AIA Instructions to Bidders, is to

   A. protect the general contractor and base the bid estimate on the more expensive carpet.

   B. base the bid estimate on the less expensive carpet and assume it is probably the carpet the architect wants.

   C. call the architect and base the bid estimate on the verbal instructions given by the architect over the telephone.

   D. write to the architect requesting formal written clarification by means of an addendum.

4. The low bid on a project is $2,220,500. The architect's final preliminary estimate of construction cost is $2,000,000. Which of the following statements is true?

   A. The architect is liable to the owner because he implicitly guaranteed the $2,000,000 estimate.

   B. The architect is liable to the owner because the low bid is more than 10 percent above the architect's estimate.

   C. If the owner suffers financial loss, the architect may be liable, if he neglected to include the general contractor's overhead and profit in the final preliminary estimate of construction cost.

   D. The architect is liable only if the owner has insufficient funds to cover the total contract sum.

5. After the bid opening for a new hospital project, a contractor discovers that his estimator misinterpreted the intent of the drawings in the construction of the concrete work included in the contract. As a result, his bid is too low by $500,000. This error results in the contractor's bid being lower than the next lowest bid by approximately $550,000. In this circumstance, the owner and architect should

   A. allow the contractor to increase his bid by $500,000 and award the contract on the basis that it is still the lowest bid.

   B. allow the contractor to withdraw his bid without penalty.

   C. correct the contract documents to eliminate all possible misunderstandings and rebid the project.

   D. offer the erring bidder the contract for the amount of the original bid. Retain the bid security if the contractor refuses to enter into a contract for that amount.

6. When specifications are *open* and allow use of *equal* products, such products

   A. must be identical to the specified products.

   B. must be approved by the architect before they can be used by the contractor.

   C. are *equal* if the contractor claims they are equal.

   D. must be named in the specifications along with the original product.

7. If an owner anticipates potential economic losses if the project is not completed on schedule, he or she may make provisions in the construction contract to deduct a specific amount of money from the contract sum for each day of delay. Such deductions are referred to as

   A. retainage.

   B. cash allowances.

   C. contingency funds.

   D. liquidated damages.

8. Subcontractors may generally file liens against an owner's project

   A. if they have not been paid by the general contractor for labor and materials supplied for the project.

   B. if they have not been paid by the general contractor for labor and materials supplied for the project, and if the owner has not paid the general contractor.

   C. even if the general contractor forced them to sign a waiver of liens.

   D. for the value of labor only, not materials.

9. Using the AIA Instructions to Bidders and General Conditions, a private developer solicits bids for an apartment project. On the basis of the architect's preliminary estimate of construction cost of $1,200,000, the owner has mortgage commitments for $1,300,000. When the project is bid, there is little competition among contractors because of a surplus of available work. Based on a low bid of $1,425,000, which of the following statements is correct?

A. Having solicited bids, the owner must accept the low bid.

B. The owner may reject all bids and compensate all contractors for their costs in preparing bids.

C. The owner may reject all bids without penalty.

D. The architect is liable for the difference between the available construction funds and the low bid.

10. Select the INCORRECT statement about sequencing of construction activities.

A. Sequencing is the responsibility of the contractor.

B. Sequencing refers to the logical order for performing required activities.

C. Sequencing involves the duration of required activities.

D. Sequencing is associated with the critical path that must be followed to complete a construction project on time.

11. If conflicting or inconsistent information appears in different parts of the contract documents, the AIA General Conditions specifies which of the following methods for deciding which will prevail?

A. By order of precedence (high to low): Agreement, Supplementary Conditions, General Conditions, Specifications, Drawings.

B. By order of precedence (high to low): Agreement, Supplementary Conditions, General Conditions, Drawings, Specifications.

C. In accordance with the judgment of the architect considering the intent of the contract documents as a whole.

D. By order of precedence (high to low): General Conditions, Supplementary Conditions, Agreement, Specifications, Drawings.

12. Which of the following statements concerning the AIA General Conditions is FALSE?

A. It contains legal ground rules generally applicable to private projects.

B. It is sometimes referred to as General Requirements and included in Division One of the specifications.

C. It is usually modified by Supplementary Conditions to respond to local conditions.

D. It is part of the Owner-Contractor Agreement and should not be used by an architect for incorporation into the contract documents unless approved by the owner and/or the owner's attorney.

13. When HVAC equipment must be ordered well in advance of the date required for installation, which of the following is the most appropriate action for the architect to take?

A. Establish a construction period of sufficient length to allow the contractor to order and take delivery on such equipment.

B. Make no special provisions, since this is strictly a construction situation and solely the contractor's responsibility.

C. Work with the HVAC consultant and the owner to order the equipment in the owner's name at the appropriate time and then assign the order to the HVAC contractor when selected.

D. Inform the owner of the situation and advise that a construction manager be hired to handle such matters.

14. Upon opening one of the bids, the architect discovers that the contract sum proposal required to be stated both in words and figures reads as follows:

Twelve million, seven hundred twenty-five thousand, four hundred twenty-five dollars, and no cents ($12,775,425.00)

On the basis of the written words the contractor is the low bidder. However, on the basis of the figures, he is the second low bidder. Which of the following statements applies?

A. The contractor is the low bidder.

B. The contractor's bid is rejected because the words and figures are contradictory.

C. The contractor is the second low bidder.

D. The contractor must submit his estimate sheets to demonstrate which contract sum applies, and the award made accordingly.

15. The bids for a project are due to be submitted to the architect at noon on April 12. At 9:00 AM on April 12, one of the bidders delivers his bid to the architect. Upon returning to the office, the bidder realizes that his bid contains a miscalculation. He then prepares a corrected version of the bid and rushes to the architect's office. Upon arriving at 11:50 AM, he asks to replace the first bid with the second submittal. Based on the provisions of the AIA documents, the architect should

A. reject both bids and disqualify the bidder.

B. retain the first bid and hold the bidder to its terms.

C. return the first bid and accept the second bid.

D. advise the owner to rebid the project because of the possibility that one of the bidders has obtained an unfair advantage over the others.

16. In the event that liens are filed against a project by a contractor or subcontractor and they are not satisfied, the owner may not be able to

A. occupy the premises.

B. file a notice of completion.

C. transfer clear title to the property.

D. obtain a certificate of occupancy.

17. The bidding documents for a project include five additive alternates. A contractor intends to bid on the project, but calculates that if more than the first three additive alternates are accepted and added to his base bid, the contract sum will be in excess of his bonding capacity. On his bid, the contractor specifically states that he is bidding on the base work and the first three alternates only. Which of the following courses of action should the architect recommend?

A. The contractor's bid should be accepted since the owner can get another contractor to construct the work included in the remaining additive alternates.

B. The contractor's bid should be rejected.

C. The contractor's bid should be considered unless the owner decides to select one of the additive alternates in addition to the first three.

D. The contractor's base bid should be considered, but his bid on all additive alternates should be rejected.

18. On most projects, architects have more than one engineering consultant. In this regard, which of the following statements is correct?

A. Each consultant uses his or her own individual instructions to bidders, general requirements, and specifications format.

B. Each consultant is ultimately responsible for integrating his or her work with the work of all other consultants to produce a coordinated project design.

C. Because of his or her specialized knowledge, each consultant establishes the design criteria applicable to his or her portion of the project.

D. Each consultant should establish a quality control program similar to the architect's to evaluate design decisions and check documents for internal consistency.

19. Because of the size of the project, the owner and architect decide to divide the project into three distinct bid packages: General, Mechanical, and Electrical. Contractors are invited to bid on any one or all three parts. Contractor A submits a bid showing a separate price for each part. The bid is qualified and states that the contractor will enter into a contract for the entire project but not for the work of the individual parts. When all the bids are opened, it is clear that the lowest total price consists of Contractor A's electrical bid combined with other contractors' general and mechanical bids. Contractor A cites the qualification in his bid and refuses to sign a contract for the electrical work only. Under the provisions of the AIA documents, which of the following statements is true?

A. The owner may require Contractor A to sign a contract for the electrical work or forfeit his bid security.

B. Contractor A is within his rights and may refuse the electrical contract without penalty.

C. The owner must reject all bids and rebid the project.

D. The owner must award the entire project to Contractor A.

20. *Fast-tracking* is a term that refers to one method of project delivery. Which of the following statements is correct with regard to fast-tracking?

   A. It developed in response to inflation and the high cost of borrowing money to finance projects.

   B. There is usually only one general contractor for the construction of the project.

   C. Architects' fees are normally lower than for a traditional project delivery system because design time is shorter.

   D. Final design decisions for the super-structure must be made prior to beginning construction of the foundation.

21. For the construction of a new office building, the owner has a budget of $1,000,000. The project is bid, and a contract is signed for $979,000. After the start of construction, unforeseen subsurface conditions are discovered, requiring a change order for an addition to the contract of $25,000. Which of the following statements is true?

   A. The owner will be required to pay $21,000 of the $25,000 and the architect the difference because the construction cost exceeds the budget by that amount.

   B. The contract sum is increased by $25,000.

   C. The owner should not agree to the change order, because the contractor bears the risk of unforeseen conditions on projects with a stipulated sum contract.

   D. The architect must issue a change order to reduce the cost of some other aspect of the project by at least $25,000.

22. If the time schedule and the date of substantial completion for a project are established, all of the following circumstances may justify time extensions for the contractor, EXCEPT

   A. more inclement weather than was anticipated.

   B. occurrence of *force majeure* or an act of God.

   C. delays by an independent trucking company in delivering necessary materials ordered by the contractor.

   D. excessive time used by the architect to review product data.

23. With regard to sections of an architect's specifications, which of the following is the correct statement?

   A. They are normally written so that the work described in each section of the specifications is let to one subcontractor.

   B. In the CSI format, each section is comprised of 16 divisions.

   C. They are written to form an integrated set of specifications for the project as a whole. The contractor determines which trade performs the work described in individual sections.

   D. Although the format for specifications may vary, each section must be consistent and contain either performance criteria or proprietary names.

24. On many projects, 10 percent of each progress payment to the contractor is retained by the owner. When and under what circumstances may retained moneys be paid to a contractor who has been required to provide a performance bond?

A. When the architect deems that the project's progress is satisfactory, the contractor's work is acceptable, and the owner no longer needs the protection.

B. If it is in accordance with the terms of the owner-contractor agreement, and the contractor's surety company consents.

C. As soon as the project is more than 90 percent complete, the architect may reduce retainage in proportion to the amount of work completed.

D. At final completion only.

25. To try to obtain the best project for the least cost, the owner and architect should seek bids from

A. two of the best known contractors.

B. three of the largest general contractors.

C. four to six of the most competitive contractors.

D. all the contractors within a 25-mile radius.

26. The owner and architect request bids comprising a base bid and two additive alternates. Three contractors submit bids as follows:

Contractor A

Base	$547,000
Add. Alt. 1	13,000
Add. Alt. 2	11,700
	$571,700

Contractor B

Base	$528,000
Add. Alt. 1	28,000
Add. Alt. 2	20,000
	$576,000

Contractor C

Base	$534,000
Add. Alt. 1	12,200
Add. Alt. 2	27,200
	$573,400

The owner decides to proceed with the project on the basis of the base bid plus alternate 2 only. Which of the following statements applies?

A. Contractor C should be awarded the contract because his base bid combined with the first alternate is low.

B. Contractor B should be awarded the contract because he submitted the lowest base bid.

C. Contractor A should be awarded the contract because he submitted the lowest total bid.

D. Contractor B should be awarded the contract because his base bid and alternate 2 is low.

27. Architects generally inform their engineering consultants about applicable codes related to the project under consideration. However, their list may be incomplete and architects may not be able to verify that the consultants have complied with all applicable codes. Under these circumstances, which of the following statements is true?

A. An architect cannot expect his or her consultants to comply with codes unless he or she has specifically pointed out those which are applicable.

B. Consultants are directly liable to the owner if they negligently fail to comply with codes.

C. Because it is difficult for an architect to check consultants' work in detail, an architect will not be liable for consultants' errors.

D. Architects are responsible for lack of code compliance. They may in turn place responsibility on their consultants for errors within their area of responsibility.

28. Architects and consultants are frequently asked to estimate the operating costs of alternative designs or systems for a project. Which of the following factors affect such estimates? Check all that apply.

A. Site location and orientation

B. Management policies of the owner/tenant

C. Energy costs

D. Quality of the maintenance staff

E. Number of occupants in the project

29. Under the AIA General Conditions,

A. property insurance for coverage during construction is purchased by the contractor.

B. builder's risk insurance for coverage during construction is purchased by the contractor.

C. property insurance coverage is voided if the owner occupies all or part of the project without the insurance company's consent.

D. required builder's risk insurance covers named perils.

30. Under the provisions of standard AIA documents, an owner must provide the architect with a program and budget for the proposed project. Which of the following statements are true?

I. The architect is entitled to rely on the fact that the project as proposed can be built for the budget established by the owner.

II. The architect must determine if the program and budget are reasonably related, and inform the owner if they are not.

III. The architect is entitled to rely on the fact that the budget established by the owner is for construction costs only, and that there are additional funds to cover other project development costs.

IV. The budget provided by the owner is a fixed limit of construction cost, and the architect is obligated to design the project within that limit.

A. I, III, and IV

B. I and IV

C. II only

D. II and IV

31. At the completion of all construction work, contractors receive final payment due under their contracts for construction. If the AIA General Conditions was used for a project and five months later the owner discovers that a portion of the work does not comply with the contract documents, which of the following statements is true?

   A. Final payment by the owner constitutes acceptance of the project and the owner has no recourse.

   B. The owner may be able to sue the contractor for damages, but cannot require the contractor to correct the non-complying work.

   C. The owner can require the contractor to return to the project and correct the non-complying work.

   D. Unless the architect noted the non-complying work during the construction phase, the owner has no recourse against the contractor.

32. When reviewing a contractor's application for payment, an architect must

   A. check the calculations only, since the owner provides all accounting services necessary for the project.

   B. verify that the project has reached the overall percentage of completion claimed by the contractor and that the total amount applied for equals the same percentage of the total contract sum.

   C. verify that the completion of construction of each element of the schedule of values is as claimed and that the amount requested for each line item is in proportion to its percentage of completion.

   D. determine that the contractor has used all previous payments for the sole purpose of compensating subcontractors and others working on the project.

33. An owner's budget for a project is $1,000,000. The architect subsequently agrees with his engineering consultants on budgets for their parts of the project as follows: Mechanical = $350,000, Structural = $110,000. The low bid subsequently turns out to be $1,100,000, consisting of: General = $540,000, Mechanical = $480,000, Structural = $90,000. The architect should

   A. ask the mechanical consultant to redesign his part of the project to reduce its cost by $130,000.

   B. ask the mechanical consultant to redesign his part of the project to reduce its cost by $100,000.

   C. redesign the architectural and general elements to reduce their cost by $100,000.

   D. require that all design team members redesign their parts of the project to reduce their respective costs by 10 percent each.

34. If the architect visits the construction site and observes a dangerous condition that could endanger the life of a worker, the architect should take which one of the following actions?

   A. Take no action, since safety is the responsibility of the contractor.

   B. Immediately call the local office of the Occupational Safety and Health Administration (OSHA) to report the situation.

   C. Inform the contractor's superintendent of the problem and tell the superintendent how to correct it.

   D. Describe the situation to the contractor's superintendent and send a written report to the owner and contractor, without suggesting a solution.

35. Performance bonds and labor and materials payment bonds are generally written for which of the following limits?

A. The limit for each bond is the full amount of the construction contract.

B. The projected amount of the general contractor's overhead and profit is the limit for the performance bond, and the value of all subcontracts is the limit for the labor and materials payment bond.

C. They may be divided into any proportion as long as the total of their limits equals the construction contract amount.

D. Seventy-five percent (75%) of the value of the construction contract is the limit for the performance bond and twenty-five percent (25%) is the limit for the labor and materials payment bond.

36. Contractors submit shop drawings for the architect's review and *approval* or *other appropriate action*. In reviewing these, architects

A. are liable for any mistake shown on the shop drawings unless corrections are marked for the contractor's attention.

B. approve shop drawings for conformance with the design concept and the information in the contract documents.

C. should request that shop drawings be submitted directly from subcontractors to save time and expedite the review process.

D. will be liable for delaying the project if they require the contractor to revise and resubmit shop drawings.

37. During the course of construction of a new elementary school, the contractor periodically complained that the architect was slow in reviewing and returning shop drawings and that this might affect the project's construction schedule. The contractor made no written claims, but the project representative recalled some of the conversations concerned with this matter. Completion was actually delayed by two months; however, the contractor accepted final payment. The following week, the contractor sued the owner to recover costs for the two-month delay. Which of the following statements is true?

A. Since the architect is under contract to the owner and did not return shop drawings in a timely manner, causing the delay, both owner and architect are liable to the contractor.

B. Since the contractor made no claims in writing at the time he or she accepted final payment, all claims were waived.

C. The fact that claims were not made in writing is irrelevant, since the architect's project representative remembers the conversations concerning delays.

D. The contractor will be able to recover only if it can be proven that the delay was caused by *force majeure*.

38. Life-cycle costing is an economic evaluation of architectural elements that includes which of the following factors?

I. First cost

II. Maintenance and operational costs

III. Repair costs

IV. Replacement cost

A. I

B. II, III, IV

C. II, IV

D. All of the above

39. LEED, the name of a program that environmentally evaluates sustainable projects, is a checklist that is concerned with which of the following? Check all that apply.

A. Indoor air quality

B. Storm water

C. Innovative energy systems

D. Aesthetic design

40. Sustainable design may require research and education that is beyond a normal architectural project. Which of the following is part of this process?

I. Energy modeling

II. Education of the client

III. Art selection

IV. Selection of energy efficient appliances

A. I, IV

B. I, II

C. I, II, IV

D. All of the above

41. The site survey for a proposed hospital building indicates an existing easement and sanitary sewer line that is located in an area designated for the construction of a multistory parking structure. Special instructions regarding this sewer line and the easement should be included in which of the following documents?

I. General Conditions

II. Instructions for Bidders

III. Supplementary Conditions

IV. Special Conditions

V. Technical Specifications

A. I and III

B. II and V

C. IV only

D. V only

42. Which of the following is NOT a valid reason for the architect to withhold payments to the contractor?

A. Contractor's request for an extension of time

B. Contractor's failure to submit a monthly progress report

C. Damage to another contractor

D. Defective work not remedied

43. To expedite construction, the ceiling subcontractor installs a portion of the suspended ceiling prior to the mechanical engineer's inspection and approval of the ductwork above. The engineer requests that the ceiling be dismantled in order to examine the mechanical work for conformance with the drawings and specifications. The cost to remove and reinstall the ceiling must be borne by the

A. ceiling subcontractor.

B. architect.

C. owner.

D. contractor.

44. Which of the following are NOT considered to be a part of the contract documents?

I. Technical specifications

II. Instructions for bidders

III. Addenda

IV. Supplementary conditions

V. Punch list

A. I and V

B. II and IV

C. III and V

D. II and V

45. Four months after the final completion and acceptance of a building, it is determined that the heating system has not been installed in accordance with the specifications and is therefore unable to maintain inside design temperatures under normal outside temperature conditions. Under these circumstances, the owner should

A. ask the mechanical subcontractor to remedy the situation under the warranty of the work.

B. hire an independent contractor to solve the problem and charge the general contractor for the cost of the work.

C. ask the general contractor to make the system perform in accordance with the specifications of the construction contract at no cost to the owner.

D. hire an independent mechanical contractor, and absorb the cost to amend the system because the problem was detected more than 90 days after final completion and acceptance of the project.

46. During the course of construction, the electrical engineer discovers that a transformer vault is constructed several feet to the east of its intended location. Upon investigation, it is found that an addendum had been issued to revise the location from that shown in the original drawings. The job superintendent claims he did not receive a copy of the addendum to indicate the revised location and therefore placed the vault in accordance with the original drawings. Several months earlier, however, at the bid opening, the successful bidder acknowledged receipt of all addenda. Under these circumstances, which of the following is correct?

A. The architect is liable, since it is the architect's responsibility to verify that the general contractor's job superintendent has a complete set of all documents prior to the start of construction.

B. The engineer is liable since his agreement with the architect concerning the construction administration phase makes him responsible for work incorrectly performed.

C. Neither architect nor engineer is liable since their only responsibility, in this case, is to determine whether pertinent contract documents have been furnished to the contractor at the appropriate times.

D. The architect is not liable for any errors committed during the course of construction regardless of their origin.

47. The purpose of the performance bond is to assure that the contractor

A. employs skilled workmen and uses first quality building materials.

B. can be bonded for projects in excess of five million dollars.

C. guarantees that his bid amount is a fixed limit of construction cost.

D. will perform all the terms and conditions of the contract for construction.

48. A period during which an architect can be subject to legal action resulting from injury or damage caused by deficient design is called (the) ______________.

49. During the course of construction, it becomes apparent that special emergency precautions must be taken by the contractor to prevent possible damage to an adjacent building on the site. Under these circumstances, the contractor

A. must obtain special instructions or authorization from the architect.

B. may act at his own discretion to prevent such threatened loss.

C. would probably be compensated for such emergency work.

D. is permitted to act with discretion only in situations where there is a threat to human life.

50. Seven months after the start of construction, the project is four weeks behind the approved construction schedule because of delays attributable to the general contractor. It becomes apparent that unless drastic steps are taken, the project completion will be delayed by at least four weeks. Under these circumstances, which of the following is a correct statement, according to the AIA General Conditions?

A. The contractor will be entitled to request an increase in the total contract price if the work is accelerated.

B. The contract time will be extended by change order as determined by the architect.

C. The contractor may request an extension of time at any time prior to substantial completion.

D. The owner may be eligible to recover damages caused by the delay in completion of the work.

51. During the construction of a speculative office building, one of the subcontractors informs the architect that he has not been paid by the general contractor for certain work required by the construction documents and that, as a consequence, he plans to file a lien against the project. Under these circumstances the architect should NOT

A. advise the subcontractor that by law he cannot file a lien against the project.

B. inform the surety of the subcontractor's intentions.

C. inform the contractor that final payment may be withheld until the claim is satisfied.

D. inform the owner of the subcontractor's intent to file a lien.

52. Which of the following has the right to order a temporary suspension of all or part of the construction work?

    A. The architect

    B. The owner

    C. The contractor

    D. The structural engineer

53. During the structural engineer's inspection of excavation prior to placing concrete footings, he notices that an area of excavation approximately ten feet deep is without barricade or protection and poses a potential hazard. In the absence of the contractor's superintendent, the engineer informs the architect of this condition. The architect should

    A. suspend all work on the job and instruct the superintendent to erect a guard rail.

    B. prepare an Architect's Supplementary Instruction form delineating the barricade while work proceeds.

    C. notify both owner and contractor in writing of his observations on the job.

    D. inform the contractor's insurance company of the condition as it exists on-site.

54. What is the minimum number of principal entrances to a public building that should be accessible to the handicapped?

    A. At least one

    B. One per 10,000 gross square feet of building

    C. One for each two floors of a multistory building

    D. One for each 200 lineal feet of building frontage

55. A large addition to a state prison includes a facility to accommodate a gas chamber. One of the contractors submitting a bid has publicly opposed capital punishment, and therefore, does not want his name to be associated with this job. Under these circumstances, the State Architect's Office would probably

    A. suggest that the contractor refrain from submitting a bid.

    B. respect the contractor's wishes and not disclose his name.

    C. reveal the contractor's name only to those who attend the bid opening.

    D. disclose the contractor's name as part of the public record of this project.

56. In an effort to save money, an owner instructs the architect to delete the specified waterproofing of exterior masonry walls. In response, the architect should

    A. resign from the project after submitting the proper written notice.

    B. advise the owner in writing that leaks in the walls may occur.

    C. refuse to guarantee the water resistance of the walls.

    D. suggest that the contractor take responsibility for the walls.

57. As a result of several complaints of nonpayment by subcontractors, it appears that the contractor may have mismanaged funds paid to him during the course of construction. An audit is ordered to determine where the money has been spent, and the cost of this investigation is paid for by the

    A. architect.

    B. contractor.

    C. owner.

    D. bonding company.

58. The preparation of record drawings is

I. often requested by the contractor.

II. usually done by the architect.

III. occasionally included in the basic architectural services.

IV. normally compensated as a change in services.

A. II only

B. I and III

C. II and IV

D. I, III, and IV

59. In submitting a bid, a contractor includes several amounts that are added to, or deducted from, the base bid if corresponding changes in project scope are accepted by the owner. These amounts are known as ____________________.

60. Tactile warnings on door operating hardware would be appropriate in all of the following locations EXCEPT

A. stages.

B. loading platforms.

C. mechanical rooms.

D. emergency exits.

61. The roofing on a project is installed prior to the architect's inspection of the roof sheathing nailing. This inspection, however, is not called for by the contract documents, nor was it previously mentioned by the architect. Nevertheless, the contractor removes part of the roofing, which discloses general compliance with the project specifications. Consequently, the contractor sends a bill to the owner for the cost of reroofing, and as a result, the architect

A. advises the owner to pay the bill.

B. advises the owner to ignore the bill.

C. offers to pay the bill, recognizing the negligence in not requesting the inspection at the proper time.

D. advises the contractor to back charge the roofing contractor.

62. If a contractor fails to complete the work within the time stipulated and agreed to, the owner might take which of the following actions? Check all that apply.

A. Dismiss the contractor

B. Replace the contractor

C. Arbitrate the matter

D. Collect liquidated damages

E. Collect a monetary penalty

63. If changes in the work result in an increase in the project cost, the contractor would be paid an amount

A. equal to the actual cost of materials and labor.

B. equal to the actual cost of the work plus overhead and profit.

C. negotiated after the work has been performed.

D. negotiated through the process of arbitration.

64. Several months after occupancy of a new building, large areas of hardwood flooring begin to buckle and warp. Because all payments to the contractor have been made, the architect's best course of action would be to

    A. advise the owner that it is now her responsibility to correct the defect.

    B. advise the flooring contractor to correct the defect at his own expense.

    C. advise the general contractor to correct the defect at his own expense.

    D. advise the general contractor to correct the defect and send the bill for extra services to the owner.

65. The period of time during which a claim against the owner's property may be brought for alleged nonpayment of labor is referred to as the ________________.

66. The carpet specifications require the contractor to furnish proof that similar installations using the material meeting the performance specification have been in continuous use for at least five years. The contractor's submittal of carpet samples for approval does not include this required information. The architect determines that the material submitted has been manufactured for less than three years, and although it meets all other requirements, there are no existing installations to meet the five-year requirement. The contractor claims his bid is based on the material submitted and that if a substitution is required, it will result in an addition to the contract. In response, the architect should do which of the following?

    A. Reject the carpet, request a substitution, and inform the owner of the anticipated cost

    B. Waive the clause in the specifications calling for the five-year performance demonstration

    C. Reject the carpet sample and insist that the contractor meet the performance requirements at no extra cost to the owner

    D. Offer to split the difference with the owner between the cost of the proposed installation and that specified

67. During the construction phase, the owner decides to call for bids for a separate contract for an electronic security system. The general contractor objects, claiming he must be allowed to do this work as an extra item, because the conduit to carry the wiring for the system was installed under the contract. In such an event, the owner

A. has the right to let additional work not be included in the basic contract for construction.

B. may ask for bids for the additional work but allow the general contractor first right of refusal.

C. has the right to let such a contract, but only with the general contractor's permission, who may refuse on the grounds of potential interference with his own work schedule and access to the project.

D. must negotiate an extra payment to the general contractor or delay installation of the security system until after final completion of the project.

68. An architect decides to employ outside consultants for the design of both mechanical and electrical systems. When an agreement is reached between the architect and the consultants, the responsibility to the owner for the adequacy of the mechanical and electrical systems rests with which of the following?

A. The architect alone

B. The consultant alone

C. The architect and consultant equally and jointly

D. The mechanical and/or electrical contractors

69. In regard to submittal, review, and approval of shop drawings, which of the following is an incorrect statement?

A. The contractor must verify catalog numbers.

B. The architect should require shop drawings only for those items that truly need them.

C. The contractor should furnish a submittal schedule to the architect.

D. The architect reviews submittals and checks for compliance with field measurements.

70. Where the provisions of the AIA General Conditions, Document A201, are incompatible with the competitive bidding laws in the state where a public works project is to be built, what is the architect's best procedure in writing the conditions of the contract for the project?

A. Call the bidder's attention to the specific differences in the invitation to bid

B. Amend the specific paragraphs that are at variance in the Supplementary Conditions

C. Request the State Public Works Board to prepare its own general conditions

D. Rewrite the AIA General Conditions and issue them as her own

71. Additional professional services performed by the architect, made necessary by the default of the contractor, must be paid for by the owner in accordance with the Owner-Architect Agreement (AIA Document B201). The owner can claim reimbursement for these fees under which of the following?

A. Liquidated damages clause

B. The bid bond

C. The performance bond

D. Labor and material payment bond

72. Under the AIA General Conditions, which of the following statements concerning a subcontractor is correct? Check all that apply.

A. The subcontractor's relationship to the owner is the same as that of the general contractor.

B. The subcontractor is subject to approval by the architect.

C. The subcontractor has a direct contractual relationship only with the general contractor.

D. The subcontractor is bound to the general contractor under the same general conditions that the general contractor is bound to the owner.

73. The *owner* is generally considered to be the person or entity who

A. provides the construction funds.

B. has the authority to make construction decisions.

C. signs the owner-contractor agreement.

D. is identified as such in the owner-contractor agreement.

74. During a prolonged period of unexpected freezing weather, the contractor is forced to provide temporary enclosures and heating devices to complete the concrete work on a project. The contractor later submits a claim for increased costs, which the owner considers excessive and refuses to pay. The project architect is therefore obliged to

A. determine the amount due and advise both parties.

B. advise the two parties to arbitrate the matter.

C. suggest that the owner pay the amount requested.

D. recommend that the contractor file a mechanic's lien.

75. Contingent liability refers to an architect's legal responsibility for the acts of the

A. architect.

B. architect's family.

C. architect's employees.

D. architect's consultants.

76. Before issuing a final certificate for payment, an architect should receive evidence that the contractor has paid for all project costs, including

I. workers' wages.

II. building materials.

III. soils engineers' fees.

IV. licensed surveyors' fees.

V. rental of construction equipment.

A. I and II

B. III and IV

C. I, II, and V

D. I, II, III, IV, and V

77. According to the AIA General Conditions, the contractor normally forwards communications to the owner

A. directly.

B. periodically.

C. through the architect.

D. by registered mail.

78. During the finishing phase of construction, the architect notices that the toilet room mirrors, which were designed to fit wall-to-wall in special alcoves, are actually eight inches narrower than shown on the drawings. The architect's recommended solution should be to

A. center the mirrors in the alcoves and paint a black, four-inch wide strip on either side.

B. center the mirrors in the alcoves and add a four-inch wide strip of matching mirror on either side.

C. furr out the alcove sides to accommodate the existing mirrors.

D. reject the mirrors and request that they be replaced in the required width.

79. If an owner wished to employ a specific contractor, and their contract was negotiated prior to completion of the construction documents, their agreement would probably be based on

A. the architect's estimate of probable construction cost.

B. a percentage of the construction cost.

C. a stipulated sum.

D. a cost plus fee arrangement.

80. According to ANSI Standard 117.1, an *accessible route* is one that is safe and usable by a

I. person in a wheelchair.

II. blind person.

III. person on crutches.

IV. person who is mentally retarded.

V. person who does not speak English.

A. I and III

B. II, IV, and V

C. I, II, III, and IV

D. I, II, III, IV, and V

The examination answers and explanations will be found on the following pages.

Do not look at the answers until you have completed the exam.

EXAMINATION ANSWERS

1. C Answer C is correct because errors are less likely to be copied or repeated, and changes may be made more easily since only one drawing has to be changed. These are the disadvantages with answers A and D. Answer B improperly relies on contractors to make design decisions. Shop drawings are intended to show installation details, not the design intent.

2. A Statement A is true. Statement C is false, since it is common practice to use more detailed estimating methods during the later phases, such as those involving subsystems. Statement D is also incorrect since architects generally prepare detailed estimates as a change in services. Statement B is incorrect because many factors other than the architect's ability to prepare schematic estimates could be involved, such as changes requested by the owner, increased engineering input, and changes in market conditions.

3. D The correct answer is D because of the specific requirements of Paragraph 3.2 of the Instructions to Bidders. This procedure allows the architect to correct the inconsistency and instruct all bidders to bid on the same basis. Answer A protects the general contractor but, if allowed, would inflate bid prices, and not all contractors would necessarily be bidding on the same basis. Incorrect answer B has disadvantages similar to answer A and may result in a potential claim for extras if the more expensive carpet is subsequently selected. The Instructions to Bidders specifically states that contractors may not rely on verbal interpretations (answer C is incorrect).

4. C Under the provisions of the AIA documents and normal practice, architects neither explicitly nor implicitly guarantee their estimate of construction cost (incorrect answer A). However, they may not be negligent in making such estimates. Answer C is the only answer that mentions negligence, and is therefore the only true statement. A bid more than 10 percent higher than the estimate does not suggest liability, unless the 10 percent figure is agreed to in the owner-architect agreement (B is incorrect). Answer D is incorrect because liability is not a result of one party's lack of funds.

5. D Strictly speaking, contractors whose bids contain clerical or computation errors of this magnitude may be allowed to withdraw their bids without penalty, but not for mistakes involving judgment. Since there is no indication that the documents were unclear or in error, or that interpretations could not have been provided by the architect under normal procedures, there is not a valid reason to rebid the project (correct answer D).

6. B *Equal* products do not have to match those specified in every detail. Generally, some details and characteristics may not be required. Thus, answer A is incorrect. Equal products do not have to be named in the specifications; the intent is to allow products with which the architect may not be familiar. Therefore, answer D is also incorrect. Answers B and C are mutually exclusive. Only B is correct; otherwise, design decisions would be made by the contractor.

7. D Retainage (A) is not specifically related to construction completion delays. Owners retain a percentage of progress payments to the contractor as partial security to insure completion of the construction work. Cash allowances (B) are used in specifications to provide for situations where scope or quality cannot be fully determined at the time bids are solicited. Contingency funds (C) are established by the owner to cover unforeseen expenses. Liquidated damages are a method of fixing in advance a good faith estimate of damages to the owner if construction is not completed on time (correct answer D).

8. A Liens are designed to protect the interests of those who supply labor and/or materials for construction projects. Thus, answer D is incorrect. Submitting proper waivers of liens will preclude subcontractors from filing liens. Therefore, answer C is also incorrect. B is also incorrect; whether the owner has paid the general contractor is irrelevant. Only answer A is correct. To avoid the filing of liens, the owner must require the general contractor to make payments to subcontractors.

9. C Under the provisions of the AIA documents, private owners may reject any or all bids. The instructions also state that the owner *intends* to award a contract to the low bidder, depending on the availability of sufficient funds. Thus, bidders are put on notice that the project might not proceed and that they must be willing to accept that risk. An architect is not liable for bids exceeding available funds.

10. C The incorrect statement we are looking for is C, which describes *scheduling* as opposed to *sequencing*. All of the other statements accurately relate to or describe the sequence of construction activities.

11. C The AIA General Conditions states that all contract documents are complementary. Discrepancies or conflicts are to be brought to the attention of the architect who will interpret the documents as a whole. Systems of precedence are sometimes included in the Supplementary Conditions, not the AIA General Conditions.

12. B The AIA General Conditions is generic and applicable to most private projects. Consequently, it must be adapted to the specific requirements of each project by the Supplementary Conditions. Only attorneys are permitted to prepare contracts to which they are not a party. Architects are not parties to the owner-contractor agreement, and they cannot unilaterally prepare a portion of that contract (statements A, C, and D are true). Division One of the specifications is called the General Requirements, but it is distinct from the General Conditions. The first is administrative in nature, the second is legal (statement B is false and is therefore the correct answer).

13. C Answers A and B do not expedite the project nor contribute to its orderly progression. They both effectively abandon the owner, who looks to the architect for advice on completing the project promptly and efficiently. Although construction managers perform this type of service, this situation does not necessarily warrant hiring a construction manager (answer D is incorrect). An architect and the engineering consultants can work with the owner to place orders. When a contractor is selected, the order can be assigned to the contractor and the equipment can be received and installed in normal fashion (correct answer C).

14. A The written words prevail in any discrepancy between words and figures. In this case, the words make the contractor the low bidder (A is correct, C is incorrect). Because of this rule of interpretation, answer B is incorrect. A bid is a binding offer without regard to the process leading to its submittal (answer D is incorrect).

15. C Up until the advertised deadline for bid submittals, contractors may withdraw any previously submitted bids and replace these with revised or new bids (C is the correct answer). Furthermore, since the bid replacement occurred prior to the bid opening deadline, the bidder in question cannot obtain an advantage over the other bidders. If the second bid had been submitted *after* the deadline, however, the architect would be justified in recommending the action in answer B.

16. C Liens provide security that those who supply labor or material for a construction project will be paid. If they are not paid in the normal course of business, they have the right to have the property sold in order to be paid from the proceeds. This affects the ownership of the property (answer C is correct). None of the other choices are correct because they are unaffected by whether or not work has been paid for.

17. B According to subparagraph 4.1.5 of the AIA Instructions to Bidders, all alternates (additive or subtractive) must be bid; otherwise, the owner will have inconsistent bids (correct answer B). Any other course of action could lead to legal action taken by any or all of the other bidders and is not in accordance with bid instructions.

18. D Each project manual and set of construction documents must be an integrated whole, not merely a collection of disparate parts. Cooperation is essential, but the *architect* is the prime professional who must produce an integrated set of contract documents (A and B are incorrect). Answer C similarly implies too large a shift of responsibility. Consultants advise architects on design criteria, but the criteria must be determined by the architect to meet the owner's program needs (answer D is correct).

19. B The AIA Instructions to Bidders, in subparagraph 4.1.6, allows bidders to qualify their bids in the circumstances described. Contractor A has responded to these provisions and acted appropriately (answer B is correct).

20. A Fast-tracking was developed to reduce the total time required to design and build a project. The prime impetus was the high cost of borrowing money, along with the price increases associated with inflation (answer A is correct). Because many construction contracts are awarded as the project develops, there is usually a construction manager on fast-track projects, instead of a single general contractor (incorrect answer B). Normally it is the construction period, not the design period, that is shortened. Furthermore, since the architect's coordination is more time-consuming and difficult, design fees are generally higher than normal on fast-track projects (answer C is incorrect). To begin foundation construction, only column locations and loads are required. Many other design decisions can be made as construction proceeds. Thus, answer D is also incorrect.

21. B The contract sum must be increased by the total sum of all change orders (answers A and D are incorrect, B is correct). Answer C is incorrect, since unforeseen subsurface conditions that increase the contractor's cost generally result in an increase in the contract sum.

22. C The AIA General Conditions specifies many of the general and specific events that entitle a contractor to extensions of time for completion of construction. The key factor is that the event must be outside of the responsibility, control, or reasonable anticipation of the contractor. All of the factors mentioned meet those criteria except answer C. While neither the owner nor the contractor controls the actions of the trucking company, delivery of materials is the contractor's problem, and delays of this type do not justify time extensions under the provisions of the contract.

23. C C is correct and A is incorrect, because the contractor determines which trade or subcontractor performs work, rather than having this determined by the format of the specifications. Answer B is incorrect because in the CSI format, there are 16 Divisions, comprised of many Sections, each with three Parts. Answer D is incorrect because it *is* acceptable to use various methods of specifying within sections, so long as each method is appropriately used.

24. B Answer B is correct. It is critical that the surety company consents to the release of retained sums. Without the surety's consent, the provisions of the performance bond may be voided, increasing the owner's risk should the contractor default. Answer A is incorrect because an architect should not unilaterally determine the extent of protection required by the owner. Answer C is incorrect because it ignores the role of the surety and solely depends on the architect for a decision. Answer D is incorrect because retainage may, under certain circumstances, be reduced prior to final completion.

25. C The correct answer is C because four to six bidders will generally assure adequate price competition without making a bidder's possibility of being successful too remote. If only two contractors are invited to bid, regardless of reputation, the lack of competition will result in a higher cost to the owner (answer A is incorrect). Three contractors may be appropriate for small projects where the number of qualified contractors is limited. For major projects, this number is too small to create sufficient competition (answer B is incorrect). Well-qualified contractors may be discouraged by the large number of bidders and the resulting low probability of getting the bid award. If contractors don't have a realistic chance to succeed, they are unlikely to bid (answer D is incorrect).

26. D Subparagraph 5.3.2 of the AIA Instructions to Bidders states that the owner may select any alternates (in any order) and that the contract award will be made on the basis of the combined cost determined by the base bid and the accepted alternates (answer D is correct).

27. D Answer A is incorrect because, under AIA documents, architects rely on the expertise of their consultants. The architect's consultants do not have a contract with the owner, and thus are not usually responsible to the owner for their professional mistakes (incorrect answer B). Answer C is also incorrect because liability and difficulty are not directly related; an architect may be liable whether a task is easy or difficult. Answer D is the only correct statement.

28. A, B, C, D, and E Operating costs are difficult to predict accurately. All assumptions must be clearly stated and consistent so that alternatives may be compared on a relative basis. Site location and orientation (A) affect the microclimate at a site and therefore the energy consumed. Energy costs must be considered (C). If the maintenance staff is experienced and does not defer maintenance, equipment will operate more efficiently (D). The number of people using a building (E) affects the demand for services, as does management policy concerning conservation (B). Thus, all the factors listed affect predictions of operating costs.

29. C Under the theory that the owner of property should insure it, the AIA General Conditions provides that the owner, not the contractor, must purchase property insurance. Builder's risk insurance *is* the property insurance during construction, and thus answers A and B are both incorrect. Answer D is incorrect because the AIA documents specify that *all risk* coverage be purchased, which covers all risks *except* named perils. Answer C is correct. Architects and owners must be aware of this requirement when all or any part of a project is substantially complete.

30. C The duties and responsibilities of both the owner and architect for cost estimates are found in the AIA Owner-Architect Agreement, Document B201. It specifically negates the intent of statement IV. It also states that the architect will balance program and budget as in statement II. Since statement II is true, statement I is false. The documents do not explicitly describe what is included in the owner's budget. Generally, however, the owner's project budget includes development costs, such as furniture, equipment, testing, inspections, etc., in addition to the basic costs of construction. To balance program and budget, the architect must be informed of which items are covered by the budget (statement III is incorrect). Therefore, only statement II is true (correct answer C).

31. C The AIA General Conditions states that, by making final payment, the owner does not waive his or her claims against the contractor for work that does not comply with the contract documents, regardless of whether the architect noted the non-complying work during construction (answers A and D are incorrect). During the first year after substantial completion, the owner may require the contractor to return to the project and correct or replace defective work. See subparagraph 12.2.2, AIA General Conditions, Document A201 (answer C is correct, answer B is incorrect).

32. C The AIA documents specifically provide that the architect has no duty to verify how the contractor has used money paid by the owner (answer D is incorrect). Because answer A does not equate money to the level of completion, it is also incorrect. The total percentage of completion is not directly related to the percentage of completion for each of the various elements and subcontracts of work. Payment to the contractor is made on the basis of completion of each item shown in the schedule of values (answer B is incorrect). Only answer C corresponds to the requirements of the general conditions. Furthermore, it represents good practice because it is the most specific and accurate way for an architect to evaluate a contractor's application for payment.

33. B The architect and each consultant are committed to a budget and, in accordance with AIA documents, to redesign if the cost of the work exceeds the budget (answer C is incorrect). The architect has achieved the goals for the architectural part of the project (Budget = $1,000,000 less $350,000 less $110,000 = $540,000, and the cost = $540,000). If the architectural elements were required to bear the burden of reduced costs, they would be out of proportion with the remaining elements. Answer D might be appropriate only if no specific budgets had been agreed upon with consultants. Although the architect would, in accordance with the agreement, be justified in asking the mechanical consultant to reduce the cost of the mechanical work by $130,000 ($480,000 – $350,000), it is not necessary to bring the project within budget. A reduction of $100,000, as described in answer B, will bring the project into compliance with its overall fixed limit ($1,100,000 – $100,000 = $1,000,000).

34. D Safety on the job is the responsibility of the contractor. However, an architect should not ignore potentially dangerous situations. He or she should inform the responsible party (the contractor, via the superintendent) of the potential danger and let them rectify the situation. A written record should be made to document the incident (correct answer D). Answer A could result in the architect being held partially liable if the architect fails to report what he or she observed and someone is hurt. Answer B suggests a rash approach that may not necessarily correct a dangerous situation in a timely manner. Answer C is incorrect because, by suggesting or ordering actions, the architect may become liable for consequences that would otherwise be the contractor's responsibility.

35. A The performance bond and payment bond are usually written together, though they do not cover the same risks. Each is normally written for the full amount of the construction contract and there is generally no reduction in premiums if lesser limits are purchased. Although this is somewhat arbitrary, experience has shown this practice to be reasonably effective and economical (answer A is correct).

36. B According to the AIA General Conditions, architects review shop drawings only for the purposes stated in correct answer B. Shop drawings may contain some information that the architect does not review, and, thus, answer A is incorrect. Answer D is incorrect, since requiring shop drawings to be revised is one of the *other appropriate actions* contemplated by the General Conditions. Answer C is incorrect because the AIA General Conditions requires the general contractor to review and approve shop drawings prior to submitting them to the architect. Architects reviewing shop drawings without a contractor's stamp of approval may assume the contractor's potential liability.

37. B The AIA General Conditions is very explicit in stating that acceptance of final payment by a contractor constitutes a waiver of all claims except those previously made *in writing* (answers A and C are incorrect). Answer D is incorrect because *force majeure* involves events beyond the control of either party (such as floods or other acts of God), and not relevant to this situation (answer B is correct).

38. D I is correct. While first cost is not the primary concern of life cycle costing, it is one of the economic factors considered. II is also correct. The cost of maintenance is part of the evaluation. III is correct as well. The durability of a product or system is considered in the cost of repair and part of the overall evaluation. IV is correct because the comparison of product or system life is one of the factors evaluated in life cycle costing.

39. A, B, and C A is correct. LEED has several options for improving IAQ (Indoor Air Quality) including filtering the air system and installing low-VOC (Volatile Organic Compound) paints and caulking. B is also correct. Methods to store, recirculate, and locally distribute rainwater are encouraged. C is correct as well. Innovative solutions to energy conservation, such as fuel cells, photovoltaic panels, and gas turbine energy production, are encouraged in the LEED accreditation system. D is incorrect. The LEED system awards no points for designs with strong aesthetics.

40. C I is correct. Computer programs that allow energy modeling of design options allow the architect a quick method of evaluating numerous different solutions. II is also correct. It is extremely important that the client be able to understand the value of sustainable design solutions. III is not correct. Art selection is at the client's discretion. IV is correct. Locating the most energy efficient appliances, plumbing fixtures, and office equipment will improve the energy efficiency of the entire project.

41. D Information of a technical nature dealing with conditions that affect the development of the site and building would be included in one of the divisions of the technical specifications, correct answer D, since the other choices are concerned with non-technical aspects of the project.

42. A All of the choices, except correct answer A, are grounds for withholding payments to contractors. A request for an extension of time, if granted, will result in a change order and a change in the contract time, but does not constitute a valid reason for withholding payments to the contractor.

43. D Paragraph 12.1 of the AIA General Conditions requires inspection and approval by the architect/engineer of all work following its installation and prior to its being covered or enclosed. Any work covered up without approval or consent by the architect/engineer must be uncovered at the contractor's expense (correct answer D).

44. D The Instructions for Bidders (II) is an element of the bid documents provided for information only, and does not form a part of the contract. The punch list (V) is a memorandum of work to be done and corrected prior to final completion. It is compiled during the latter stages of construction and does not constitute a part of the contract documents. All of the other documents do, however, form a part of the construction contract. The correct answer is therefore D.

45. C Paragraph 12.2 of the AIA General Conditions states that if any work is found to be defective or not in accordance with the contract documents within one year of final completion and acceptance by the owner, the contractor must correct such work promptly upon receipt of a written notice from the owner (correct answer C).

46. C The architect's responsibility in this situation is limited to issuing the addenda and ascertaining the contractor's receipt of same. By acknowledging their receipt at the time of bid opening, the contractor agrees that all addenda are part of the contract documents, thereby assuming responsibility for incorporating any change shown on the addenda into the work. The AIA Instructions to Bidders states that all addenda issued prior to execution of the contract are a part of the contract documents. The correct answer is therefore C.

47. D The performance bond provides for contractor compliance with the terms of the contract and, in the event of his or her default, protects the owner against loss up to the amount of the bond (correct answer D).

48. Statute of limitations A statute of limitations is a law that imposes time limits upon the right of action by one party against another for alleged damage or injury.

49. B The AIA General Conditions, Article 10, defines protection of life and property at the construction site. It is the contractor's responsibility to act expeditiously in order to prevent loss or injury to adjacent property (correct answer B). Extra compensation for such work would be appropriate only if determined by previous agreement and included in the contract for construction.

50. D Article 8 of the AIA General Conditions defines the contract time and the conditions under which time extensions may be granted. The contract time may be extended only if the project's completion is delayed for specific reasons, such as changes ordered in the work, labor disputes, fire, unusual delay in transportation, unanticipated adverse weather conditions, or any other causes beyond the contractor's control. Delay authorized by the owner pending arbitration, or for any other reasonable cause as determined by the architect, may justify a time extension. In this case, the delay is attributable to the contractor, and therefore, an extension of the contract time is not warranted (B and C are incorrect). A is likewise incorrect. If the contractor's delay damages the owner, he or she may be able to recover the damages from the contractor (correct answer D).

51. A Under the terms and conditions of AIA Document A201 and the lien laws in effect in most states, any subcontractor or material supplier is entitled to file a lien for non-payment of completed work. A is therefore the correct answer. The other choices define actions that the architect must take in order to protect the owner's interests in regard to the satisfactory completion of the work under the terms of the contract.

52. B The General Conditions (A201) states that if the contractor fails to correct defective work, etc., the owner may order the contractor, in writing, to stop the work until the cause for such an order is eliminated. See Subparagraph 2.3. Under no circumstances should the architect or any of his or her employees or consultants take on this responsibility.

53. C The contractor has the sole responsibility for the protection of work and property as outlined in the AIA General Conditions, Article 10. However, if the architect or any of his or her consultants becomes aware of any potentially hazardous condition, the architect is obliged to notify the owner and contractor (correct answer C).

54. A A reasonable number, but always at least one, of the principal entrances to a public building should be part of an accessible route. The number of such entrances must be sufficient to accommodate the disabled users of a building facility.

55. D All aspects of bidding procedures on public work are public information and are subject to public scrutiny. With almost no exception, public building construction is bid competitively, and requests to submit proposals are openly advertised. The date, place, and time of bid opening are also published, so that anyone may attend the opening. The names and bids of the contractors are read aloud, and thus, they become part of the permanent public record. The contractor in question can be true to his principles or his business, but in this case, not both.

56. B One must assume that the architect would not have specified the waterproofing unless it was necessary to avoid future leaks. Therefore, if the architect agrees to the owner's deletion, there must be a letter on file warning the owner that leaks might occur. The deletion should be stated in a change order, which should also explain the reason for the change. This kind of precautionary measure will eliminate the architect's responsibility for potential design deficiencies. Incidentally, it would not be ethical for the architect to either resign or suggest that the contractor guarantee poor-quality work.

57. **C** According to the Owner-Architect Agreement, the owner agrees to pay for accounting and auditing services that may be required by the execution of the work. In this case, there is sufficient suspicion to justify an audit, the results of which would be made available to the owner, architect, and all concerned subcontractors.

58. **C** If the owner authorizes the architect to prepare record drawings, this is normally considered a Change in Services, with additional compensation.

59. **Alternates** The correct term is *alternates,* which are specific items added to or deducted from the base bid. Alternates are generally recommended by the architect and approved by the owner before they are included in determining the low bid.

60. **D** The ANSI standards specify that doors leading to areas that might prove dangerous to a blind person must be made identifiable to the touch by a textured surface on the door operating hardware. Textured surfaces, however, should not be provided for emergency exit doors or on doors other than those leading to hazardous areas.

61. **A** The AIA General Conditions provides that if any work has been covered, which the architect has not specifically requested to observe prior to being covered, the architect may request to see the work and it must be uncovered by the contractor. If the work is found to comply with the contract documents, however, the cost of uncovering and replacement must be borne by the owner.

62. **D and E** If the contractor failed to complete the work as promised, the construction contract might contain a liquidated damages clause or a penalty clause that would enable the owner to collect a fixed sum of money. If the job were going well, but behind schedule, the owner would probably not dismiss or replace the contractor. Penalty clauses, to be legally enforceable, must be accompanied by bonus provisions that reward contractors for completing the work earlier than the date originally stipulated.

63. **B** According to the AIA General Conditions, the cost or credit resulting from changes in the work may be determined in a number of mutually acceptable ways: by lump sum, unit prices, or cost plus fixed or percentage fee. If none of these methods is agreeable, the cost is determined by the architect on the basis of reasonable direct expense, plus an allowance for overhead and profit. The method used should always be agreed upon before any work begins.

64. **C** Under the AIA General Conditions, Subparagraph 12.2.2, defective workmanship discovered within one year after completion of the work must be corrected. Therefore, the general contractor must correct the defect at his own expense. The architect should not deal directly with the flooring contractor, nor should the owner be expected to repair or pay for any such work.

65. **Lien period** The lien period is the time limit during which workers or contractors may file a mechanic's lien for labor or material expended on a project. Conversely, a *statute of limitations* is the time limit for

legal action of one party against another for alleged damages.

66. C The architect checks and approves samples for compliance with information shown in the contract documents. As the owner's agent, she cannot reduce the quality of materials and workmanship required by the documents without the owner's written consent. The bid price is based on the specifications, and any deviation must be in the form of a change order. The architect's only option is to reject the sample and insist on conformance with the specifications.

67. A Article 6 of the AIA General Conditions (Document A201) permits the owner to enter into separate contracts at his or her discretion. In that event, the general contractor must cooperate with the work of others.

68. A The architect is responsible to the owner for all work, including that of his consultants, since there is no contractual relationship between the owner and the architect's consultants.

69. D The requirements for submittal, coordination, review, and approval of shop drawings are defined in Paragraph 3.12 of the AIA General Conditions (A201). In approving shop drawings, the architect states that reviewing is only for conformance with the design concept of the project and compliance with the information contained in the contract documents. The contractor must check for compliance with field measurements.

70. B In all cases where the General Conditions are at variance with state laws or procedures, the most efficient means to accommodate such situations is to amend the specific paragraphs that are at variance in the Supplementary Conditions. To rewrite the AIA General Conditions (D) and assume the responsibility for its legal adequacy is dangerous. To allow the client to assume that responsibility (C) is even more dangerous. The invitation to bid (A) is not part of the legal contract between owner and contractor, and therefore, anything incorporated into it is difficult, if not impossible, to enforce legally.

71. C The bonding company insures the faithful performance for completion of the contract. This assurance is provided by the performance bond.

72. B, C, and D All of the statements concerning subcontractors are correct except A. A subcontractor has no contractual relationship to the owner. At the time of bidding, the architect, with the owner's approval, may, however, reject any subcontractor and ask that he or she be replaced.

73. D As defined by the AIA General Conditions, Subparagraph 2.1.1, the owner is the party identified as such in the owner-contractor agreement. Depending on the circumstances, persons other than the owner may provide funds, make decisions, and even sign agreements, if authorized by the owner to perform these duties.

74. A Under the circumstances described, the architect should determine the fair cost of the additional expenses claimed by the contractor. The architect is obligated to resolve disputes to keep the job going and avoid litigation. If after all this, there is still no agreement, the parties may arbitrate the matter as a last resort.

75. D An architect's liability for her own acts and those of the architect's employees is known as direct liability. Contingent liability, on the other hand, refers to the architect's liability for the acts of independent consultants who are retained to perform professional services.

76. C Before receiving the final payment, the contractor must furnish affidavits stating that all indebtedness connected with the project has been satisfied in order to avoid the possibility of liens being filed against the property. Among the costs for which the contractor is responsible are wages, materials, and equipment. Payment of professional fees, however, is generally the direct responsibility of the owner.

77. C Subparagraph 4.2.4 of the AIA General Conditions states that the owner and contractor shall endeavor to communicate through the architect.

78. D One of the architect's responsibilities is to insure that the intent of the contract documents is carried out. In some cases, this requires professional judgment. In this case, all of the suggested solutions would change the proportion and character of the toilet room detailing. Therefore, the architect should reject the mirrors and insist that the drawings be followed as originally conceived.

79. D The cost plus fee arrangement is normally employed in negotiated contracts and is especially useful when construction work precedes the completion of construction documents. A cost plus fee contract is fair, flexible, and allows the owner to benefit from fluctuating market conditions by modifying the program or materials as work progresses. A percentage of construction cost is generally used for owner-architect contracts; architects' estimates are never used as a basis for construction contracts, and a stipulated sum is almost always used when the work is competitively bid.

80. C An *accessible route* is a path that is safe and usable by disabled people in wheelchairs, as well as those with other disabilities. The term *disability* is defined as *a limitation or loss of use of a physical, mental, or sensory body part or function.* Therefore, all the choices would require an accessible route, except the person who is unable to speak English. That person is considered to have a social, rather than a physical or mental, handicap.

INDEX

Notes

Notes

Notes

Notes

Notes

Notes

Notes

Notes

Notes

1. Italian seasoning
2. Garlik salt
3. Pasil
4. origono

34000
3000
37000
24000
61000
10000
71000
20000
91000
1250
92250
24000
116250

1 May
2 June
3 July
4 August
5 septem

5 x 6000 = 30000

32000
30000
62000
Bonus 10000
72000
Tax Ret. 23000
95000